W0079322

HUMAN ACTION AND ITS EXPLANATION

SYNTHESE LIBRARY

MONOGRAPHS ON EPISTEMOLOGY,

LOGIC, METHODOLOGY, PHILOSOPHY OF SCIENCE,

SOCIOLOGY OF SCIENCE AND OF KNOWLEDGE,

AND ON THE MATHEMATICAL METHODS OF

SOCIAL AND BEHAVIORAL SCIENCES

Managing Editor:

JAAKKO HINTIKKA, *Academy of Finland and Stanford University*

Editors:

ROBERT S. COHEN, *Boston University*

DONALD DAVIDSON, *University of Chicago*

GABRIËL NUCHELMANS, *University of Leyden*

WESLEY C. SALMON, *University of Arizona*

VOLUME 116

HUMAN ACTION
AND ITS EXPLANATION

A Study on the Philosophical Foundations of Psychology

RAIMO TUOMELA

Department of Philosophy, University of Helsinki

D. REIDEL PUBLISHING COMPANY

DORDRECHT–HOLLAND/BOSTON–U.S.A.

Library of Congress Cataloging in Publication Data

Tuomela, Raimo.
    Human action and its explanation.

    (Synthese library; v. 116)
    Bibliography: p.
    Including indexes.
    1. Psychology—Philosophy. I. Title.
BF38.T86   150′.1   77–24522

ISBN-13: 978-94-010-1244-7   e-ISBN-13: 978-94-010-1242-3
DOI: 10.1007/978-94-010-1242-3

Published by D. Reidel Publishing Company,
P.O. Box 17, Dordrecht, Holland

Sold and distributed in the U.S.A., Canada, and Mexico
by D. Reidel Publishing Company, Inc.
Lincoln Building, 160 Old Derby Street, Hingham,
Mass. 02043, U.S.A.

All Rights Reserved
Copyright © 1977 by D. Reidel Publishing Company, Dordrecht, Holland
Softcover reprint of the hardcover 1st edition 1977
No part of the material protected by this copyright notice may be reproduced or
utilized in any form or by any means, electronic or mechanical,
including photocopying, recording or by any informational storage and
retrieval system, without written permission from the copyright owner

CONTENTS

To Sirkka

PREFACE

This book presents a unified and systematic philosophical account of human actions and their explanation, and it does it in the spirit of scientific realism. In addition, various other related topics, such as psychological concept formation and the nature of mental events and states, are discussed. This is due to the fact that the key problems in the philosophy of psychology are interconnected to a high degree. This interwovenness has affected the discussion of these problems in that often the same topic is discussed in several contexts in the book. I hope the reader does not find this too frustrating.

The theory of action developed in this book, especially in its latter half, is a causalist one. In a sense it can be regarded as an explication and refinement of a typical common sense view of actions and the mental episodes causally responsible for them. It has, of course, not been possible to discuss all the relevant philosophical problems in great detail, even if I have regarded it as necessary to give a brief treatment of relatively many problems. Rather, I have concentrated on some key issues and hope that future research will help to clarify the rest.

A remark on my syntactic conventions is due here. Generally, symbols may be used autonymously whenever this cannot be expected to cause confusion. As usual, the single quote-operation is used to form names of expressions. Numbered formulas and statements are referred to only by their numbers, when they occur in the same chapter (e.g. formula *(18)*); otherwise they are referred to by the number of the chapter and the formula number (e.g. by *(7.18)* when formula *(18)* of Chapter 7 is meant).

I would especially like to thank Professors Robert Audi, Richard Rorty and Zoltan Domotor, as well as Dr. Lars Hertzberg for discussions concerning some of the topics dealt with in this book. Professor Ilkka Niiniluoto made some helpful remarks on Chapter 11 when I presented the material in my seminar. I wish to thank Mr. Farrell Delman for checking the English of most of this book.

I am very much indebted to the Academy of Finland for research grants which made it possible for me to be on leave from my professorship during 1974 and the latter half of 1976 to work on this project.

With appropriate permission, I have in this book used some passages of the following papers of mine:

'Psychological Concepts and Functionalism', in Hintikka. J. (ed.): 1976, *Essays On Wittgenstein, Acta Philosophica Fennica*, vol. 28, North-Holland, Amsterdam, pp. 364–393
'Causality and Action', forthcoming in Hintikka. J. and R. Butts (eds.), *Logic, Methodology and Philosophy of Science V*, Reidel, Dordrecht
'Purposive Causation of Action', forthcoming in Cohen, R. and M. Wartofsky (eds.), *Boston Studies in the Philosophy of Science*, vol. 31, Reidel, Dordrecht
'Dispositions, Realism, and Explanation', *Synthese* **34** (1977), 457–478.

Helsinki, December 1976

Raimo Tuomela

CHAPTER 1

SCIENTIFIC REALISM AND PSYCHOLOGY

Wittgenstein once said that in psychology there are experimental methods and conceptual confusion. This still seems largely true. As evidence for the claim about conceptual confusion it suffices to take a look at almost any psychological work on cognitive psychology (perhaps excluding some of the most recent).[1] It is hoped that this book will bring at least some clarification into the conceptual framework of cognitive and motivation psychology.

Our basic aim in this book is to develop a causal account of complex actions (cf. especially Chapter 10). We shall also discuss the related matter of how to explain actions causally. This latter task again presupposes that we have a relatively good grip on explanatory psychological concepts (e.g., proattitudes, intentions, beliefs). Thus, we shall devote relatively much space to the investigation of the nature of inner psychological states and events, especially those to be called conceptual states and events.

Our development of a causal philosophical theory of complex actions will naturally involve a conceptual clarification of various types of action and of various psychological concepts for inner psychological states and episodes, and most importantly, of the various interrelations between these groups of concepts. The resulting system should give a kind of philosophical and methodological foundation for cognitive and motivation psychology, i.e. roughly for all psychological theories which deal with human actions and their explanation by reference to wants, intentions, beliefs, and related factors. Thus, what we are going to do should be of interest not only to philosophers and methodologists but to psychologists as well. Philosophy of psychology should, and can, be seen as intimately related to the science of psychology.

The general philosophical view that this book espouses could be called *critical scientific realism* (see Tuomela (1973a) for a brief exposition). One way of viewing our present work is then to regard it as an attempt to try and see how critical scientific realism applies in detail to some of the central philosophical problems of psychology (and other related social sciences, too). Even if most of our arguments and developments will be more or less independent of any particular realist assumptions, perhaps at least some kind of "spirit of realism" is likely to be felt throughout the book. Especially our view of the nature of explanatory psychological concepts as theoretico-reportive concepts (Chapter 4) and our treatment of causality (Chapters 9, 12) will be seen to have a clearly realist flavor.

Our version of critical scientific realism primarily involves a certain view of what there is and of how knowledge of what there is can be obtained. Roughly and concisely speaking, critical scientific realism says that science is about a reality that exists independently of observers. In principle, and contra Kant, this reality is knowable, though normally only in a symbolic, partial and distorted way.

Knowledge about this reality is primarily obtained by means of scientific theorizing and observation, and this knowledge always remains corrigible. A central feature of this view is that there exist objects and properties which are not observable or experiencible at all. We need special theoretical concepts and theoretical discourse to speak about this non-sensible (non-empirical) part of reality.

Our scientific realism can be partially and crudely summarized by the following concise epistemic-ontological thesis:[2]

(R)   All non-sentient physical objects and all sentient beings have exactly the constituents and properties ascribed to them by the theoretical scientific terms that are required for the best scientific explanations of the behavior of these entities.

What (R) really comprises can of course only be seen after its philosophical key terms have been explicated. Especially much depends on what the 'best scientific explanation' involves and on how much content is given to 'behavior'. What we understand by the first phrase will roughly become clear later in this book (also see Tuomela (1973a)). 'Behavior' is to be taken in an inclusive sense in our idealized (R). In the case of *persons* (or agents), our primary objects of investigation in this book, this term will be taken to cover, in addition to actions and other overt behavior, also mental states and episodes. Thus the "sensuous" features of the world (e.g. the qualitative features involved in seeing a red rose and in sensing the smell of the rose) will also become included. (This is something, e.g., Sellars (1963) has strongly emphasized.)

Thesis (R) involves the idea that what explains best describes best. As it is scientific theories which explain best we can almost take as our general motto: *scientia mensura*. As Sellars puts it, "in the dimension of describing and explaining the world, science is the measure of all things, of what is that it is, and of what is not that it is not" (Sellars (1963), p. 173). In all interesting cases the best explaining theories (R) speaks about are yet to be found.

Sometimes and for some purpose it is useful to philosophize by means of highly idealized concepts ("ideal types"). We shall occasionally below employ Sellars' idealized notions "manifest image" and "scientific image" when discussing broader vistas (cf. Sellars (1963)). The manifest image of the world consists, roughly, of "Strawsonian" entities such as tables, trees, stones, colors, sizes, and so on, and of the framework of persons or agents capable of knowing, believing, wanting, hoping, saying, promising, acting, obeying norms, building societies, and, generally, bringing about cultural products. The manifest image can be characterized as the conceptual framework within which (Western) man became aware of himself. Moreover, man is essentially to be regarded as that being which conceives of itself in terms of the manifest image.

In addition to the above epistemic-ontological criterion one can also characterize the manifest image epistemically by saying that within it only correlational inductive methods (say, Mill's methods and many ordinary statistical multivariate methods of data analysis) are used. No postulation of unobserved objects nor of properties "lying behind" the observable ones and explaining their behavior is involved. An example of the use of the postulational method is given by the explanation of the macro-properties of gases (e.g., the Boyle–Charles law) by means of the kinetic theory of gases.

The scientific image can be characterized as consisting of those objects and properties which scientific theories truly postulate to exist. Epistemically (or methodologically) the scientific image can be described by saying that it makes extensive use of the postulational method.

Which of these images is primary? Obviously all scientific realists will say that ontologically or in "the order of being" the scientific image is primary. How about the "order of conceiving", viz. from the point of view of concept formation and gathering knowledge? Here scientific realists differ. Radical ones want to make the scientific image primary even in the order of conceiving, whereas, e.g., Sellars considers the manifest image primary in this respect. When discussing psychology and the social sciences in this book we shall agree with Sellars' view of the primacy of the manifest image in the order of conceiving. However, we emphasize the social and conventional nature of this primacy and the possibility of changing linguistic conventions and habits.

We said earlier that Sellars' notions of the manifest and the scientific image are really ideal types. In fact, it is easy to show that they have several unrealistic features.

First of all, these notions are vague. Concerning the manifest image, we may ask whether, e.g., a distant star which can be seen only through a powerful telescope belongs to the manifest image. How about bacteria seen only through microscopes? They are certainly not postulated, however.

Next, it would be easy to argue that there are several incompatible but "equally possible" refinements of the manifest image (cf. e.g. the incompatibilist versus the reconciliationist explications of the freedom of the will). So, we may ask, what is the use of the definite article in the connection of this notion.

Still another criticism against Sellars' notion of the manifest image is that theorizing within the social sciences in fact typically employs the postulational method even when it is otherwise within the conceptual framework of the manifest image (e.g. statistical factor analysis is often used in a postulational way).

The scientific image is supposed to be an asymptotic unification of the scientific pictures different scientific theories give us at the limit. Now this Peircean "limit science" which is supposed to tell us the final truth about the world is of course something mythical. We do not a priori know how it will look like (and how the final scientific image will be like); we do not even know whether science will grow towards any such limit at all. Furthermore, our present ununified science probably gives a very bad estimate of the final scientific image.

I would also like to point out that there are other ways of forming theories in science than by starting from observable entities and by postulating unobservable entities to explain the behavior of these observable ones. Many physical theories simply do not speak about any observable entities at all and hence do not have any observational content (cf. Tuomela (1973a)).

What my above criticisms show is at least that the notions of the manifest image and the scientific image are problematic and cannot be used in serious philosophical argumentation without qualifications. Still we are going to occasionally employ the truths involved in these notions. Our basic setting will be the following.

We think that psychological theorizing (at least within cognitive and motivation psychology) has to start within the manifest image and to essentially employ the concepts of the framework of persons. However, on our way to fusing the manifest and the scientific image into a "three-dimensional" picture, we employ the postulational method to introduce inner psychological entities (states, episodes, etc.) to explain overt actions. Future science (presumably neurophysiology and neuropsychology) will then tell us more about these inner psychological entities. It should also be emphasized that this method makes the manifest and the scientific image in part *conceptually* connected. Within a setting like this, the above criticisms against the vagueness and ambiguity of the notions of the manifest and the scientific image can be avoided, it seems. (If necessary, we will, so to speak, relativize these images to research paradigms and programmes, as will in effect be done in Chapters 3 and 4.)

As mentioned earlier, we shall in this book try to create a causal theory of action which applies to arbitrarily complex actions. The various notions of action that we are going to deal with are typical elements of a refined framework of persons (agents). But in our treatment it will turn out that all an agent ever does (in our achievement sense of doing) is to move his body, while the rest is up to nature. Movements of the body seem naturalistic enough not to present great obstacles for unifying the manifest image with the scientific image.

Our causal theory of actions basically claims that intentional action-events are causally brought about by the agent's effective intendings, and, more generally, any action token conceptually involves reference to some kind of antecedent causal mental event or state (see Chapter 10).

It will, not surprisingly, turn out that the acceptability of our causal theory of action is going to depend strongly on the acceptability of a certain kind of *functionalist* account of inner mental events. In general, if one wants to construe an action theory so that it at least does not conflict with an attempt to fuse the manifest and the scientific images, special attention must be paid to the conceptual construal of mental states and episodes. Thus, in fusing the manifest and the scientific images together into a "stereoscopic" view one must give an account of both how "raw feels" (non-conceptual mental events and episodes such as sense impressions, sensations, and feelings) and how "thoughts" (i.e. conceptual events and episodes like perceivings and desirings) are to be characterized. (We assume here that raw feels and thoughts are really understood as inner episodes, etc., and that they are taken to involve some categorical features over and above their various overt-dispositional and overt-categorical ones.)

We are not really going to discuss the philosophical problems related to raw feels in this book. But thoughts (conceptual events and episodes, inner actualizations of propositional attitudes) must be discussed in some detail because of their importance. As said earlier, we are going to construe the concepts representing wantings, believings, perceivings, etc., as theoretical (or theoretico-reportive) concepts. That they can be so construed will be shown in Chapter 4. One essential condition for our construal is that we reject the Myth of the Given (cf. Sellars (1956)). According to this myth words for mental entities are somehow causally and independently of learning given their meanings, due to the impact of some suitable extra-linguistic mental entities (cf. Sellars (1956)). Therefore mental entities (sense data, etc.) would be (externally) fixed and they could not be theoretical entities in any sense.

We are going to construe thoughts (thinkings) *analogically* on the basis of overt speech (cf. Chapters 3 and 4). This account, if successful, will also account for the *intentionality* or intentional aboutness of thoughts. For one can then claim that intentionality is reducible to the *semantical* metalinguistic discourse (for semantic "aboutness") pertaining to a language for overt public entities. This accounts for the conceptual side of intentionality. In the real world (*in rerum natura*) there will be very few exactly statable "criteria", such as necessary conditions, for the non-metaphorical application of the categories of intentionality and thinking. One such elementary criterion is that any "entity" to which one can adequately ascribe thoughts must have some suitable representational sensory mechanism for "picturing" the "facts" of the world; furthermore, it must be capable of self-controlled activity (even if it is hard to be very exact about this).

There is another criterion for intentionality and thinking, which, though vague, is central. It is that the entity can be taken to be "one of us" and to share our intentions and norms and be capable of fulfilling and obeying them. This idea also involves the important insight that what is peculiar to and distinctive about persons is their "rule-following" nature, which can only be accounted for by means of a *normative* or prescriptive language, and it is in this sense that social science and the humanities most basically differ from the physical sciences. On the other hand, in the dimension of describing science can in principle tell us everything there is to be told about persons, societies and cultures. It is roughly in this sense that a nomological social science is possible.

NOTES

¹ To support our claim consider this recent statement:
"Our conceptual framework has grown out of our own research in the attitude area. It emphasizes the necessity of distinguishing among beliefs, attitudes, intentions and behavior; four distinct variables that have often been used *interchangeably* in the past." (Fishbein and Ajzen (1975) p. vi, my italics).

² The kind of scientific realism espoused by this work on the whole comes close to the views of Wilfrid Sellars (see, e.g., Sellars (1963), (1967), (1968), and (1974). The thesis (R) below would probably be acceptable to Sellars, too, although I perhaps want to emphasize its *regulative* and *idealized* character more than Sellars does.

My (R) can be interpreted to amount to the *Sellarsian scientific realism* of Cornman (1971). Note that (R) should be taken to concern only *contingent* descriptive properties.

CHAPTER 2

HUMAN ACTION

1. ACTIONS AS ACHIEVEMENTS

As we all know human behavior can be described and classified in a variety of ways. We speak about activities, actions, achievements, habitual and automated behavior, reflexes, and so on. It is plausible to think that our explanations of behavior are strongly linked with our ways of describing behavior. We shall in fact argue later that one's general views concerning the nature of man and the sources of man's activity strongly affect one's way of conceptualizing and explaining behavior.

In this book we shall concentrate on behavior as action and on some of the various ways one can explain actions. This chapter is in part devoted to a preliminary clarification of some key concepts related to action.

Among actions we customarily include behavior such as the following: raising one's arm, opening a door, crossing a street, paying a debt, killing somebody, saying or promising something, refusing or forbearing to do something, and so on. These actions can be done rationally, intentionally, voluntarily, deliberately, etc.

Actions must of course be distinguished from bodily movements (and reflexes), even if normal actions in a sense consist of bodily movements. Paying a debt by signing a check, say, certainly involves more than my hand and body moving in a certain way. But what is this more? This is one of our basic problems in this book.

Generally speaking, actions can be primarily characterized either in terms of their *antecedents* or in terms of their *consequences.* In the first type of characterizing action, reference can be made to such things as the agent's intentions, acts of will, purposes, reasons, wants, desires, beliefs, emotions and feelings. For instance, in our view an action may be conceived as behavior caused or brought about by effective intentions (states or episodes of intending).

When actions are characterized in terms of their consequences the emphasis often is on the idea of an agent intervening with and changing the external world. Or, with a slightly different twist, an action can be considered as a kind of "response" to *task* specifications in such a way that they essentially involve a public element—a result by which the correctness of the action performed can be judged.

In any case, to act (in this achievement or task sense) is to bring about public *results* or changes (or to prevent these change-events from occurring) in the world, which, so to speak, contains an opportunity for this type of intervention. Thus, an agent's opening the window presupposes that the window was not open before his acting and that it did not open without the agent's intervention. The change in the world brought about by the agent's action is, of course, the window becoming open. Furthermore, such results, which are conceptually involved in this concept of action, may themselves have various intended and unintended causal and other consequences (in this case, e.g., the ventilation of the room and the agent getting rid of his headache).[1]

In our view, to be developed later in detail, to give an adequate account of the notion of action one has to refer both to the antecedents and the consequences of a behavior process. The primary antecedents will in our approach be effective intentions (intendings) and beliefs, while wants, desires, duties and obligations, etc. are to be construed as secondary intention-forming antecedents.

We distinguish between action tokens, which are singular events, and action types, which correspondingly are generic events (universals). We shall later define and classify both action tokens and action types in detail.

A singular action token is a complex event brought about by an agent. It is complex in the sense of being process-like: an action token consists of a sequence of events (see especially Chapter 10). (We are going to technically call actions events, even if in many cases it would accord more with ordinary usage to call them processes.)

When discussing events seriously, one has to give account of how they are to be described or otherwise characterized and "picked out". This means that one must give a criterion of identity for events (and especially actions). Only then can one speak of "redescriptions" of actions. To get a glimpse of some of the problems involved let us quote a well-known passage from Davidson: "I flip the switch, turn on the light, and illuminate the room. Unbeknownst to me I also alert a prowler to the fact that I am home. Here I do not do four things, but only one, of which four descriptions have been given" (Davidson (1963), p. 686). A rather similar view of the identity of actions is held by Anscombe (1957) 'and von Wright (1971). Against these "radical *unifiers*" we have the "*fine grain*" approach held, for example, by Goldman (1970). According to this approach we can say that there are four actions in question here. Intermediate positions are possible as well (see next section).

Let us draw an "action-tree" for Davidson's example in order to see what is involved. To make the example a little more interesting assume in addition that the agent, call him A, hurts his finger when flipping the switch and that the lighting of the bulb, due to a causal mechanism unknown to A, explodes a bomb. We get the simple diagram in Figure 2.1.

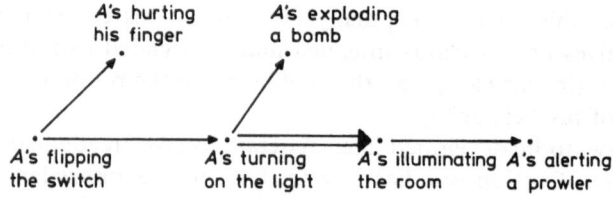

Figure 2.1

In this "action tree" the nodes represents different actions (for a fine grain theorist) or different descriptions of the same action (for a unifier). Single arrows represent a causal (or, more generally, factual) nomic relationship whereas the double arrows represent a "conceptual" relationship (relationship based on semantical or conceptual connections). In this example all the actions (or action descriptions) are (or have to do with) singular actions. Thus we can read from the diagram, for instance, that the singular action of A's flipping the switch causally or *factually* generated A's turning on the light. This is based on the fact that the flipping of the switch caused the light turning on. We may understand that in our present case the light's turning on (semantically) *means* the room becoming illuminated; thus A's turning on the light *conceptually* generates A's illuminating the room. Notice also that A's flipping the switch in this tree generates two otherwise independent actions. The possibility of branching gives our diagram its tree-like character (or, more generally, a graph-structure).

For radical unifiers our action tree still represents only one action, whereas for a fine grain theorist there are now six different actions represented. Our fine grain theorist might perhaps concede that all the actions on the tree are the same or equivalent in the sense of being generated by the same primitive or basic action of flipping the switch, though these actions are not strictly identical. One reason why they might not be considered strictly identical is simply that they hold different "generational" relationships to each other; another reason is that they have different results, as we shall argue later.

A difficulty for the fine grain theorist is this: the action tree is presumably supposed to represent what the agent did in a certain situation; but it is not restricted to representing previously intended actions, and may well include unintended or non-intentional actions; but there are, then, for a fine grain theorist an indefinite number of actions an agent performed in that given situation. One may invent all kinds of causally and conceptually generated actions to enlarge the tree: *A* flips the switch forcefully, *A* flips the switch forcefully by his left thumb, *A* flips the switch forcefully with his left thumb by standing firmly on his feet, *A* causes a bomb to explode through a mechanism connected to the switch.

It seems that in the debate between the radical unifiers and the fine grain theorists the distinction between actions (as extralinguistic events) and action descriptions has not been kept clear enough. The same holds for the notions of a ("bare") singular event and an exemplification of an action type. Let us now make a brief digression into this difficult topic, thereby shaping the groundfloor of our conceptual framework.

2. ACTIONS AND EVENTS

Our discussion has clearly suggested that one important task in characterizing actions is to give them philosophically illuminating criteria of identity: When do two actions count as the same? Or better (as we, strictly speaking, must answer the previous question by "never"): When is an identity statement '$a = b$', representing the identity of two actions *a* and *b*, true? For a number of reasons, it is important that such a criterion of identity be given to *singular* or tokened actions in the first place. One reason for this is that when we describe what an agent actually did, we of course have to start with action tokens. Another reason is that actions are so context-dependent that what is said about the relationship between action tokens very often does not hold true for the corresponding action types. This is true especially of the generational relationships between actions; and we think that any theory of action must give an account of them.

We claimed that actions are events; and events are here understood in a broad sense (including, e.g., short-term states). Before discussing identity criteria for events, we must obviously say something about the nature of events. Our emphasis in this book will be on *singular* events as contrasted with *generic* ones. Singular events can be construed either as *unstructured* or as *structured* particulars. Furthermore, events may be regarded as either *extensional* or *intensional*, although we shall be less concerned with this distinction here. We shall in this book adopt a view according to which events can in a certain sense be treated as "bare" particulars while in another sense they are structured. In the sense in which they are bare entities, they are extensional, and in the sense in which they are structured, they are intensional.

Before discussing events in detail a few words have to be said about the reasons for considering events an ontological category on a par with, e.g., "substances". Let us note here that in this book the class of singular events (including, e.g., explosions, switch flippings, rememberings) will be taken as a rather broad category so as to include what might be called short-term states (i.e., perceivings and wantings). Thus, when we later in Chapter 3 speak about Jonesian states and episodes, they (except for process-episodes) will fall under the broad category of events as discussed here and elsewhere in this book.

What kind of reasons are there for taking singular events seriously? One important reason comes from explanation and causality. We often explain singular events and speak of causation between singular events (see Chapter 9). This also makes it natural to construe actions as events, as will be seen later in this book.

A closely related reason is that we want to speak about redescriptions of events (e.g., actions) and to distinguish between events and their different descriptions. (A satisfactory account of actions cannot be given without speaking about redescriptions of actions, and for this it is best to require that actions be construed as events.)

There are also more direct reasons, based on linguistic considerations, for the necessity of an autonomous category of singular concrete events. Davidson (1969) forcefully argues that the logical forms of some sentences require event-ontology as strongly as substances are required by some other statements. Thus, 'Jones buttered a piece of toast in the bathroom' seems to require event-ontology in its logical form in order to account in the *most economical* way for the fact that it entails 'Jones buttered a piece of toast' (see next section). A good example of a sentence which requires event-ontology is the so-called numerically restrictive sentence (see Wallace (1971) and next section).

Let us now go on to consider the event of somebody's raising his arm. We may regard it as a concrete event. But is it to be treated as a "bare" event or as a structured event? We can conceptually and epistemically think of this event as a "substratum" or a "bracketed" event. But from an ontological point of view, one has to be careful in speaking of substrata and unstructured events. Substrata without any qualities or properties are only *conceptual abstractions* or "bracketings" of real events. Any event that occurs in reality has factual properties. The point we are trying to make is indeed that all singular events do exemplify universals; but still a datable singular event is not to be regarded essentialistically as something which is somehow *constituted* by its instantiating a certain property (universal) rather than another one. Singular events are not reducible to universals.

What we perceive is an event conceptualized in a certain way. In our example we may perceive, for instance, a bodily movement of the arm going up or, alternatively, an intentional arm raising by the agent. These conceptualized events are structured, for they are events seen as exemplifying some specific universal (i.e., some generic events or properties).

It is unstructured events (i.e., events whose structure has not been displayed) that we are going to quantify over in our formal developments. Such events can be regarded as extensional entities and discussable within standard predicate logic. My conception of unstructured events seems to correspond fairly closely to Davidson's view of singular events (cf. Davidson (1969)), although I do not know exactly how far the agreement goes.

Structured events as we shall discuss them in this book are events as conceptualized within the "manifest image" (see Chapter 1). My view of conceptualized events (events with explicitly displayed structure) can be explained partly by reference to the rather well-known account by Kim (see, e.g., Kim (1973)). Kim thinks of a singular event as a concrete object (or n-tuple of objects) exemplifying a property (or n-adic relation) at a time. This exemplified property is to be intensionally construed. To take a simple dyadic example, the event of John's greeting at time t_0 is represented by [$\langle$John, $t_0\rangle$, greeting], or, more formally, by [$\langle a, t_0\rangle$, P]. The property P is the *constitutive* property of this event. That is, each singular event is assumed to be uniquely associated with a certain property which determines the *generic* event (or event type) that the singular event exemplifies. (The constitutive property P is, presumably, not really an *element* of the event, but it is a *fact* that the event exemplifies P.) The generic event in question can be represented by [$\langle x, t \rangle$, P] where x and t are variables ranging, respectively, over objects (including agents) and times. Each singular event can, however, *"merely exemplify"* a number of different properties. Thus, the event 'John's greeting' may exemplify the property 'uttering the sound *hello*' but also 'standing in front of the university'.

An identity condition for monadic events might now be formulated as follows (cf. Kim (1973), p. 223):

(K)   Two singular events [$\langle a, t \rangle$, P] and [$\langle b, t' \rangle$, Q] are identical if and only if $a = b$, $t = t'$, and $P = Q$.

Identity conditions for n-adic events become slightly more complicated, although they, of course, are based on the above idea, too (cf. Kim (1973)).

Goldman (1970) has applied the above Kimian kind of analysis of events to action contexts. In the above definition of identity, we may take a and b to be agents and P and Q to be action properties. Thus, two singular actions are said to be identical if and only if they involve the same agents, the same action properties, and the same times (cf. Goldman (1970), p. 10). What results is the (or a) fine grain approach to identifying actions we discussed earlier. There are, however, difficulties with Kim's and Goldman's views, and some of them must be taken up here.

The constitutive properties in Kim's account are assumed to represent generic events. Therefore singular events are really analyzed as being merely exemplifications of generic events. But now the problem arises what the set of constitutive properties (or generic events) really is, and that question Kim does not answer. Nor has anybody else answered it (fully) satisfactorily. Still, it is clear that this approach is a fine grain approach, and too much so, I think. Thus, for Goldman and Kim individual stabbings can never be killings, nor can singular arm wavings be greetings.

A related problem is what the identity conditions for properties exactly are. Linguistic criteria have often been tried. Thus, for example, Goldman (1970) suggests that two properties P and Q are identical if and only if they are designated by synonymous predicates. But that at least presupposes an analysis of synonymity which is independent of property identity. That seems hard to give. What is more, the suggested criterion seems clearly false. Let me indicate why.

Goldman's criterion is this: "if there are phrases in a language which express [properties] ϕ and ϕ', then their being synonymous is a necessary and sufficient condition for their expressing the same property" (Goldman (1970), p. 12). But as even Goldman himself notices one can use non-synonymous descriptions to express one and the same property. Thus, 'blue' and 'the color of the sky' gives one such pair and 'temperature' and 'mean kinetic energy' serves as another. Thus, synonymity is not necessary for property identity.

Synonymity is not even sufficient for property identity. I do not have a very good scientific example to show this, but in any case my argument goes as follows. Assume as true that theoretical terms in science are often "observationally" open (cf. Tuomela (1973a) for arguments). To use our later terminology, if $T(\lambda \cup \mu)$ is a theory with λ as its set of observational predicates and μ its set of theoretical predicates (see Chapter 4 and Tuomela (1973a) for this dichotomy), then the model-theoretic interpretations of the μ-predicates are not, normally at least, completely fixed by the interpretations of the λ-predicates. Assume, furthermore, as acceptable the *semantic* principle of *Conceptual Functionalism* (see (CF) of Chapter 4) according to which, roughly, the λ-predicates determine the meanings of the μ-predicates. This principle seems good and tenable at least for cognitive and motivation psychology. Then it can be argued that, in general, the μ-predicates *multiply designate*, i.e., one predicate will simultaneously designate several properties (which model-theoretically are represented by different and nonisomorphic extensions). Thus, going back to Goldman's criterion, 'ego strength' could stand for the two synonymous property-designating expressions (which are identical!) in a personality theory. Still 'ego strength' designates a number of different properties. This result is in good agreement with what has been found out in personality psychology by means of experimental work, although no sufficiently exact personality theories exist to really prove this point in full detail.

As we have noticed, Kim's approach is faced with the difficult and so far unsolved problem of property identity. Before more can be said about this problem his account cannot be seriously evaluated. A related feature about Kim's approach that we noticed is that it is essentially a fine grain approach: each singular structured event is associated with exactly one constitutive property. How fine grained the approach really is of course depends on how the "final" set of generic events will look.

A. Rosenberg (1974) argues that in face of problems like the above Kim's approach has to be modified by allowing for more than one constitutive property for each singular event (call this criterion (K')). This modification perhaps makes the approach a little less fine grained, but it is still faced with the difficulties of selecting the total set of constitutive properties and the set of constitutive properties for each particular event and of giving identity criteria for properties. Furthermore, Rosenberg claims that Kim's modified approach now is equivalent to Davidson's approach (as far as event identities are concerned), given a suitable backing law account of singular causation. If so, it could be said that Kim's modified criterion of event identity gives the *meaning* of identity statements while Davidson's causalistic approach (see below) gives (the) *epistemic* grounds for them. But Rosenberg's claim seems incorrect. It is instructive to see why.

Davidson's (1969) well known criterion is as follows:

(D)   Two singular events e_1 and e_2 are identical, viz. $e_1 = e_2$, if and only if

(e) (e caused $e_1 \rightarrow e$ caused e_2)   and
(e) (e_1 caused $e \rightarrow e_2$ caused e)

Let us postpone a discussion of the adequacy of (D) for a while in order to discuss immediately Rosenberg's thesis.

The type of backing law account of causation Rosenberg (and presumably Kim) have in mind is simply this:

(CC) e_1 caused e_2 if and only if the constitutive properties of e_1 and e_2 are "constantly conjoined" or lawfully correlated.

Presumably (CC) is to be understood so that for e_1's causing e_2 it suffices that *one* of e_1's constitutive properties be lawfully correlated with *one* of e_2's constitutive properties. We shall, later in this book, indicate that, as it stands, (CC) is inadequate in several respects, and we are going to replace it by a much stronger condition (see Chapter 9). But, for the sake of argument, let us here consider only (CC).

It is obvious that if two events e_1 and e_2 are identical in the sense of (K') then, as their constitutive properties coincide, these events have got to satisfy the right hand side of (D). But how about the converse? It seems that it does not have to be true. Assume that e_1 is associated with some finite set $\{P_1, \ldots, P_m\}$ of constitutive properties and e_2 with the set $\{Q_1, \ldots, Q_n\}$ (if the sets really turn out to be identical then of course $m = n$). If we now go through all the events and find that the right hand side of (D) comes out true, what do we have? The idea here would be to try and guarantee that each P_i, $i = 1, \ldots, m$, could be correlated with a Q_j, $j = 1, \ldots, n$, so that P_i and Q_j are found to play exactly the same causal roles with respect to every other property. But this is not entailed by the fulfilment of the right hand side of (D) without further qualification. The qualification needed is just that we really have to go through *all* the properties (whatever their set really is). If we had worked with Kim's original view, which associates a unique constitutive property with each event, that problem obviously would not have arisen.

Perhaps we may accept the above qualification, after all. There is, however, still another difficulty here. For, as said, in order for (D) to entail (K') what we essentially need is the truth of this criterion:

(D*)   Two properties P and Q are identical, viz. $P = Q$, if and only if P and Q have identical causal relations with respect to every other property.

However, even if the entailment goes from right to left, the converse implication can hardly be accepted as true (irrespective of whether only some kind of scientifically "good" and "nonartificial" properties are admitted in the class of all properties here).

First, I think non-causal nomic connections should be accepted as affecting the identity of properties, and criterion (D*) ignores them. (An example of a non-causal law would be, "The formation of the respiratory system in a human embryo always succeeds the formation of the circulatory system".)

Secondly, it seems that we cannot rule out the possibility that even if P and Q play identical causal (or nomic) roles they may at the same time have different, either "accidental" or conceptual connections to some properties. As an example of a conceptual connection, which should matter when discussing property identity, consider "Extending one's arm when cycling means signalling for a turn". (When discussing examples like this we admittedly run into problems concerning the ontological status of universals like properties and types, but we cannot go into those problems here.)

We have thus found that Kim's modified criterion (K') and Davidson's criterion (D) are different. Furthermore, in our approach neither of these criteria will be considered acceptable as such. We have already discussed the criteria (K) and (K'). Now a few words about (D) are in order.

The most basic fault with (D) is that it is *circular* as a meaning analysis of event identity. This is simply due to the fact that the right hand side of

(D) makes use of quantifiers which are intended to range over events (there is no way of interpreting (D) without admitting events into the range of the quantifiers). But this clearly requires that these events have already been identified (in some sufficiently clear sense). Therefore I would not take (D) as a meaning criterion but rather as an *epistemic* criterion.

There is also a technical fault in (D). It is due to the use of the material implication $\rightarrow$ in it. Suppose that e_1' is an event which has no causes and no effects. Then if we substitute e_1' for e_1 in (D) we obtain the result that e_1' is identical with any arbitrary event. This fault can be removed by using another explicate for the if-then relation in (D). For instance Lewis' variably strict conditional $\square\!\!\rightarrow$ might be considered as a good candidate here (cf. Lewis (1973a)).

We shall, in addition, use our causal predicate $C(-, -)$ (to be defined later in Chapter 9) to explicate singular causation. Let us call the resulting criterion (D'). (D') can be used as an, or perhaps "the", epistemological criterion for event identity in "normal" contexts, i.e., where events like e_1' above are not concerned. For even (D') does not give us anything constructive and useful with respect to such "random" events, although it technically removes the mentioned paradox connected with (D).

In the case of unstructured singular events (i.e. events whose structure has not been displayed) there is yet another epistemological criterion of identity which must be taken up here. If singular events are viewed as such unstructured events, then intuitively they are spatiotemporal "worms", or "chunks", or "slices", or even sequences of such. This suggests that they can be given a *spatiotemporal* criterion of identity.

It is clear that neither same place nor same time alone gives more than a necessary condition for identity. There can clearly be events that happen at the same place at different times, and there can be simultaneous events occupying different places. But jointly these factors give an identity criterion for two singular events e_1 and e_2:

(PT) $e_1 = e_2$ if and only if e_1 and e_2 occur
          at the same time and at the same place.

This criterion may seem either too obviously true or too obviously false. It seems obviously false for conceptualized events (cf., e.g., arm-movement and arm-raising or the ball warming example below).

In the case of nonconceptualized events it again may seem so obviously true as to be unilluminating. Still, for instance, Davidson seems sceptical about (PT). Let us consider why. First, we have to make an agreement concerning location. We take as the location of an event the location of the smallest material object a change in which is identical with the event in question. (Otherwise it can be argued on the basis of the Frege–Quine–Davidson theorem that every simultaneous event takes place everywhere; cf. Davidson (1969).) Given this convention for location (accepted by Davidson) we consider Davidson's example: "If a metal ball becomes warmer during a certain minute, and during the same minute rotates through 35 degrees, must we say these are the same event?" (Davidson (1969), p. 230). Davidson himself suggests the possibility that the warming of the ball during this minute is identical with the sum of the motions of the particles that constitute the ball during this time; and let us construe rotation similarly. This (or something like this) must indeed be said here. But it should be emphasized that when making statements (like the above) about this situation we of course cannot help introducing and using conceptualizations of the situation. This is what makes us think that two events took place. But, I claim that we are here speaking of the same event which, however, was conceptualized in two different ways.

One trouble with (PT) is that the location of mental events is problematic. In the case of wantings, believings, etc. the person presumably is the substance which determines the location (cf. our convention). But how about, e. g., sensations and impressions? What about aches in phantom limbs? It seems to me that the *adverbial* account of these cases works best, and hence the location again will be the whole person, irrespective of how unilluminating that may sound (cf. Chapter 4).

To sum up, we have not found damaging counterexamples against (PT). It seems acceptable as an epistemological criterion for (unstructured) events. I do not think, however, that (PT) qualifies as a meaning analysis of identity statements, for that in effect would require going through all the descriptive information there is in principle to be had about the entities in question.

How is the meaning of identity statements then to be analyzed, if neither (K), (K'), (D') nor (PT) qualify in this task? What comes closest to a meaning criterion of identity is simply Leibniz's criterion, of course:

(L) $e_1 = e_2$ if and only if $(F)[e_1$ has F if and only if e_2 has $F]$.

In (L) F ranges over properties. It should be emphasized that the events e_1 and e_2 here should be taken as nonstructured events (in my sense). Otherwise, difficulties connected to the independent semantical ascribability of properties to e_1 and e_2 will arise.

As we have indicated there is something important in Kim's type of account. However, as we have seen, it connects singular events to generic events somewhat "too closely" (no matter whether (K) or (K') is accepted). We prefer to approach the matter as follows.

Singular events' having structure can be regarded as a matter of one's pragmatic interests. Consider thus, for instance, a monadic singular event $e = [\langle a, t \rangle, P]$. Here a represents the "locus" and t the time of occurrence of e. P can be termed the aspect property of the event e. It should be emphasized however that P should be taken to represent the aspect from which we, so to speak, view the particular in question. It reflects and represents our pragmatic interests rather than being an a priori *essential* property of the event in something like the sense of traditional essentialism.

There is no prefixed set of aspect properties for a given singular event. In principle it can be viewed from any aspect. Thus the structured event $e = [\langle a, t \rangle, P]$ can be taken to be just an "unstructured" spatiotemporal event viewed from the aspect P. In other words 'e' and 'e as a P' represent one and the same event. (Any event can have many names and descriptions, of course.)

The aspect properties are universals. We often speak about types as well. Thus, we shall frequently speak of singular actions u as X'ings, where X is an action type. In this book we shall not make any strong assumptions about the precise metaphysical status and nature of properties and types (and other universals).

In our language (which may be a first order extensional language or a more complex one) we may pick out a singular event e by singular terms. Perhaps the event has a proper name (individual constant) 'e' by which the event can be registered. Below we shall in fact for convenience assume that for every singular event e we pick out for consideration there is available a singular term 'e' suitable for this referring task. Because of this correspondence it does not matter much whether we speak in the material mode or the linguistic mode about the event. We shall prefer the former way of speaking unless confusions are to be expected. Following common practice we will then not pay much attention to this difference, which of course is important in principle.

Similarly, as with singular events, let us assume that the predicate 'P' *designates* the aspect property P. While 'e' suffices to pick out e, it does not in itself contain any *descriptive* information about e. However, a statement of the kind '$P(e)$' (an "aspect statement" for e) does give descriptive information about e, viz. it says that e has P. But then it also describes e as a P, i.e. as a specified structured event.

In general, then, we have the connection between a singular unstructured event, say e, and a corresponding structured event that the unstructured e *is*, as a structured event, *e as a P* (if P is the aspect property of e); or, alternatively, e is *e under a certain description D*. Here D could be just $P(e)$.

In our language we thus have, among other things, the following semantic devices for "speaking about" events. First, we have singular terms (proper names and definite descriptions) referring to singular events. We also have singular variables running over singular events. Next, we have aspect statements such as '$P(e)$'. Such a statement represents a *fact* (e.g., the occurrence of an event) just in case '$(Ex)(x = e)$' is true, viz. if and only if 'e' indeed refers to an existing entity e. (That the existential statement captures the occurrence of e may not be beyond dispute, but here we cannot pause to discuss it.) In addition, there are of course an endless amount of other descriptive statements, both true and false, about e.

Redescribing a structured singular event does not present any special semantical problem for us. It just amounts to using different new aspect statements and other statements containing new predicates such that these statements are true of the event.

Can we say anything about *generic* events within our framework? I think we can, and even in several different ways. First, *open sentences* like '$P(x)$', where x is an individual variable, represent *generic events* or *event types*, e.g., "arm-raising". When x is replaced by a name of a singular event we have a statement describing a singular event. (I have here, as well as in the case of singular events, omitted the explicit linguistic representation of locus and time. In any case, they may be represented by variables in a rather obvious way, if needed.) Secondly, *existential statements* of the form '$(Ex)P(x)$' represent generic events in another sense, viz. they say that a singular event exemplifying the generic event $P(x)$ exists. Thus, the existential statement '(Ex) Buttered (Jones, toast, x)' represents the non-pure generic event of Jones' buttering the toast (see Section 3). Similarly, the existential statement '(Ex) Buttered (A, p, x)' represents the pure generic event of buttering, provided A and p are variables. Furthermore, we may also quantify over predicates (if our language is a second-order one) and in that way discuss, e.g., the existence of generic events and relationships between them.

To summarize our view on event identities, the meaning of identity statements is given by Leibniz's criterion (L); (D') and (PT) can be taken to give epistemic grounds (at least an important part of them) for such identity statements. Criterion (K) will also prove to be useful in determining the truth of identity statements (especially in the case of establishing non-identity).

Let us once more consider Goldman's criterion of identity, according to which two singular actions are said to be identical (in the above sense) if and only if they involve the same agent, the same action-property, and the same time. One may now think of the following kind of counter-example to this. An agent might sign a check simultaneously with both of his hands. Wouldn't there then be two different singular acts of signing? In answer to this, one might have to require that the specific *manner* or *style* of doing the action has to be taken into account. Then I would think the criterion has a better chance of being an acceptable epistemological identity criterion, though not yet a meaning criterion, in the sense we have been discussing. But there is another use of Goldman's original, un-amended criterion. It seems to be an acceptable *semantic* criterion for the notion of *exemplifying the same action type*. (For that notion the manner of exemplification is irrelevant.) We shall later in effect accept this suggestion. One qualification is needed, however, and it is that the singular actions, whose identity is in question, be action tokens$_1$ in our later technical sense (see Chapter 10).

Now we can finally return to the switch flipping example of the last section. A fine grain theorist, like Goldman, claims that the following events are all different:

$a = $ A's flipping the switch
$b = $ A's turning on the light
$c = $ A's illuminating the room
$d = $ A's alerting the prowler.

Presumably the same would hold for the rest of our list:

$e = $ A's hurting his finger
$f = $ A's exploding a bomb

In other words, for a fine grain theorist $a \neq b \neq c \neq d \neq e \neq f$. These non-identities are primarily due to the non-identity of the constitutive properties of these events.

A radical unifier would presumably say here that all the actions are identical, i.e. $a = b = c = d = e = f$. How should we react to these different views on the basis of our discussion in this section?

First, we notice that a fine grain theorist is obviously making his claims with respect to structured events, whereas a unifier has in mind something like our unstructured events. A unifier would say that the multiplicity concerns the descriptions or conceptualizations of actions, not the actions themselves. This is what we are willing to maintain, too. To clarify our own position a few additional remarks are still needed.

As we have emphasized and argued above, we regard actions as events, viz. as events brought about by the agent in question. To anticipate our later discussion let us point out that we take actions to be complex events, i.e. sequences or processes. In acting an agent causally brings about, by his effective intendings (or tryings), whatever behavior is needed to satisfy the purpose expressed in the agent's intending. Combining this with our achievement-view of action, we get as the "structure" of an intentionally performed action token u this: $u = \langle t, \ldots, b, \ldots, r \rangle$, where u is regarded as an unstructured event, t is an effective intending, b is the bodily behavior involved in the action, and r is the result of the action. (See Chapter 10 for more details.)

Now, clearly, two action tokens cannot be identical if their results (taken to be events in the spatiotemporal order) differ. Let us now look at our switch flipping example. As I construe the events in that example, *all* the results r_j, $j = a, b, d, e, f$, are different, except that we may perhaps take $r_b = r_c$. To be a little more specific, I am claiming, for instance, that the result r_a of the switch being flipped is an event which is nonidentical with $r_b =$ the light going on, and so on.

This position yields a moderately fine grained view, even if it starts from considerations and assumptions almost opposite to Goldman's. It is easy to see how this view relates to the radical unifiers' view. Consider, for example, the action of A's alerting the prowler, which on my construal involves the result event r_d of the prowler's becoming alerted. Now, the bodily action involved in A's alerting the prowler names an action that a radical unifier presumably is speaking about. This bodily action is identical with the corresponding bodily actions involved in a, b, c, e, and f (see Chapter 10 for related discussion).

What was just said naturally fits well together with our criteria of identity for events. In examples like this it is primarily (D') and (PT) that are put into use, and nonidentities of action-events are usually determined on the basis of their result-events.

3. ACTIONS AND ACTION STATEMENTS

In order to get a better grip on action statements (and on action-explanations), we shall now introduce a formal (or semiformal) framework. Our main purpose here is not to give a theory of the logical form of psychological statements or anything like that but rather to introduce a framework for the logical characterization of psychological statements within which philosophical problems can better be discussed.

We start by considering the following action statement about Jones' buttering a piece of toast which has often been discussed in the literature (cf. Davidson (1966)):

(*1*)   Jones buttered the toast slowly, deliberately, in the bathroom, with a knife, at midnight.

One may say that the task of formalizing action statements is to exhibit their logical structure. Furthermore, one may require that this be done so that it is shown how the meanings of action sentences depend on their structure and indeed so that Tarski's semantical theory of truth be respected (cf. Davidson (1966) and (1975)).

If one accepts a strict "bringing about"-analysis of actions such as Chisholm (1966) or von Wright (1968) and (1971) one might think of the formalization of action statements as follows. An action consists of an agent's bringing about an event (or state). To analyze this we, for simplicity, first consider the statement

(*2*)   Jones buttered the toast.

Here, the agent Jones is then said to have brought about the state of affairs that the toast is buttered. Speaking in logico-linguistic terms, on analogy with the case when we represent intentional attitudes (e.g. belief) by sentential operators, we may now think of the agent functioning as a sentential operator, which operates on the sentence describing the result of the action. More exactly, following Pörn (1974), we may represent (*2*) by a formal statement '$E_A p$', which reads 'A brought it about that p'. The bringing about-operator E_A can still be analyzed further:

$$E_A p = D_A p \ \& \ C'_A p$$

where '$D_A p$' reads 'it is necessary for something that A does that p' and '$C'_A p$' reads 'but for A's action it might not be that p'. '$D_A p$' thus stands for the necessity aspect of action according to which an action is necessary for its result (see von Wright (1968)). But $C'_A p$ does not guarantee the satisfaction of p, which it should. (This may be called the sufficiency requirement.) But we are not interested here in patching such flaws but rather in the general feasibility of this type of approach. For there are many difficulties attached to this approach to formalizing action statements. I shall now briefly take up some of the most troublesome of them (cf. Davidson (1966)). The problems to be mentioned below seem to me to apply to the entire approach and not only to some particular version of it.

In order to get the agency aspect correct, (2) must be rendered not, e.g., as 'Jones brought it about that the toast was buttered' but rather as

(3)   Jones brought it about that Jones buttered the toast.

But now we can see that no philosophical illumination and hardly any logical gain is to be obtained from such an analysis. First, what does (3) tell us about agency (and about bringing about) that the contained sentence 'Jones buttered the toast' does not? Does the difference between these two statements represent something? If it does, it has to be clarified so that the "standard" objections to agent causation do not apply (cf. Sellars (1969a)). If it does not, it is hard to see how (3) clarifies the meaning of 'Jones buttered the toast'. Naturally, a logical analysis of action sentences should somehow clarify their meaning.

Secondly, the above kind of "bringing about"-analysis at least does not show how the meanings of action statements depend on the meanings of their contained parts. This is seen for instance from the fact that this analysis gives no account of the so called *variable polyadicity* problem (cf. Kenny (1965)).

Consider again statement (1) and compare it with (2). The approach we are criticizing gives no solution to the problem of how to handle polyadicity of action verbs. (2) is represented in the bringing about-theory by means of the two-place predicate 'Buttered (Jones, the toast)'. But in order to account for the fact that Jones buttered his toast in the bathroom a new argument apparently has to be introduced into the predicate, and so on. So it would seem that there is no end to adding new arguments and hence to increasing the polyadicity of the action predicate. Thus, the logical form of action statements seems to be strongly affected by rather arbitrary additions. One problem that follows from this is that it does not seem possible to account for such obvious logical inferences as that (1) entails (2).

There are also difficulties connected to problems of action identity and the modification of action verbs, which affect the above kind of approach, although we shall not discuss them now.

In view of our criticisms it seems that we are well motivated to look for a completely different type of approach. We thus turn to an interesting approach advocated by Davidson (see, e.g., Davidson (1966)).

Let us consider statement (2). Does it describe a singular action? It does not in the sense of picking out one particular action. What Jones did in the bathroom that night (and what is more fully described by (1)) is something that makes (2) true, to be sure. But still (2) does not describe only that particular action, for it is satisfied by all singular butterings of the toast by Jones. Thus (2) seems to describe something like a generic (or partially generic) event, but not a singular event. It should be noticed that if, as a matter of fact, Jones buttered the toast only once in his life and if hence (2) is, as a matter of fact, satisfied by only one singular event, this does not change the semantic (or logical) situation. As to its logical form, (2) is *existential*, and satisfiable *in principle* by several singular events.

Now Davidson, having made this observation, makes an interesting suggestion for exhibiting the logical form, or "deep structure", of (2). Its philosophically interesting content is that the ontology of statements should be made explicit in the formalization. In this case the logical form of (2) becomes:

(4)   (Ex) Buttered (Jones, the toast, x).

Thus the idea is to consider the action verb 'butter' as a three place predicate rather than as a dyadic one.[2] The symbol 'x' is a variable which runs over singular events (actions). This variable can be replaced by singular terms naming singular events. In fact, the standard semantical device we have for referring to singular events is just singular terms. Whether or not a statement not containing such singular terms succeeds in picking out a singular event will always be a contingent matter.

Singular actions are thus to be named by singular terms such as proper names or definite descriptions (see Chomsky (1970) for argument that nominalizations occur in linguistic deep structures). For instance, nominalizations such as 'Jones' buttering the toast' can be construed as singular terms by the stipulation that 'Jones' buttering the toast' names just that singular action we have been speaking about.

It should be emphasized that Davidson's idea of exhibiting singular ontological entities in the formalization does not apply only to events. It can be applied to any kind of entities that one's language treats as particulars. Thus, it applies to singular states (also those not included here in the class of events), situations and processes. Thus, whenever singular entities are somehow considered philosophically and methodologically important Davidson's approach can be applied. As we remarked in the previous section, this technique has proved to be interesting and even necessary, for instance, in the treatment of causation and determination (cf., e.g., Davidson (1967), Berofsky (1971), Tuomela (1974b)). Furthermore, it seems to show new interesting applications for the notion of quantificational *depth* when the quantifiers range over, possibly complex, singular events. (Cf. the philosophical applications in Hintikka (1973) and the methodological importance given to the notion of depth in Tuomela (1973a) and Chapter 4 below.)

Let us now return to see how Davidson's formalization works in the case of the more complex action statement (*1*). Davidson omits from his treatment such locutions as 'slowly' and 'deliberately'. The phrase 'slowly' is omitted in his analysis mainly as it does not introduce a new entity in the formalization and as it is not only a problem for action sentences but for many other kinds of discourse as well (cf. 'Jones is a good actor'). Phrases like 'deliberately' (or 'intentionally', etc.) are omitted by Davidson because they are tied to the agent rather than to the action.[3]

Let us also forget about the phrase 'with a knife' in (*1*) for a moment. With all these omissions we are left with

(5)   Jones buttered the toast in the bathroom, at midnight.

I take it that Davidson's formalization (formal translation) of (5) would be:

(6)   (*Ex*) Buttered (Jones, the toast, x) & In (the bathroom, x) & At (midnight, x).

Verbally (6) reads: There is a singular event such that this event is a buttering of the toast by Jones and such that this event occurs in the bathroom and at midnight.

The Davidsonian way of formalizing action statements may more generally be represented by

(7)   (*Ex*) Verbed (Agent, Object, x) & T(Object, x).

This represents a simplified case, however, for in general the predicate 'Verbed' may involve several agents and objects, but that does not affect the logical and philosophical situation, it seems. The predicate T is assumed to take care of the place (in the bathroom) and time (at midnight) of the event as well as all the other relevant attributes of the event (e.g., the instrument). I have simplified the account here in that there should in fact be several conjuncts with predicates like T and with different objects (cf. statement (6)). But our simplification does not affect the central theoretical issues involved.

It is easy to see that, given the formalizations (6) of (1) and (4) of (2), statement (1) entails statement (2). Notice also that the formalizations have been performed by means of standard first-order logic only. Davidson's approach is in fact a continuation of the programme for formalizing natural language advocated (in a sense) by Frege, Tarski, and Reichenbach.

Let us here still mention two more illustrations of the Davidsonian type of formalization. We first consider

(8)   Shem kicked Shaun before Shaun kicked Shem.

We can understand the problematic temporal 'before' as a predicate between singular events and get this as the logical form:

(9)   (Ex)(Ey) (Kicked (Shem, Shaun, x) &
         Kicked (Shaun, Shem, y) & Before (x, y)).

For another illustration of the usefulness and flexibility of this approach, consider the following "numerical" sentence:

(10)   Shem kicked Shaun at least twice.

The logical form of (10) can be taken as given by:

(11)   (Ex)(Ey) (Kicked (Shem, Shaun, x) &
         Kicked (Shem, Shaun, y) & x ≠ y).

It is again easy to see how (11) explicitly entails that Shem kicked Shaun at least once. This fact and such related facts as that (1) trivially entails (2) within Davidson's approach can be taken to show something important. Namely, this indicates that the approach indeed is able to recursively account for the meanings of sentences on the basis of their parts. In other words, in the manner of Tarski's well known Convention T ('T' for 'true' here) inferential relationships between sentences are accounted for recursively on the basis of the truth conditions of these sentences. An adequate theory of truth for a natural language must, according to Convention T, be stated using only a *finite* number of non-logical axioms and must entail all sentences (roughly) of the form '"Grass is green" is true if and only if grass is green'.[4]

Giving an account of the conditions under which a sentence is true, i.e., giving the truth conditions of the sentence, can be taken to be intimately related to giving an account of the meaning of the sentence. Davidson has even gone so far as to claim that to give an analysis of the meaning of a sentence *is* to give its truth conditions in the Tarskian sense.

Davidson's programme then can be summarized by saying the following. The meaning of a statement in a natural language is given by showing how the Tarskian truth conditions of the statement recursively depend on the truth conditions of its atomic parts. But the meanings of atomic sentences are not really accounted for in the Tarski–Davidson theory. It is conceivable that Davidson would accept a meaning account ultimately in terms of truth-independent *picturing* (i.e., language-world picturing) for them. In any case such an account is assumed below, although we shall not further clarify this notion of picturing here (see Sellars (1968) and J. Rosenberg (1974) for accounts of it).

We may still put Davidson's programme in slightly different terms by saying this. The idea is to exhibit the deep structures or (what is taken to be identical) the logical forms of natural language statements so that the inferential relationships between sentences are laid bare (cf. our discussion in Chapter 3). As the inferential relationships are displayed within standard quantification theory, Tarski's Convention T should be accepted. Hence, we get the above mentioned connections to truth conditions (recall that the validity of rules of inference is just a semantical matter of truth). Therefore, too, giving the logical form for a sentence is for Davidson an analysis of the meaning of the sentence.

Below we shall accept Davidson's programme as a kind of working hypothesis, except that we surmise that only *representational* meaning is analyzable thus. There are other richer aspects of meaning which may escape this analysis (cf. meaning in the contexts of communication and speech acts). We shall briefly return to these problems later in Chapter 3.

Even if Davidson's account were to capture all the important meanings of 'meaning', it seems too austere in several other respects. One potential difficulty is connected with the treatment of adverbs and with the related problem of predicate modification. To discuss these we return to the phrases 'slowly' and 'with a knife'. It would seem odd to account for 'slowly' by an additional conjunct 'Slowly (x)' or even 'Slow (x)'. 'Slowly' may thus seem to modify the verb rather than the event spoken about in the statement. Therefore, we would have to deal with something like 'Slowly-buttered' instead of 'Buttered' in the formalization. As to the manner phrase 'with a knife', Davidson's suggestion would presumably be to include it in the (complex) predicate T. But I would rather think of this phrase as modifying the verb. Thus, we would get 'With-a-knife-buttered' by augmentation from 'Buttered'.

Clark (1970) and Sellars (1973a) have criticized Davidson's approach for neglecting verb-modification. They accordingly concentrate heavily on verb-modification. To see what is involved here, let us consider Sellars' approach concerning the logical form of (1). Sellars considers a statement almost identical with (1) (see Sellars (1973a), p. 199). From it we can easily extrapolate that Sellars' (and probably Clark's) formalization of (1) would be:

(12)   [In-the-bathroom [At-midnight [With-a-knife
        [Slowly]]]] Buttered (Jones, the toast).

Before discussing the details of (12), a few general remarks are in order. First, one might even think of going further and formalizing 'the toast' as a modifier of the verb (though that certainly would not be accepted by Sellars). But that depends on what is to be said about verb modification in the "last" analysis. Clark has tried to sketch a theory of verb modification, which we seem to need now (cf. Clark (1970)). We can see that such a theory would be a higher-order theory. Thus the statement $O_n[O_{n-1} \cdots O_1]]]$ Verbed (. . .) belongs prima facie to a $n + 1$'st order logic. It seems to me, however, that such a theory of verb- or predicate-modification can be reduced to second-order logic. (But that may depend on the final form of the theory.) It is, however, important to notice that the above kind of operators can be conceived of as functors which yield as their values *structured* predicates. Thus, I think we may treat, for example, the operator, 'Slowly', simply as follows:

(13)   Slowly [Buttered] = Slowly-buttered.

Here, 'Slowly' is a functor which applied to 'Buttered' yields the first-order predicate, 'Slowly-buttered', as its value.

It should be possible to construe a general theory of predicate modification in the above manner by means of just a finite number of "core" predicates and only a finite number of recursive modifying operators. Then and only then can we have a recursively learnable language (cf. Davidson (1965)). Such a theory should of course account for inferences in the earlier discussed manner. Thus, a statement containing a context . . . slowly-buttered . . . should entail another, otherwise the same, statement, except that the latter contains the replaced context . . . buttered. . . .

For our present purposes we shall still make another assumption or rather a restriction. We assume that a theory-to-be of predicate modification has generated a finite amount of structured predicates which suffice for the purposes of psychological theory formation. Thus below, when we use the placeholder 'Mod-Verbed ( )' for action predicates, we assume that the verb in question is either a core predicate or a modified predicate obtained by means of admissible transformations from core predicates. Thus, we can leave most of the problems involved in verb-modification to linguists and logicians.

Let us now return to the Sellars–Clark formalization (*12*). The first important thing to notice about it is that it is incorrect. Even if some adverbs in the original statement (*1*) might be taken to modify the verb 'Buttered', not all of them do. I think that place and time do not modify the verb but are properties of the singular event itself. Thus 'Slowly' and 'With-the-knife' modify 'Buttered' and do not introduce a new entity in the formalization, whereas 'In-the-bathroom' (place) and 'At-midnight' (time) in a sense do. So we would get (leaving the nature of the verb-modification implicit):

(*14*)   (*Ex*) Mod-Buttered (Jones, the toast, *x*) &
          At (midnight, *x*) & In (bathroom, *x*).

But have we really proved that predicate modification is necessary? I think not. That is very hard to do. The matter is more a question of convenience until we arrive at a broad linguistic theory of logical form and meaning telling us what is right.

If someone has strong nominalistic and anti-essentialist intuitions it is very hard to convince him of the necessity of predicate modification. It should be kept in mind that the event variable *x* can in principle be taken to refer to the whole behavior situation, i.e., a complex event including more than a bodily movement.

One of the best pair of sentences for showing the need for (at least) predicate modification would be of the following type:

(*15*)   Clumsily, John kicked the ball.

(*16*)   John clumsily kicked the ball.

One might claim that on a certain occasion (*15*) could be truly uttered, whereas (*16*) would be false (the kick itself was perfect). But, if this is granted, perhaps we must include both 'clumsily$_1$' and 'clumsily$_2$' in our deep structures. However, that seems to make deep structures (logical forms) strongly dependent on linguistic (and perhaps extralinguistic) context.[5]

Another example apparently showing the need for predicate modification is given by:

(*17*)   John moved quickly but raced slowly.

John might be a runner left last in the final heat of 100 m dash at the Olympic games. A mechanical application of Davidsonian formalization might seem to yield (after rather obvious steps):

(*18*)   (*Ex*) (Moves (John, *x*) & Quick (*x*) & Races (John, *x*) & Not-Quick (*x*)).

Of course, (*18*) is a logically inconsistent sentence. However, we can easily see that 'quickly' and 'slowly' in (*17*) implicitly contain *different* standards of reference. Thus, 'quickly' should presumably be taken relative to (something like) the class of human beings and 'slowly' relative to the class of competitors in the final heat. Making these relativizations explicit obviously blocks the inference to (*18*). Still we may ask, as above, should we make deep structures dependent on background information such as comparison standards for the adverbs?

The list of examples of the above kind can be continued and similar remarks and counterremarks can be made. In this book we shall not discuss this problem further, but allow for the possibility of predicate modification as sketched earlier. Notice, that our way of treating the situation in no way destroys the general and most basic features of the Frege–Davidson programme; it essentially involves only a new selection of basic predicates.

For the time being it is too early to evaluate the possibilities and (final) adequacy of the Frege–Davidson programme. Theoretically viewed, linguistic theory is still, in spite of the wealth of data produced, almost in its infancy. (For a good, recent discussion of the status of the Frege–Davidson programme see Harman (1972)).[6]

We shall end this chapter by noticing that the modern approaches to intensional logic and the Davidsonian approach are not formally very far from each other (even if they may differ interpretationally as to their ontologies). For one thing, Cresswell (1974) has shown that Davidson's approach can be embedded in Montague's approach under one condition. The condition is that two distinct events never occur in exactly the same set of possible worlds. However, for this condition to be maintained for singular events such as our unstructured events, the set of possible worlds would become extremely large, since it would have to correspond to all the possible ways of conceptualizing substratum events. The situation thus depends very much on what one accepts as one's set of possible worlds.

One can also approach the same matter from a slightly different angle. (Here I am indebted to Prof. Zoltan Domotor.) In Kripke–Montague-type logic one usually interprets an n-place predicate P by an interpretation function (I) as follows, using ordinary set-theoretic notation:

(a) $I(P) : D^n \rightarrow 2^S$

Here D is a set representing the domain of interpretation, $2 = \{0, 1\}$ represents the truth values, and S represents one's set of possible worlds, situations, events, or indices (or whatever one calls them). Finally, 2^S stands for the set of functions from S to 2. Now, in Artificial Intelligence S is usually taken to be the set of situations and the interpretation of the predicate P is given by:

(b) $I(P) : S \rightarrow (D^n \rightarrow 2)$.

Within Davidson's approach the interpretation clearly becomes this:

(c) $I(P) : S \times D^n \rightarrow 2$.

But it is easily seen that the interpretations (a), (b), and (c) are set-theoretically equivalent as long as just ordinary functions are allowed.[7] Furthermore, they remain equivalent under homomorphic mappings if instead of S we use a structure $\langle S, R \rangle$, where R is an "alternativeness" relation (partial ordering) familiar from modal logic.

Thus, purely formally, Davidson's approach is embeddable into the "possible worlds" approach. For a more general argument that corresponding to any intensional language there exists an extensional (perhaps somewhat artificial looking) language see Parsons (1970) and especially

Lewis (1974). This extensional language is "Ramsey-equivalent" (i.e., after quantifying existentially over metalinguistic semantic talk) with the intensional language. A price to be paid for extensionality is a somewhat complicated ontology. Formally speaking, one may, however, always "go extensional" it seems; the extensional language in question being in the general case a suitable predicate logic, which *most comfortably* (although perhaps not logically necessarily) is allowed to quantify over at least two types of entities and hence to be at least second-order (cf. Tuomela (1973a) on the scope of first-order formalization).

NOTES

[1]   One of the best analyses of the "bringing-about" aspect of action is in von Wright (1968). In this work the following four elementary action types are distinguished: 1) producing a change (or result), 2) preventing a change, 3) forbearing to produce a change, 4) forbearing to prevent a change. However, as we shall accept and operate with a somewhat different notion of action than that of von Wright, we shall not here go into the details of his theory.
    Let us remark here that we shall in this book concentrate on "positive" actions and thus leave out such "negative" actions as forbearances and negligences. In principle, however, our treatment covers them as well. Basically, such negative actions are to be analyzed in terms of suitable negated action descriptions. In our ontology, however, we do not have any *properly* negative actions any more than other negative events.
[2]   One can handle the tenses of verbs either as in (4) or one can, alternatively, restrict oneself to present tenses and add in contexts like (4) that x is past.
[3]   Such adverbs as 'intentional' can be formalized by using the locution 'It is intentional of A that p', where 'A' names the agent and p says what the agent did (cf. Davidson (1966)). It is essential that in p we use a name of A under which he recognizes himself (e.g. we may use 'A' or a pronoun). Thus we may in our logic use a *sentential* intentionality operator '$I_A^*(p)$' which reads 'It is intentional of A that p' to account for the intentionality of the action. Thus, if we in our example, formalized by (6) below, want to add that Jones performed his action intentionally, we just add the conjunct '$I_{Jones}^*(p)$'. Of course this logical method is not meant to be a *philosophical* solution of intentionality at all (cf. Chapters 3, 7, and 10).
    Another, and perhaps more satisfactory way of formalizing that-clauses in contexts like 'It is intentional of A that p', 'A intends that p', 'A believes that p' is provided by Harman's operator '$\doteq$' (see Harman (1972)). This is an operator that converts any logical part of a logical structure (or deep structure or "proposition") into the corresponding part of the *structural* name of the proposition in question. In other words, $\doteq$ will grind out the syntactical deep structure when operated with on a surface sentence. E.g. '$\doteq$' applied to '$p \ \& \ q$' gives the logical form '$\doteq p \doteq \& \doteq q$'. As this operator clearly changes the logical properties of whatever it applies to it can be used to clarify issues such as the opacity of belief-contexts. We shall later occasionally employ "dot-quotes", originally introduced by Sellars, basically in the same role as Harman uses his $\doteq$.
[4]   We shall understand Convention T in a strong sense here. Thus we assume that it, in particular, entails all the appropriate instances of 'x is true if and only if p' where 'x' is replaced by a name *descriptive of the logical form* of a sentence s of the object language

and '*p*' is replaced by *s* itself. (This version of Convention T escapes the counterexample proposed in Hintikka (1975).)

⁵ Parsons (1970) gives a somewhat similar example for the necessity of predicate modification. Consider

    (a) John painstakingly wrote illegibly.
    (b) John wrote painstakingly and illegibly.

Here it may seem clear that 'painstakingly' modifies the illegible writing in (a), whereas in (b) it only modifies the writing. But somebody with strong nominalistic intuitions may object.

⁶ Harman (1972) discusses various approaches to formalizing natural language statements and finding their logical forms. He defends a somewhat strict, nominalistic, version of what we have called the Frege–Davidson programme against, e.g., various operator-approaches.

Harman proposes that the following five criteria of adequacy should be imposed on any theory of logical form: 1) A theory of logical form must assign forms to sentences in a way that permits a finite theory of truth for the language (in our above sense). A theory of logical form should minimize 2) the number of new rules of logic (i.e., standard quantification theory) and 3) the number of new logical and nonlogical axioms. 4) A theory of logical form should avoid unnecessary ontological commitments and 5) it should be compatible with syntax.

Even if we subscribe to these principles, we have above argued for the need of predicate modification, contrary to Harman. At the present state of art there is relatively much room for disagreement within the scope of the above five conditions of adequacy.

⁷ To give an example, consider again Jones' buttering the toast. To keep consistent with the above treatment we now, instead of (*4*), consider the translation '(Ex) Butters $(x$, Jones, the toast)', and we consider the interpretation given by functions of kind

$$(*)\qquad I \text{ (Butters)}: S \times D_1 \times D_2 \to 2,$$

where S is the event domain, D_1 the domain of agents (including, e.g., Jones) and D_2 the domain of objects (including, e.g., the piece of toast). Our (*) is an instance of (c), which is in one-one correspondence with (b). (b) again is merely a notational variant of (a). This reduces the present example of Davidsonian formalization to Kripke–Montague semantics.

CHAPTER 3

MENTAL EPISODES

1. THE STREAM OF CONSCIOUSNESS AND THE MYTH OF THE GIVEN

One of our main concerns in this book is a study of the intention-belief-action triangle of concepts or, perhaps better, actions and their (causal) determination by activated conative and doxastic states. Therefore, we obviously have to discuss the nature of conative and doxastic attitudes at some length. We shall see below that these concepts can be construed as *theoretico-reportive* concepts so that there will be conceptual connections between attitude concepts and action concepts. What is more, it will be argued that in the case of wants and intentions it is a *conceptual* truth that they are, among other things, states *causing* behavior in suitable circumstances.

It must be added immediately, however, that only *activated* or *effective* states of wanting or intending can "productively" cause behavior (e.g., actions). The activation of a wanting or an intending involves something *episodic* (e.g., an event or a process) to take place in the agent (independently of whether this episode is a direct causal effect of something external, as is often the case in perception, or not). To account for such episodic mental events, we must discuss the agent's "stream of consciousness" (and, in fact, any *actual* mental events and processes, conscious or not). The main task of this chapter is to construe the stream of consciousness in a kind of behaviorist-flavored fashion, which is meant to avoid the Scylla of logical behaviorism (and other extreme reductionist forms of behaviorism) and the Charybdis of Cartesian mentalism.

We divide mental episodes (short-term states, events, and processes) into "*raw feels*" and *representational* episodes. Among raw feels we include, for example, impressions, sensations and feelings, construed as non-representational (or non-conceptual). A visual impression of a brown table, for instance, can be construed as a raw feel episode. Roughly, it is a non-conceptual episode taking place in a person when he sees a brown table, or when it looks to him that over there is a brown table. This raw feel is something in the natural order, which can be taken to explain the person's conceptual episode of seeing a brown table (or its looking to the person that there is a brown table). Another example of a raw feel would be the feeling of pain, which is to be distinguished from a person's representational belief or awareness that he has that pain.

It is important to make the distinction between raw feels and representational states. A failure to make it has brought about much confusion not only in contemporary discussion but also in the case of many classics of philosophy (notably the British empiricists). This was in effect pointed out clearly already by Green when, discussing Locke and Hume, he spoke about "the fundamental confusion, on which all empirical psychology rests, between two essentially distinct questions — one metaphysical, What is the simplest element of knowledge? the other physiological, What are the conditions in the individual human organism in virtue of which it becomes a vehicle of knowledge?" (Green (1885), p. 19). We shall not now pause to argue for the necessity of making this distinction (see Sellars (1956) and Rorty (1976) for excellent discussions of that).

Let it be noted here, too, that the problems that raw feels and representational episodes, respectively, pose for mind-body theories are quite different accordingly, although we shall not in this book be deeply concerned with the mind-body problem. Our focus will be on representational episodes such as wantings, intendings, believings, and perceivings. No specific philosophical problems related to raw feels will be discussed below.

In Sellars (1956) a certain kind of functionalist account of mental episodes is sketched in terms of a myth about our "Rylean ancestors". This view, which might be termed *conceptual functionalism* or, perhaps, *conceptual behaviorism*, will give a suitable starting point for our developments. Before going to Sellars' account, which also is called an *analogy theory*, a few preparatory remarks are in order.

According to our view concepts standing for inner mental episodes are to be construed as a kind of theoretical concepts. At first the idea that mental concepts are theoretical may sound odd. How can my toothache or my desire be theoretical? They may seem fully "introspectable" and in that wide sense "observable" by the "inner eye". This is erroneous, however. There are several good reasons for regarding mental concepts as theoretical concepts. (This, of course, depends on how theoretical concepts are characterized.) A more comprehensive discussion of these reasons will be left till next chapter. Here we shall concentrate on the most basic philosophical reason. It is the rejection of the empiricist "Myth of the Given" and the adoption of a realist view instead.

What is the Myth of the Given? It must perhaps be said immediately that it is a myth that there is only one specific Myth of the Given. For instance, several versions of this myth can be extracted alone from Sellars (1956); cf. Cornman (1972). On a general level we can, however, say that the Myth of the Given claims that all factual knowledge and the meanings of factual predicates are to be solidly and incorrigibly based on the information given to the senses. There is thus an "analytic", *ostensive* tie between our fundamental descriptive vocabulary and the extralinguistic world (physical objects and attributes or extralinguistic sense contents). Thus, observational terms (especially) get their meanings entirely from their causal or other strictly fixed ties to either external physical entities or, usually alternatively, to extraconceptual "sense data" or sense contents (often taken to be causally brought about by the former). Observationality is thus a *logical* (or semantical) feature of observation terms in this empiricist construal based on the Myth of the Given. Theoretical concepts then come to have an auxiliary status in a clearly instrumentalistic sense. (Cf. especially Sellars (1956) for this aspect of the Myth of the Given and see, e.g., Tuomela (1973a) for a discussion and survey of the empiricists' attempts to establish an observational-theoretical dichotomy.)

Another important idea involved in the Myth of the Given is that "given" entities such as sense data and especially thoughts (these two are in fact conflated in the doctrine) are regarded as extraconceptual entities experiencible independently of language and of learning. That is, such givens are really ghosts in the machine in something like a Cartesian sense. (See Rorty (1976) for a lucid discussion of this.)

The Myth of the Given (in both of its above versions) must be rejected, however. The notion of an immediately known particular, e.g., a sense datum, is an incoherent one. For, first, knowing a particular can only mean knowing some fact about this particular. But as knowledge of a fact requires a *justification* of one's belief in the fact, one cannot know anything without presupposing some prior knowledge.

The rejection of the Myth of the Given amounts, by and large, to the recognition that one does not come to have a concept of something because one has noticed that kind of thing but rather it is just the other way round. This leads to an epistemology without a firm "rock bottom" basis. All knowledge presupposes some "*Vorverständnis*". Thus, even observations are realized to be "theory-laden" (with respect to some prior "theory").

What interests us especially here is that the rejection of the Myth of the Given, and thus of the idea of an ostensively given conceptual framework with absolute authenticity, makes possible a realistic construal of inner mental episodes as theoretical episodes. Accordingly, it also becomes possible for the corresponding theoretical concepts to be used "evidentially", i.e., in direct "reporting", which role of course was reserved solely for observational concepts within the empiricist construal. What else is involved in regarding mental concepts as theoretical will be discussed later.

2. Sellars' Analogy Account of Mental Episodes

Let us now proceed to a discussion of Sellars' instructive analogy account of inner mental episodes, even if we have not yet perhaps given fully convincing reasons for the need of such episodes. Our, as well as Sellars', basic reason will be the *explanatory* function of these episodes, and we shall illustrate it below and especially in the next chapter. Generally, we can say here that inner mental episodes such as perceivings and desirings may, and "normally" do, *cause* overt behavior. This general idea can be employed in the explanation of the fact that people are capable of certain kinds of intelligent behavior even when no public verbal productions accompany such behavior.

We need an account of mental concepts which does justice to this explanatory idea. In Sellars' account inner episodes are introduced in the manner of an analogy: just as the observable behavior of gases is micro-explained by the kinetic theory, so is overt behavior to be explained in terms of inner episodes.

Sellars divides mental events and processes into conceptual ones (e.g., thinkings, wantings, believings, perceivings) and non-conceptual ones (e.g., sensations and impressions). (This is essentially the same distinction as our raw feel — representational episode dichotomy discussed above.) We shall concentrate here on the conceptual ones and often use the term 'thought' to generically cover all of them.

It should be emphasized that thoughts thus understood are *actualities*, some goings on in the person's "mind" (in the first place, though they ultimately will presumably turn out to be propositional brain processes). Even Ryle would admit, I think, that when a person's propensity to behave changes, we can say that a change in a state of the person is involved. But what primarily is in question here is the existence and characterization of inner episodes which, no matter how hypothetical-categorically mongrel they are, still admit of some kind of a *purely* categorical description. It is not required, however, that we *now* should be able to give anything but relational, i.e., *functional*, descriptions of these episodes. Furthermore, if functionalism of the general kind that I will be defending is correct, inner mental episodes do not even in principle admit a categorical *psychological* description (apart from our trivially and generally describing them as "episodic").[1]

A functionalist programme of the Sellarsian kind must satisfy two central demands:

(1) "It requires that a form of linguistic behaviour be describable which, though rich enough to serve as a basis for the explicit introduction of the framework of conceptual episodes, does not, as thus described, presuppose any reference, however implicit, to such episodes. In other words, it must be possible to have a conception pertaining to linguistic behaviour which, though adapted to the above purpose, is genuinely independent of concepts pertaining to mental acts, as we actually can conceive of physical objects in a way which is genuinely free of reference to microphysical particles. Otherwise the supposed 'introduction' of the framework would be a sham.

(2) It requires an account of how a framework adopted as an explanatory hypothesis could come to serve as the vehicle of direct or non-inferential self knowledge (apperception)." (Sellars (1968), pp. 71–72.)

These requirements have in fact been motivated above in our discussion. We shall later still consider separately the idea that inner mental episodes should be explanatory. This requirement of explanation can be formulated more stringently and it could in fact be regarded as a third important requirement for a functionalist programme.

In Sellars' functionalism the framework of thinkings is introduced as an *analogical* one the fundamentum of which is meaningful overt speech: thinking is (in a certain analogical sense) internalized silent speech. This is contrary to another traditional view according to which speech is to be construed as overt thinking. This latter view can be associated with such diverse theoreticians as Aristotle, the Port-Royal grammarians, Chisholm, and Katz. The view that language is conceptually primary again goes with people like Plato, Humboldt, Whorf, Wittgenstein, Quine, Geach, and, as mentioned, Sellars.

To summarize, it is important to remember that for Sellars thinking is *ontologically* prior to speech and, indeed, its *cause*. But in the order of conceiving, speech is the fundamentum. Notice, furthermore, that for Sellars to give a conceptual clarification of a piece of linguistic behavior (speech) is to characterize this piece of speech *functionally*, in terms of its linguistic role (broadly understood). Thus, given the analogy account of thinking, to say what a person thinks is to give a functional classification of his thinking and, ultimately, of his linguistic (and other intelligent overt) behavior.

To carry out his programme Sellars starts by viewing a community through the eyes of a "Rylean" behaviorist (cf. Ryle (1949)). The reader is thus invited to think of the following myth. Suppose we had Rylean ancestors, to use Sellars' terminology, who employed a somewhat limited behavioristic language in their communication. The descriptive predicates of the language concern only public properties of public objects located in space and enduring through time. This language permits the standard logical operations, and it also allows subjunctive conditionals. The only way our Rylean ancestors could speak of anything resembling our present notion of thinking was in terms of overt verbal episodes, most centrally "thinkings-out-loud". Similarly, e.g., perceiving could only be approached through "perceiving-out-loud", and so on. No "inner" episodes, conceptual nor non-conceptual, are admitted by this Rylean language.

On the other hand, it should be noticed that *action*-talk (e.g., about speech acts) is permitted. That is, the language does not consider behavior only as mere movements and responses, etc.

Given this Rylean language, Sellars' claim now is that if we add to it the resources of 1) *semantical* discourse and 2) *theoretical* discourse (i.e., discourse concerning unobservable inner episodes), that suffices to enable the language users to acquire the concept of mental episode (event, state, process) as we *now* have it.[2]

Semantical discourse will enable the members of the Rylean community to characterize verbal episodes in semantical terms and thus to give these episodes characteristics similar to the intentional characteristics we presently (in our culture) give to mental episodes. If acceptable, this construal in a sense reduces intentionality or intentional "aboutness" to metalinguistic semantical discourse. (We shall return to this problem.) The addition of a theoretical discourse about inner episodes then gives a new kind of entities, which share the aboutness aspect with verbal episodes but which still are distinct from the latter.

This introduction of inner mental episodes in the mythical community
is due to a certain genius:

"According to our philosophical myth a protoscientific member of the
[Rylean] community, Jones by name, develops the hypothesis that people's
propensities to think-out-loud, now this, now that, change during periods
of silence as they would have changed, if they had, during the interval,
been engaged in a steady stream of thinkings-out-loud of various kinds,
because they are the subjects of imperceptible periods which:

(a) are analogous to thinkings-out-loud;
(b) culminate, in candid speech, in thinkings-out-loud of the kind to
    which they are specifically analogous;
(c) are correlated with the verbal propensities which, when actualized,
    are actualized in such thinkings-out-loud;
(d) occur, that is, not only when one is silent but in candid speech,
    as the initial stage of a process which comes 'into the open' so to
    speak, as overt speech (or as sub-vocal speech), but which can occur
    without this culmination, and does so when we acquire the ability
    to keep our thoughts to ourselves." (Sellars (1968), p. 159.)

Jones then goes on to teach his fellow members to use mental language
(according to his hypothesis) not only in the third person case but also
in the first-person case (see below and Sellars (1956)).

There are several problematic aspects concerning the myth summarized
here. One is the general epistemic status of the myth. Should it be considered an anthropological hypothesis? Does it involve a factual hypothesis
about language learning? I think we can safely give a negative answer to
both of these questions. (This seems to be Sellars' own opinion, too.)
The myth does not have to be taken as a factual hypothesis. We should
rather think of the situation as analogous to contract theories in political
philosophy. We are dealing with conceptual reconstruction here, and
hence it suffices that it is merely conceptually possible that the above
historical and psychological theses are true. This conceptual reconstruction and explanation also contains a promissory note concerning the
explanation of behavior, as we noticed earlier.

A few other features about the Jones myth must be mentioned here as
well. One is that, as it stands, it is compatible both with materialism and
dualism and with any reasonable mind-body theory I can think of. More
generally, a functionalist view, of which the analogy theory is a special
version, does not, without further assumptions, entail materialism, as is

sometimes thought. In the next chapter we shall explicate the analogy account in a way which makes mental talk referential but which still leaves it a "limiting" *logical* possibility that a Jonesian theory or postulate system does not introduce (new) inner entities (e.g., events and processes) over and above the overt ones, even if it does introduce new mental predicates.

Notice that inner Jonesian entities are *inner* entities (in language using beings) roughly in the sense molecular impacts are in gases, not in the sense Rylean "ghosts" are in machines.[3]

Given an analogy account of the above kind one can see how intentional aboutness reduces to metalinguistic semantical discourse. The basic idea is that an item is intentional just in case this item is the kind of thing which makes *reference* to something. Consider, for example, Tom's belief that the earth is flat. This dispositional state may become actualized by an inner episode when we ask Tom whether or not the earth is flat. This inner episode might now, in accordance with Sellars' analogy theory, get an overt manifestation and "culminate, in candid speech", in Tom's thinking-out-loud either spontaneously or deliberately (when it is a saying) that the earth is flat. This can take place, e.g., by his uttering just the sentence 'The earth is flat'. We understand this physical utterance in terms of our semantical discourse. (We perceive it as a sentence, carve it up into words, explain the meanings of words and the structural composition of the sentence, and so on, in the obvious way.)

Mental "acts" or thinkings thus become in our analysis unmystified (from a semantical point of view, at least). Thinkings are relations between persons and linguistic entities and not between persons and some entities enjoying "intentional inexistence" in something like Brentano's sense (see our further discussion in Chapter 7).

It should be pointed out that Sellars' analogy theory emphasizes too much the "output" side, so to speak. For instance, an account of a person's seeing a tree over there has to be analyzed in terms of some suitable input (involving the tree as an ingredient of a state of affairs or event) causing the person's perceiving. We shall correct this bias in the next chapter.

Before going to discuss the fulfilment of Sellars' programme in more detail, we shall first briefly comment on the second criterion of adequacy and the self-reporting use of psychological language referring to inner episodes. There is not much to discuss here as Sellars just takes it to be a contingent fact that people learn to use psychological language on the basis of social rewards and punishments:

"For once our fictitious ancestor, Jones, has developed the theory that overt verbal behavior is the expression of thoughts, and taught his compatriots to make use of the theory in interpreting each other's behaviour, it is but a short step to the use of this language in self-description. Thus when Tom, watching Dick, has behavioural evidence which warrants the use of the sentence (in the language of the theory) 'Dick is thinking "p"' (or 'Dick is thinking that p'), Dick, using the same behavioural evidence, can say, in the language of the theory, 'I am thinking "p"' (or 'I am thinking that p'). And it now turns out — need it have? — that Dick can be trained to give reasonably reliable self-descriptions, using the language of the theory, without having to observe his overt behaviour. Jones brings this about, roughly, by applauding utterances by Dick of 'I am thinking that p' when the behavioral evidence strongly supports the theoretical statement 'Dick is thinking that p' and by frowning on utterances of 'I am thinking that p', when the evidence does not support this theoretical statement. Our ancestors begin to speak of the privileged access each of us has to his own thoughts. *What began as a language with a purely theoretical use has gained a reporting role.*" (Sellars (1956), pp. 188–189.)

As long as the kind of learning spoken about in this quotation is a conceptual possibility and not contradicted by psychological findings Sellars is safe on this point. As a consequence, even self-reporting use becomes as much dependent on social practice and linguistic conventions as one can possibly demand (cf. Wittgenstein, Vygotsky).

3. ANALOGY AND THE LANGUAGE OF THOUGHT

One very problematic aspect about Sellars' functionalist construction (and, indeed, in any analogy account) is involved in the notion of analogy. Sellars does not in his works really give an essentially better characterization of the analogy assumed to exist between thought and speech than what the passage cited earlier gives.

We have to notice first that the analogy does not concern any intrinsic (non-functional) features of thoughts, and hence thoughts are not going to have any intrinsic or contentual psychological features, e.g. introspective qualities, connected to them. But what (structural) features could then be involved in the positive analogy? How does the "language of thought" differ from the "surface language" (potential and actual speech)? This latter distinction immediately brings up the familiar linguistic distinction between "deep structures" and "surface structures" of items of linguistic discourse. It is almost part of our common "knowledge" today that *some* such distinction has to be made in an adequate theory of language.

There are lots of *linguistic* arguments for making such a distinction (cf. e.g., Chomsky (1965), also recall our discussion in Chapter 2). I will not review them except for the few examples that will be taken up below in a related context.

It may be emphasized that there is rather strong *psychological* evidence for the existence of deep structures. (See Fodor, Bever, and Garrett (1974) for an extensive review concerning the psychological reality of deep structures.) Perhaps the strongest evidence comes from studies of the perceived semantical similarity of sentences. Such similarity seems rather well predictable on the base of the similarity of their underlying structural descriptions or deep structures (in the standard Chomskyan (1965) sense).

When developing an analogy theory of thought, which presupposes a language of thought, we immediately notice some disanalogies between the language of thought and overt speech which indicate at least that the distinction between thinking and speaking is not simply a question of being in a "keeping-one's-thoughts-to-oneself state of mind".

One easily noticeable disanalogy comes from *temporal* properties, such as duration of speech. Obviously, thinking takes place much faster than speaking. Thus, thinking is not in any important sense "reciting for oneself". But this disanalogy does not have to disturb an analogy theorist in the least.

We can also see that a great many typically *grammatical* features show disanalogy. For instance, consider these active and passive forms: 'John hit Bill', and 'Bill was hit by John'. They presumably have the same logical structure but different surface structures (see, e.g., Chomsky (1965), Harman (1973) for a relevant discussion). *Indexical* statements (e.g., "It rains today" or "I am fine") are problematic, too, as they may not seem to express propositions. But I must bypass them here, and simply assume that they can be acceptably accounted for (e.g., Lakoff (1975) gives a good beginning).

We also have to consider standard cases of *ambiguity*: 'They are flying planes' is a surface form derivable from at least two different deep structures. An analogous remark holds for 'Bill is standing near a bank'. Notice here that deep structures may contain clearly semantic information in the form of indexed words such as 'bank$_1$' and 'bank$_2$'.

As said, there is much linguistic literature on the topic of deep vs. surface structure (and the related "competence" vs. "performance" distinction). This is not the place to review and comment on the literature, especially when the whole discussion is theoretically still at a very programmatic stage (despite the wealth of examples and data produced). We shall go ahead and boldly assume that, at least as a first approximation, the language of thought amounts to the same thing as the deep structure of language, which again can be identified with the logical form of language. (A weaker claim, sufficient for our purposes, would be that the sentences of the language of thought still be only structurally *analogous* to linguistic deep structures and logical forms.)

That a deep structure of a sentence equals its logical form has been claimed by various authors, especially the "generative semanticists" (e.g., Lakoff (1971), (1972), McCawley (1972)) and several philosophers and logicians (e.g., Davidson (1975), Harman (1972), (1973)). Current evidence indicates, however, that Chomskyan (1965) syntactic deep structures (base structures) cannot quite be identified with the generative semanticists' (and our) deep structures (see Lakoff (1971) and Chomsky (1971)). This issue seems to be in part empirical (in the sense linguistics is empirical).

The proposed identity between logical form (in the Tarskian truth-determining sense) and deep structure in the generative semanticists' sense can be regarded as an analytic or conceptual identity (cf. Heidrich (1975)). The identity (or structural similarity, if we accept the weaker thesis) between logical forms and statements in the language of thought again is broadly *factual* in character. This follows from our formulation of the analogy theory.

If we identify the sentences of the language of thought with the logical forms of the surface sentences of the language we imply that any (conceptual) thinking must take place in some representational system or other. If higher animals and small babies can think, as I believe they can (in a rudimentary sense), some representational system, in this case presumably different from full human natural language, must then be involved. Also, various forms of "body-talk" (e.g., "giving a fist" to somebody) involves a representational system.

The essential feature about thinking, then, is its *representational* character. Accordingly, we may propose that thoughts (as events or states) must be tokens (or at least analogous to tokens) of deep structure sentences. Given this, a functional account of inner mental events and states is deeply involved with giving an account of the language of thought. As the language of thought deals with meaning as representation (and truth), an account of thinking involves an account of meaning in this sense. However, meaning as involved in *communication* and in the theory of *speech acts* may still need an extra account, though this also presupposes and depends on a clarification of meaning in this representational sense.[4]

Here, we cannot of course go deeply into the enormously complex matter of giving a "full" account of the language of thought. I do accept most of what Harman (1973) says about the matter, but even that is very little. One feature to be mentioned here is that Harman takes mental states to be tokens of "sentences under analysis". A sentence under analysis is a surface sentence coupled with its logical form. On the other hand, I prefer to take mental states to be (at best) tokens of deep structure sentences only. I do this primarily because of the existence of "full" thinking which, nevertheless, is coupled with linguistic disability and defect. Thus, there is some evidence for the claim that children think in a much more complicated manner than they are able to express in surface form. Mental states hence do not have to involve the transformation rules for going from deep structures to surface structures.

If the view I have sketched above is even approximately acceptable, it obviously follows that thinking is to a high degree language-dependent. How much it depends on a particular (natural) language in turn is a function of how much interlinguistic objectivity and how many linguistic universals there are in the deep structures for these different languages.

What I have said above may seem to contradict Sellars' analogy account, but whether that is so depends on how liberally one uses the notion of analogy here, and thus it is not worth quarreling about.

What has been said about analogy and aboutness must of course appear in the logical forms of statements about propositional attitudes. In Chapter 2 we discussed the problem of logical forms in general and applied it to the case of action sentences. Analogously with the Davidsonian proposal, we can briefly say this. Switching now to a slightly more interesting example, consider a belief statement: 'Bill believes that John opened the window'. Its logical form is (essentially) '(Ey) Believes (Bill, (Ex) Opened (John, the window, x), y)'. In other words, believing is treated as a state represented by a three-place predicate. The above statement can be read as 'there is a singular state y such that y is a state of believing by Bill that John opened the window'. We shall later in Chapter 6 discuss in some more detail this extension of the Frege–Davidson type of formalization into the context of propositional attitudes. It suffices to remark here that the statement in the second argument of the belief predicate accounts for the "aboutness" of believing.

4. RULES OF LANGUAGE

The rest of this chapter will be devoted to a discussion concerning the non-circularity requirement imposed on a functionalist analogy account of mental events and processes. This matter is very central for the programme, even if it is somewhat peripheral from the point of view of action theory in general. However, as our particular version of a causal theory of action strongly relies on a functionalist account of mental episodes and as the matter is of deep intrinsic interest we shall make a digression into this circularity issue.

We start by a discussion of rules of language, which will be seen to be central for Sellars' account and, to almost the same extent, for our own view.

In Sellars' analogy theory the framework of thinking is an analogical one the fundamentum of which is meaningful overt speech (potential and actual overt linguistic behavior). Here speech is to be understood in terms of the uniformities and propensities which connect utterances with (1) other utterances (at the same or different level of language), (2) with the perceptible environment, and (3) with courses of action (including linguistic behavior). (See Sellars (1967), p. 310. We shall later discuss the types of uniformities involved here.) What is central for our present purposes is that in Sellars' theory a) linguistic behavior (activities consisting both of actions and non-actions) is *conceptual* activity in a primary sense and b) linguistic behavior is through and through *rule-governed* and thus in an important way involves a *prescriptive* aspect (cf. Sellars (1969b), p. 510). We shall below accept both a) and b).

To see why linguistic rules play such an important role in Sellars' theory we briefly consider his account of the central semantical notions of meaning and truth.

To say of a linguistic utterance (e.g., noun or sentential expression) that it means something is not to construe meaning as a relation between a linguistic entity and some *non*-linguistic entity (thought, proposition, etc.) but to go about "nominalistically" as follows. Consider, e.g., the Finnish common noun '*talo*'. The analysis now becomes:

(a)   '*talo*' (in Finnish) means house,

which has the sense of

(b)   '*talo*'s (in Finnish) are ·house·s.

Here '·house·' is a common noun which applies to items in any language which play the representational *role* played in our language [English] by the sign design which occurs between the dot quotes. (Accordingly, we might indeed take the dot-quote operation, when applied to the surface form of a sentence, to give its logical form.) The word 'means' indicates that the context is linguistic, and it also reminds us that in order for the statement to do its job directly, the unique common-noun forming convention must be understood, and the sign 'house' must belong to the active vocabulary of the person to whom the statement is made. There it plays the same role as the word '*talo*' plays in Finnish. Many details ought to be added, of course, before we have working theory of meaning. But here that is not needed, for we can already see that the key notion is that of *playing a role* in a language. This notion is analyzed in terms of overt speech, i.e., in terms of the uniformities and propensities of the linguistic behavior of language users. These uniformities again are based on *linguistic rules* in a strong and essential way which involves the causal efficacy of rule expressions (cf. Sellars (1967), p. 310, and below).

Similarly, in the case of the semantic notion of truth linguistic rules are central, since the truth of statements becomes analyzed roughly as follows: A sentence (in a language $\mathscr{L}$) is true if and only if it is semantically assertable according to the rules of language $\mathscr{L}$.

Sellars argues that a proper understanding of the nature and status of linguistic rules is a *sine qua non* of a correct interpretation of the sense in which linguistic behavior can be said to *be* (and not merely to express) conceptual activity. As we indicated earlier, these linguistic rules may concern 1) world-language or language-entry uniformities, 2) intra-linguistic uniformities, and 3) language-world or language departure uniformities. But what we are here mostly interested in is a different type of classification of rules, viz. their classification into *rules of action* (ought-to-do rules) and *rules of criticism* (ought-to-be rules).

Rules which specify what one ought to do are rules of action. Their prototype is the conditional rule

(*1*)   One ought to do *X*, if *C*.

Rules of criticism, on the other hand, specify how something ought to be.
An important type of ought-to-be rules is the following:

(*2*)   *A*'s ought to be in state *S*, if *C*.

It should be noticed here that the *A*'s, which are the subjects of the rule,
need not (on Sellars' account) have the concept of what it is to be in state *S*
or of what it is for *C* to obtain.

Before we go on to give specific examples of ought-to-be and ought-to-
do norms and compare them, a general point has to be discussed. This is
the problem of giving an account of what it is to *satisfy* a linguistic norm,
or, what it is to *follow* or *obey* such a rule. Analogously with the case of
laws of nature, we may look for either naturalistic or for some other kind
of criteria of rule-obeying. The other kind of criteria include at least
psychological ones and (what might be called) *conceptual* ones.

First we can notice that mere behavioral uniformities ("constant
conjunctions" or merely conforming behavior) give at best necessary
criteria of satisfaction for rules such as (*1*) and (*2*). Thus, for instance, in
the case of satisfying (*1*) we have to say more than that people invariably
but "accidentally" do *X* when *C* is the case. One commonly accepted
approach to this problem is to say something like this: To satisfy a rule
is to *obey* it with the *intention* of fulfilling the demands of the rule. This
criterion is psychological in that it refers to an agent's intentions. For
instance, it may be understood to entail the agent's *awareness* of the rule,
his ability to *apply* the rule in appropriate (including *new*) circumstances,
the agent's ability to *criticize* the rule, the possibility for him to *err* or to
break the rule, etc. Furthermore, at least in the case of objectively sanc-
tioned rules, obeying would be taken to entail the *recognition* of rewards
and punishments and reacting to them in one way or another.

But this kind of psychological approach to clarifying the notion of
satisfying a rule is not available to, e.g., Sellars, for he cannot in his
programme rely on the above kind of psychological notions on the pain of
circularity. Sellars' answer is based on the notion of *pattern governed
behavior*. This is behavior which is more than merely rule conforming
behavior (in a "constant conjunction" sense) but which is not yet rule-
obeying in the above strong psychological sense. "Roughly it is the
concept of behavior which exhibits a pattern, not because it is brought
about by the intention that it exhibit this pattern, but because the

propensity to emit behavior of the pattern has been selectively reinforced and the propensity to emit behavior which does not conform to this pattern has been selectively extinguished" (Sellars (1973b), p. 489).

A useful analogy comes from natural selection which results in, for example, the patterns of behavior which constitute the "dance" of the bees. The basic thing here is that in a sense pattern governed behavior is something performed *because of* the whole behaving system's achieving its goal. Thus, in a weak sense at least, there is a *reason* for this behavior. But it is essential to keep in mind that a piece of pattern governed behavior as such is not an action in a full sense (something one can decide or intend to do), even if actions may consist of pattern governed behavior. Therefore, pattern governed behavior is not correct or incorrect as actions are correct or incorrect. Pattern governed behavior is thus subject to ought-to-be rules and not directly to ought-to-do rules. However, we note for further reference that even if ought-to-be rules are to be distinguished from ought-to-do rules, on Sellars' account there is an essential connection between them. As Sellars puts it, "the connection is, roughly, that ought-to-be's imply ought-to-do's" (Sellars (1969b), p. 508).

One may ask whether Sellars really has succeeded in defining the notion of pattern governed behavior so that it lies "in between" mere rule-conforming behavior and rule-obeying behavior. But let us accept Sellars' characterization as successful for the time being; we shall return to some aspects of it later.

We next consider some examples of ought-to-be rules (cf. Sellars (1969b), pp. 508–511):

   (3)   One ought to feel sorrow for bereaved people.

   (4)   These rats ought to be in state S, if C.

   (5)   (*Ceteris paribus*) one ought to respond to red objects in sun-light by uttering or being disposed to utter 'this is red'.

An interesting thing about (3) is that it seems to presuppose that the subjects of the rule have the concept of what it is to be bereaved. Notice also that feeling sorrow is not an action, but it may be taken to involve some activity.

Rules (4) and (5), on the other hand, do not presuppose that the subjects have the concept in question, not at least in a strong sense of having a concept. But, for a number of reasons, these rules are not very good for illustrating Sellars' basic idea. Rule (*4*) is about rats, not about people, and this fact makes it only indirectly relevant. Rule (*5*) again is very problematic in that the ceteris paribus clause must contain something which amounts to requiring that the subject is in a "responding-to-colours-of-environmental-objects-state". (If one is about to be run over by a red car, say, one ought to jump aside and ought not to utter 'this is red'!) But, if such a psychological qualification is needed, rule (*5*) is of no use in clarifying Sellars' account just because of its reliance on an inner mental state.

The basic problem about (*5*) is that it relies on an untenable idea of the conditioning of verbal behavior. To avoid such stimulus-boundedness of linguistic behavior, we should operate rather with weaker rules like:

(*5**)   (*Ceteris paribus*) one ought not to respond to red objects in sunlight by uttering or being disposed to utter 'this is not red'.

As 'this is not read' can be considered equivalent to 'it is not the case that this is red' (*5**) equals (assuming the "usual" connection between may's and ought's, viz. 'may' equals 'not-ought-not'):

(*5***)   (*Ceteris paribus*) one may respond to red objects in sunlight by uttering or being disposed to utter 'this is red'.

(Sellars (1969b) also mentions the possibility of using may-rules instead of ought-rules in contexts like this.) We shall not further discuss may be-rules any more than may do-rules. Let me just point out that (*5***) is rather uninformative and uninteresting as contrasted with (*5*). This indicates to me that the machinery consisting of ought-rules and may-rules is too rough to do justice to the various delicate and complex features involved in thinking.

In what sense does an ought-to-be rule imply an ought-to-do rule? Sellars gives a couple of examples of this. It suffices to consider the following ought-to-do rule corresponding to, and entailed by, (*3*) (cf. Sellars (1969b), p. 509):

(*3'*)   (Other things being equal and where possible) one ought to bring it about that people feel sorrow for the bereaved.

We shall later return to the acceptability of claims like the above that (*3*) entails (*3'*).

5. CONCEPTUALITY AND MENTAL EPISODES

5.1. Let us now go on to discuss whether Sellars' analogy account satisfies the non-circularity requirement. Marras has recently criticized Sellars' analogy theory on two grounds (see Marras (1973a), (1973b), (1973c)). First, Sellars' analogy theory is claimed to be circular in that Sellars' analysis, after all, relies on inner episodes. Secondly, Sellars' attempt to explain conceptuality (the having of concepts) in terms of linguistic rules also is circular, as the latter notion presupposes conceptuality. Because of their importance concerning any analogy account and, more generally, any functionalist account of thinking, we shall go into a fairly detailed discussion of these arguments.

Let us thus consider the first alleged circle in some more detail. The bare bones of the argument can be reconstrued as follows (cf. Sellars (1973b), p. 486 and Marras (1973b)):

(P1)   If the concept of the conceptuality of inner episodes (i.e., thoughts) is to be analyzed by means of the concept of a rule of language, it is essential that the concept of a rule of language does not presuppose the concept of the conceptuality of inner states.

(P2)   The concept of a rule of language presupposes the concept of the conceptuality of inner episodes.

(C)   The concept of the conceptuality of inner episodes cannot be analyzed by means of the concept of a rule of language.

(P1) is of course accepted by Sellars (and by Marras). But Sellars, contrary to Marras, naturally does not accept (P2). Notice that (P2) has to be accepted as soon as rule-following is analyzed in terms of rule-obeying and awareness of rules (see our earlier discussion). Here Sellars' move is, not unexpectedly, to accept only the following weaker postulate (cf. Sellars (1973b), p. 487):

(P2*)   The concept of an ought-to-do rule of language presupposes the concept of the conceptuality of thinkings-out-loud (in a Rylean sense).

But Marras argues that this will not save Sellars as it immediately leads to a dilemma. To put the matter briefly, the dilemma arises as follows. The Rylean thinkings-out-loud of (P2*) are items of linguistic behavior. But they are not merely behavioral items in some physicalistic sense (movements of the body, sounds uttered, etc.). They can be described as having semantical (or conceptual) properties. Thus as Sellars himself argues in so many words, they exist in "an ambience of semantical rules". We are not dealing with items like utterances but with sayings (activities, which need not, however, be actions, viz. action tokens$_2$ in our later terminology of Chapter 10), and this difference is due to the embedding of the latter in a system of linguistic rules. But the semantical rules explaining the conceptuality of thinkings-out-loud are either ought-to-be rules (at the level of pattern governed behavior) or ought-to-do rules (at the level of linguistic actions or rule-obeying behavior). In the latter case we immediately seem to get into a vicious circle, because our starting point was just to analyze rule-obeying behavior in terms of Rylean thinkings-out-loud. The former alternative, Marras argues, leads to a vicious circle, too, as soon as one accepts, as Sellars does, that ought-to-be rules entail a related ought-to-do rule. Therefore also the assumption (P2*) leads to a vicious circle.

Furthermore, just because of the mentioned connection between ought-to-be and ought-to-do rules, it also follows that ought-to-be rules presuppose the concept of the conceptuality of thinkings-out-loud. Thus, by the argument just given above, Sellars seems forced to accept (P2), after all, and hence the ensuing paradoxical conclusion.

Marras' second charge of circularity simply says that Sellars' analysis of conceptual activity in terms of rules of language leads to a vicious circle. We construe it as follows:

(P1′)   If the concept of conceptual activity is to be analyzed by means of the concept of a rule of language, it is essential that the concept of a rule of language does not presuppose the concept of conceptual activity.

(P2′)   The concept of a rule of language presupposes the concept of conceptual activity.

(C′)   The concept of conceptual activity cannot be analyzed by means of the concept of a rule of language.

Sellars of course does not accept the conclusion (C′) of this conceptuality paradox. How that is possible, if it is, will be intimately connected with the notions of ought-to-be rule and pattern governed behavior. But before we enter these deep waters we make a couple of remarks on the circularity paradoxes.

We first notice that for Sellars (C′) is stronger than (C). That is, if conceptual activity (including especially verbal behavior and speech) cannot be analyzed in terms of rules of language then neither can the conceptuality of inner mental episodes be so analyzed, given the assumptions of the analogy theory.

There are two ways of blocking the first paradox, which are available for at least some analogy theorists, but not for Sellars.

(i) One can deny that an ought-to-be rule entails an ought-to-do rule (in the sense, e.g., rule (3) has been supposed to entail rule (3′)).

(ii) One can abandon the idea of analyzing the conceptuality of inner episodes in terms of rules of language.

There is also a third way out, which perhaps also Sellars would like to use:

(iii) If we employ may-be and may-do rules (as in (5**)) instead of ought-to-be and ought-to-do rules the paradox is at least technically blocked. Furthermore, I don't think one is forced to assume that a may-be rule entails any may-do rule for reasons similar to those relevant to point (i). Notice here, too, that even if an ought-to-be rule would be taken to entail an ought-to-do rule in the discussed manner, the analogous may-be rule does not entail the corresponding may-do rule (although the converse holds true) assuming the "usual" connection between may and ought.

Of these three ways out I find (i) and (iii) clearly acceptable. (We shall discuss (ii) later.) Concerning (i), ought-to-be rules do not entail any ought-to-do rules unless some kind of deontological Kantian type of ethics is accepted (as Sellars does). Only in (something like) a deontically perfect world must we accept (absolute) ought-to-do rules concerning all the means for achieving a goal which "ought-to-be".

The second way out of the first paradox will be commented on later, as said, and solution (iii) should be clear enough as it stands.

Now we proceed to a discussion of Sellars' (1973b) way out which concerns primarily the conceptuality paradox, but which, by entailment, also qualifies as a fourth way out of the first paradox, if successful. Let us cite Sellars: "Essential to any language are three types of pattern governed linguistic behavior

1. Language Entry Transition: the speaker responds to objects in

perceptual situations, and to certain states of himself, with appropriate linguistic conceptual episodes.
2. Intra-linguistic Moves: the speaker's linguistic conceptual episodes
   tend to occur in patterns of valid inference (theoretical and
   practical), and tend not to occur in patterns which violate logical
   principles.
3. Language Departure Transitions: the speaker responds to such
   linguistic conceptual episodes as 'I will now raise my hand' with an
   upward motion of the hand, etc.

It is essential to note that not only are the abilities to engage in these types
of linguistic conceptual activity *acquired* as pattern governed activity,
but they *remain* pattern governed activity. The linguistic conceptual
activities, which are perceptual takings, inferences and volitions *never*
become *obeyings* of *ought-to-do* rules." (Sellars (1973b), p. 490.) I think
that herein lies much of a solution to our conceptuality paradox, even if
Sellars perhaps does not quite successfully spell that out in his reply to
Marras.

To clarify Sellars' proposal let us consider a simple case of inference.
An agent is supposed to have two premises 'p' and 'if p then q'. He then
infers 'q'. This is a piece of pattern governed behavior, which is not an
action and hence not an obeying of any ought-to-do rule. Yet it can be
taken to be a piece of conceptual activity. I think we can say that it is
normally governed by the ought-to-be rule: If 'p' and 'if p then q' are true,
then one ought not to infer '$\sim q$', and under epistemically normal conditions one ought to infer 'q'. Let us accept that this rule (or something
closely similar to it) has been internalized by the agent so that it indeed is
causally efficacious (cf. e.g., Tuomela (1975) and Chapter 8 below for an
argument concerning the causal character of inference).

Now this kind of inference represents a type of activity which a) has no
action tokens (in our full sense of action tokens$_2$; see Chapter 10), b) has
at least some overt tokens (inferrings-out-loud, which are action tokens$_1$
in our later sense of Chapter 10), and c) is (more or less clearly) conceptual. Is that not enough to prove Sellars' point? Surely we can also add
to point a) that there is no corresponding ought-to-do rule (in the sense
($3'$) corresponds to (3)) that the agent would be obeying when engaged in
inferring.

There are two counterobjections an intentionalist like Marras can still
make (and he has indeed made them). First there is the objection that,

even if there are no ought-to-do rules which directly correspond to ought-to-be rules governing perceptual takings, inferences, and volitions, still the *understanding* of these concepts relies on ought-to-do rules and thus on the conceptual framework of agency. The second objection says that the ought-to-be rules applying to these pattern governed behaviors still presuppose that the agent has certain conceptual capacities, which are exercized in these behaviors. This would entail that Sellars' answer is circular, after all.

To evaluate the first objection we consider rule (5) (assuming, for the sake of argument that (5), instead of (5**), can be used). There are two ought-to-do rules related to (5):

(5′)   (*Ceteris paribus*) one ought to say 'this is red' in the presence of red objects in sunlight.

(5″)   (*Ceteris paribus*) one ought to bring it about that people respond to red objects in sunlight by uttering or being disposed to utter 'this is red'.

Now it is (5″) that really corresponds to (5) in our earlier sense, for in (5) the uttering 'this is red' is a non-action (i.e., not an action token$_2$) and does not presuppose that the subject or trainee has the concepts of red, object, sunlight, and saying 'this is red'. On the contrary, (5″) requires this of the trainer (though not of the trainee), and (5′) requires it even of the trainee. But (5′) is not entailed by (5) primarily because an utterance 'this is red' is not always an action. Rule (5′) is applicable "only" to those cases (that is, the normal situations) when 'this is red' is a *saying*.

The intentionalist argument now relies on the validity of the following hypothetical conditional (cf. Marras (1973b), p. 481): (S) For a token expression to have a certain sense is for it to be a member of a class of expressions of a given type which *may* be used (and *would* actually be used, *ceteris paribus*, in a situation appropriate for linguistic action) by competent members of the linguistic community at large, to convey a corresponding thought.

According to this semantic principle a token expression 'this is red' would have a certain sense (meaning) even when it "belongs to" a non-action (i.e., something not an action token$_2$) such as a perceptual talking-out-loud or to a "mere" response; which clearly does not require any actual exercise of concepts by the person uttering it.

How can a naturalist answer this intentionalist argument? First, I think Marras here speaks about meaning or sense from the point of view of an account of *communication* rather than from the point of view of meaning as *representation*, which thinking primarily involves. These are different senses of meaning and different aspects of "languaging". An utterance can have meaning in some representational sense without being intimately related to possible or actual uses of the expression type which that utterance tokens (see our remarks on action tokens$_1$ and their type-independence in Section 3 of Chapter 10). Of course, I don't want to deny that sharp distinctions are hard to make here and that, quoting Sellars, "One isn't a full fledged member of the linguistic community until one not only *conforms* to linguistic ought-to-be's (and may-be's) by exhibiting the required uniformities, but grasps these ought-to-be's and may-be's themselves (i.e., knows the rules of language). One must, therefore, have the concept of oneself as an agent, as not only the *subject-matter* of ought-to-be's but the agent-subject of ought-to-do's." (Sellars (1969b), p. 513.)

Nevertheless, one can accept this general *ideal* and still claim that most of us are not full-fledged members of our linguistic community in this sense (who, e.g., knows all the rules of language?), although we still are able to engage in various conceptual activities. Thus, we may argue, from a linguistic point of view the conceptuality expressed in verbal activities comes in *degrees*. To illustrate this, consider, e.g., having the concept of table. A rudimentary concept of table is already possessed by a person at the conceptual level of a few months old child. For such a being a table is probably some rather diffuse perceptual entity with some shape, size, color and texture plus the property of hurting when getting kicked, etc. This is far from, for example, a scientist's concept of table which contains both of Eddington's famous tables as its representings and probably everything needed for solving Eddington's puzzle.

5.2. Perhaps we must, however, agree with the intentionalist that full-fledged linguistic ability (at least) *indirectly* presupposes language as an instrument and the framework of agency (and hence ought-to-do or at least may-do rules). But, as said, the other side of the coin is language in thinking (and representation).

Marras' thesis (S) may work as a meaning-criterion for expressions but it is insufficient for characterizing *conceptuality*, I argue. Various pattern governed behaviors, such as inferences and "volitings", which never are full actions can be regarded as more or less fully conceptual. These cases may be taken to show the plausibility of a *naturalistic* account (cf. Section 3 of Chapter 10).

Let me now make some programmatic claims on conceptuality: The non-metaphorically understood conceptuality of (human) activities is in them (in a sense difficult to make explicit) rather than merely put there, we may argue. Thus it belongs to the goal-directedness of behavior, meaningful self-control and adaptivity, the concept of work, as well as the various forms of interaction between people.

I will not here try to formulate a coherent naturalist account. But let me still make some remarks. First, it has to be emphasized that the conceptuality of human activities in an important sense involves something non-naturalistic. This has to be discussed before the naturalistic aspect can be analyzed.

In view of our earlier discussion it comes as no surprise that this non-naturalistic element is the prescriptive element of conceptual activity. The question really amounts to the well known and accepted irreducibility of 'ought' to 'is'.[5]

The above point can be elaborated as follows. We have above discussed naturalistic conditions for the applicability of rules of language in describing conceptual activity. In doing this we have played down the distinction between descriptive (or "theoretical") and prescriptive (or "practical") discourse. However, granted that prescriptive discourse (involving essentially irreducible token-reflexives such as 'I' and 'we') is indeed irreducible to descriptive discourse, naturalistic (and purely descriptive) criteria are not of course going to exhaust what is interesting in analogy accounts of conceptual thinking. Rules of language play their role in a certain community and, as said, they have an irreducible "we"-character. We may decide to apply agent-discourse, e.g., to amoebas, compasses and pens, and hence anthromorphize practically anything. But there is an irreducible element essentially involved here (cf. our discussion of "incorrigibility" in Chapter 4).

However, naturalistic facts are still relevant (in a way difficult to specify exactly) in that by means of them we come to roughly distinguish between which beings genuinely can share community intentions and standards with us and which cannot, and hence between "proper" conceptual thinking and thinking in a metaphorical sense (such as a compass needle "knowing" where the north is). What these naturalistic facts are like, however, is a difficult matter.

Roughly speaking, conceptual activities would seem to involve goal-directedness of behavior, meaningful self-control and adaptivity, as well as various forms of interaction between the members of a society. Perhaps one might functionally characterize conceptuality by saying that it is something in virtue of which the framework of agency and of linguistic rules are applicable to the description of human activities. Perhaps one could also say that conceptuality *in re* ontologically precedes conceptuality explicated in terms of linguistic rules. If so, linguistic rules may be taken to *express* conceptuality, but conceptuality cannot be *explained* in terms of linguistic rules. Rather, the *applicability* of linguistic rules is, in principle, to be explained in terms of naturalistically (and non-intentionally) characterized verbal and non-verbal conceptual activities.

It is to be emphasized, furthermore, that linguistic rules (of the kind discussed above) seem too coarse and clear-cut for expressing the various forms and degrees of conceptuality exhibited by human (and also non-human) activities. Naturalistic "criteria" might do better in this respect, too. I shall not here make any attempt to formulate such naturalistic conditions. They are bound to be complicated as historical and evolutionary considerations will have to be taken into account. In addition, such "criteria" will to a great extent be factual and hence not knowable a priori. Science may and presumably will tell us that amoebas and worms, for example, do not have at all the same kind and as powerful representational and semantical capabilities as, e.g., apes, dolphins, human beings, and so on.[6]

These promissory notes for a naturalistic account of conceptuality will have to suffice here. We now return to the mentioned criticism of Sellars that even those pattern governed behaviors which are a) thinkings-out-loud and b) non-actions, require conceptual capacities by the subject. I think that in the case of full-fledged members of a linguistic community that is indeed the case (think of, e.g., 'this is red' as a saying contra a mere utterance). But, if we accept, as we did above, that there are rudimentary forms of linguistic activity so that in some case, e.g., only (5) but not (5') is satisfied (or "internalized") by a speaker, then the charge can be dismissed.

Before going on to summarize our lengthy and complex discussion concerning circularity paradoxes, we will make a comment on the rule-boundedness of conceptual activity. What we have so far said in favor of a naturalist account of conceptuality (i.e., especially what goes beyond Sellars' naturalism) is still compatible with Sellars' thesis that a) linguistic activity is conceptual activity in a primary sense and even with that b) linguistic behavior is through and through rule-governed, if these theses are sufficiently *liberally* understood. To emphasize this last point we may still consider one more difficulty for a strict Sellars-type-rule-analysis of conceptuality. Think of a mathematician who intends (intends-out-loud, say) to prove Fermat's last conjecture. I think it is completely legitimate to speak of such intentions, which cannot be realized recursively. Carrying out this intention cannot thus be conforming to or obeying a *recursive* rule. So we must either say, contra Sellars, that intending is not intimately bound up with linguistic rules or we must accept non-recursive rules among our linguistic rules. But if the latter alternative is adopted, we have to explain in a non-question-begging way what kind of non-recursive rules for instance the above kind of creative and innovative activities could conform to.

5.3. Now we are ready to give a (partial) summary of our results concerning the circularity paradoxes. First, Sellars' line of argument based on pattern governed non-actions blocks *direct* dependence on ought-to-do rules and the framework of agency. In terms of our syllogistic arguments this means that Sellars does not have to accept either (P2) or (P2') if 'presupposes' is strictly interpreted. (This gives the promised fourth way out of the first paradox concerning the conceptuality of inner episodes.)

Secondly, our discussion related to the semantic principle (S) has shown that ought-to-be rules and pattern governed behavior is, however, *indirectly* related to ought-to-do rules and the framework of agency. But we must notice that the indirect dependence is on the level of linguistic competence, so to speak. We just have to refer to a conceptual framework but not necessarily to the exemplifications in the actual world of the concepts of the framework, and we do not have to account for the causal efficacy of the linguistic rules in terms of psychological criteria (cf. Section 4). We do not directly refer to, say, any particular agent's actual inner episodes, and the indirect conceptual dependence, of course, does not commit us (or Sellars) to the existence of anything like "Cartesian" inner episodes totally independent of language (or conceptual system).

As in the case of the first paradox, we can also, in the case of the conceptuality paradox, block *direct* reference to ought-to-do rules and the framework of agency either by rejecting the discussed problematic entailment relation between ought-to-be rules and ought-to-do rules or by operating with may-be and may-do rules. These moves, however, do not exclude the commitment to *indirect* reference to ought-to-do rules and the framework of agency.

Finally, we also have the possibility here to give up analyzing conceptuality *strictly* in terms of rules of language. This is in fact what we have been arguing, if by 'analysis' one means either a strict Sellarsian type of analysis discussed above or an *explanatory* analysis. Instead we have suggested a somewhat wider type of naturalistic analysis. However, we shall not press this point here as no such sufficiently developed alternative naturalistic analysis is yet available.

As a conclusion we can now see, I think, that none of the difficulties encountered by analogy accounts discussed in this chapter are fatal. Our results indicate that the conceptuality of inner episodes can be analyzed in terms of a suitable analogy account without direct circularity.

On the other hand, it has to be emphasized that problems of thinking are enormously complex and delicate. Therefore reductionistic philosophical analyses of these questions are bound to fail. But I do not think an analogy account has to be reductive in any strong sense — e.g., indirect dependence on the framework of agency, as above, must be admitted.

We can now also notice a consequence of our results for functionalist analyses of mental concepts. Consider for instance the notion of trying (or, equivalently, willing or effective intending) that I regard as a key concept in a causal type of action theory (see below Chapters 6, 8, and 10). In the simple case of arm raising, we, roughly, characterize the agent's trying to raise his arm as a mental event which, the world suitably cooperating, causes the arm to rise. Thus, there is both a conceptual and a causal connection between the concepts of trying (effective intending) and the action of arm raising. It is to be emphasized that trying exists as an ontologically "genuine" event separate from the overt action event of arm raising.

The trying event is conceptually characterized (primarily) in terms of the concept of arm raising, which is an action concept. In a causal theory of action, on the other hand, one characterizes actions by reference to suitable mental antecedents (e.g., tryings, cf. Chapter 6). As will be argued in Chapter 4 one can make some of the connections between inner and overt episodes (described in terms of suitable linguistic deep structures) analytic while the others still remain synthetic. Thus no fatal conceptual circularity arises from this semantical "reciprocity", still less do we get ontological circularity.

The considerations and results of this chapter now show that one can simultaneously accept a suitable causal theory of action and a "public" and "social" view of inner mental episodes, and this can be accomplished basically in terms of a suitable analogy account of mental episodes (cf. Chapter 4 for details). It is worth noticing that if, on the contrary, the circularity charges discussed in this chapter had remained unanswerable, this would clearly have been a serious blow not only for functionalist (e.g., analogy) analyses of mental concepts but also for causal theories of action (assuming that one has a distaste for "the ghost in the machine" or that one finds this notion inherently obscure).

It may also be remarked in this connection that analogy accounts (and, more generally, functionalist accounts) are not demolished by the famous "infinite regression" argument advocated by Ryle and Malcolm. This argument is closely related to Marras' circularity paradoxes discussed above. (It can be applied to both versions of the paradox.) Our remedy to the latter argument qualifies for the infinite regression problem as well.

Let us briefly consider Malcolm's formulation of the infinite regression problem (Malcolm (1971), p. 391):

"If we say that the way in which a person knows that something in front of him is a dog is by his seeing that the creature "fits" his Idea of a dog, then we need to ask, 'How does he know that this is an example of fitting?' What guides his judgment here? Does he not need a second-order Idea which shows him what it is like for something to fit an Idea? That is, will he not need a model of fitting? But then, surely, a third-order Idea will be required to guide his use of the model of fitting. And so on. An infinite regress has been generated and nothing has been explained.

On the other hand, if we are willing to say that the person just knows that the creature he sees fits the first Idea (the Idea of a dog), and no explanation of this knowledge is needed – if we are willing to stop here in our search for an explanation of how the man knows that this thing is a dog – then we did not need to start on our search in the first place. We could, just as rationally, have said that the man or child, just knows (without using any model, pattern or Idea at all) that the thing he sees is a dog."

Malcolm applies his dilemma to argue that the cognitive processes and structures postulated by psychologists are completely useless. Rendered useless are Chomsky's grammars (as innate structures), Piaget's and other structuralists' explanatory concepts, memory structures, and just about everything "inner" psychologists have recently constructed models about. Were Malcolm right, not only my notions of intending and trying but also the notion of conduct plan that we will strongly rely on would be useless (see Chapters 7–10). Well, who is right?

As we saw above, an analogy account of inner episodes in a sense relies on rule following. If this rule following had been explicated in terms of awarenesses or similar inner episodes the infinite regression would immediately be on. But the strategy was to go to overt pattern governed behavior instead and thus to dismiss the infinite regression at least prima facie. What Marras' argument tries to show is that this answer does not work after all. To the extent we above succeeded in answering Marras' charges we have blocked the first horn of Malcolm's dilemma. (Note: this horn of course prima facie applies to Marras and the other intentionalists we were opposing.)

How about the second horn of the dilemma? Well, an analogy theorist explains inner episodes conceptually in terms of public social practice. In the conceptual order he thus does not think that inner episodes explain behavior, rather it is vice versa. To this extent we agree with Malcolm.

But there is much more involved here. If we want to think of inner mental episodes as causes of behavior and thus as explaining behavior, we think of them in the order of being. It is in this sense and to this end that psychologists use cognitive processes, etc., in their theories. Such psychological processes and structures may be very useful for explanation and prediction of behavior even if they would turn out not to be neurophysiologically reducible (in the sense of type – type identity or something related). That is, even if psychological theories had postulated irreducible "little men" to explain behavior these little men might do a lot of work. If we learn, for example, that people recognize visual patterns by a template matching process rather than by a feature extraction process we have obtained important explanatory information. Of course, we would not have learned anything about the meanings of 'template matching' and 'feature extraction', as somebody like Malcolm might want to object. But such an objection would of course be silly.

I conclude that neither will this second horn of Malcolm's dilemma affect an analogy theorist.

NOTES

[1] Sellars does not accept this for non-conceptual episodes. See, e.g., Sellars (1956) p. 192 and (1968) p. 22, where he states a basic problem concerning the functional definability of sensations and feelings, etc. (e.g., the intrinsic quality of red and the intrinsic feeling of pain are taken as irreducible non-functional qualities). I shall not in this book discuss these qualia problems, as they do not concern wants, beliefs and other propositional attitudes that we focus on.

My own solution would essentially be to define all psychological concepts functionally (and thus not include anything about the private feelings of pain, etc., into the *concept* of pain). The "intrinsic qualities" can be accounted for by suitably restricting the admissible intended models of psychological theories (they should apply to organisms constructed of suitable biological "hardware", etc.). This choice of intended models cannot perhaps be made purely by currently available verbal means—cf. my comments in Chapter 4. (See Shoemaker (1975) for a good functionalist account of the "qualia" problem.)

[2] A terminological remark is due here. In this book we use 'episode' in a wide sense, which covers a) events, b) short-term states, and c) processes. We shall not attempt to clarify events in more detail than what was done in Chapter 2. Processes can be analyzed as sequences of events. For our purposes short-term states are like events except that they may have a somewhat longer duration than we ordinarily think events have.

Below we will often for stylistic reasons speak of events and episodes almost interchangeably, as long as this does not distort our treatment.

[3] In this book we will use 'Rylean' and 'pre-Jonesian' practically interchangeably in phrases such as 'Rylean language', 'Rylean entity', etc. Strictly speaking, a pre-Jonesian language is to be regarded as a Rylean one enriched with semantical discourse. A Jonesian language is then obtained from a pre-Jonesian language by adding the theoretical discourse as specified in the Jones myth.

[4] For an account of meaning in the communicational sense see Harman (1973) and, especially, Lakoff (1975). Lakoff's theory, with which our view is in concordance, gives an interesting account of indexicals and of "conversational implicatures".

[5] This point has in fact been made by Sellars himself in a number of places (cf., e.g., Sellars (1956), §§5, 36), Sellars (1963), pp. 39–40). Let me here express my thanks to Professor Richard Rorty for discussions concerning, among other things, this Kantian aspect of Sellars' view. In his works, most notably in the forthcoming *Philosophy and the Mirror of Nature*, Rorty has driven this idea further to a rather strong pragmatistic position, which differs at least in emphasis from Sellars' view and also from the view expounded in this chapter.

[6] It is, accordingly, no accident that chimpanzees rather than, say, rats have been taught to speak and think in American Sign Language and other comparable human or human-like languages.

CHAPTER 4

CONCEPT FORMATION IN PSYCHOLOGY

1. PSYCHOLOGICAL CONCEPTS AS THEORETICO-REPORTIVE CONCEPTS

One central idea involved in the Myth of the Given is that, according to it, there is some kind of analytic, ostensive tie between one's descriptive vocabulary and the extralinguistic world so that this descriptive vocabulary gets a privileged status. Especially, the *meaning* of an observational predicate (e.g., 'red') is taken to be determined by what is immediately "given" at the time of the acceptance of any observational sentence containing that predicate. This "given" could be something external (to the person), which causally brings about the person's acceptance (or disposition to accept) the relevant observational sentence or it could be an extralinguistic sensory event (sense datum) which leads the agent to accept the observational sentence.

Similarly, applied to predicates standing for raw feels (e.g., pain) or for representational mental states (perceivings, wantings, etc.) the Myth of the Given assumes a similar ostensive tie. The meanings of such mental predicates are taken to be fully determined by some extralinguistic inner givens (raw feels or representational episodes).

We can thus see that the acceptance of the Myth of the Given would almost make psychological predicates (representing inner episodes) *observational* in the old positivist sense, if observation is taken very broadly so as to include introspection — "observation by the inner eye". One qualification needed for this to have any plausibility at all is that people must somehow (e.g., by introspection) be able to use these predicates in reports concerning the first-person case. The other-person case would, however, pose a problem for this view. For would not 'having toothache' then be observational in my own case but theoretical when applied to somebody else?

If we reject the Myth of the Given we also reject the idea that some conceptual frameworks, or parts of such, are somehow *logically* tied to the world (rather than, for example, this tie being in part causal and in part due to social convention). Then also the first-person use of mental language is best construed as *learnt* as in the analogy theory of mental episodes.

Thus, mental predicates such as 'having a toothache' or 'perceiving a table' or 'wanting to have a meal' can be construed to have an "anybody-ascriptive" use such that both the first-person and the third-person uses are learned rather than "given". Both in the first person and in the third person case we can then regard mental predicates as *theoretical*. This is exactly what is accomplished by the Jones myth of Chapter 3 (also see Sellars (1956)).

It is also part and parcel of the Jones myth that mental predicates are *reportive*; and thus they represent *theoretico-reportive* concepts. The reporting use of mental predicates presupposes that they, suitably construed, occur in statements which can be true or false and that they can have ordinary referential semantics. This is also involved in the mythical Jonesian theory. (We shall soon discuss a special form of the reporting use of mental concepts in the context of a kind of incorrigibility peculiar to mental talk.)

There are several other ways than the rejection of the Myth of the Given and the adoption of a scientific realist construal (cf. the Jones myth) of mental predicates to defend the view that mental predicates are reportive, but we shall not here discuss them (see e.g., Cornman (1971)). It should be emphasized still that when we hypothesize below that mental predicates may be used reportively we will assume an *adverbial* construal of them (e.g., '*A* perceives *x*' becomes '*A* is in a state of perceiving *x*' or, better and more generally, '*A* Mod-Verbs').

Our rejection of the Myth of the Given also entails that we reject the observational/theoretical dichotomy in the old positivist sense. There is no such semantical or logical dichotomy, nor are there any observational terms which *in principle* have an epistemically privileged status as compared with terms called theoretical. On the contrary, in principle it is possible to use any descriptive predicates in direct commerce with reality, i.e., in direct reporting. Instead of using 'water' one may thus use 'a bunch of H_2O molecules' (assuming the identity of these expressions) and instead of 'having a toothache' one may use 'being in such and such a brain state' (or something like that).

In general, it is possible in principle to eliminate an entire conceptual framework such as the manifest image in favor of the scientific image, and this elimination does not have to involve an isomorphic predicate-by-predicate elimination. Rather one conceptual framework is holistically replaced by another conceptual framework. What the exact requirements for such a replacement are is of course very problematic; presumably it would take place by scientific reduction in general. But the important thing about it for us is that given any extralinguistic situation it is in principle possible that competing frameworks (e.g., the manifest image and the scientific image or some theories belonging to them, respectively) could be applied. That is, several frameworks can be used to directly describe or report the situation (or, if you like, the sensory stimulation brought about by that situation) in terms of suitable singular statements belonging to the respective frameworks. Thus, for instance, micro-properties (of gases, say) may in principle be used for direct reporting. Evidential use is not restricted to macropredicates or predicates commonly called observational (cf. 'temperature', 'red').

However, it has to be emphasized that such replaceability need not be possible "in practice". Indeed, it is very important from a methodological point of view to draw a distinction between observational or "data" predicates (or concepts) and theory-dependent explanatory predicates, while the "pragmatic" character of this dichotomy is emphasized.

Let us briefly consider my *methodological* observational/theoretical-characterization best elaborated in Tuomela (1973a). Let us reproduce the core of that analysis here:

(OT) A nonlogical concept P occurring in a theory T belonging to a paradigm K is called *observational* with respect to theory T if and only if every representative scientist within K can (validly and reliably) "measure" P in the typical applications of T without relying on the truth of the theory T.

A nonlogical concept P occurring in a theory T belonging to a paradigm K is called *theoretical* with respect to T if and only if (a) P is not observational (with respect to T) in the strong sense that of every representative scientist within K it is true to say that he cannot "measure" P in *all* typical applications of T without relying on the truth of T, and (b) P has been introduced into T in order to explain the behavior (i.e., those aspects of it T deals with) of the objects T is about.

In this characterization 'paradigm' refers to a Kuhnian paradigm in the sense of a constellation of group commitments. The reader is referred to Tuomela (1973a), pp. 17–20, for a detailed clarification of this definition.

It is assumed in the above characterization that theoretical concepts are concepts occurring in some scientific theory which in some sense determines their meaning and use. In fact, I have elsewhere accepted the view that theoretical concepts can be introduced postulationally by means of a scientific theory so that the meanings of these theoretical concepts are determined roughly as follows (cf. Tuomela (1973a), pp. 122–123):

(M) The meanings of the extralogical predicates of a scientific theory T are in general determined by the entire theory. That is, they are determined by the syntactical formation rules, logical and mathematical axioms as well as the rules of derivation of the language of the theory, by all the scientific axioms (and theorems) of the theory and by the adopted semantical rules.

If my claim that psychological concepts standing for inner episodes are a species of theoretical concepts is acceptable, then most things said in Tuomela (1973a) about theoretical concepts, *a fortiori*, apply to them. One aspect about the above meaning thesis (M) is in fact reflected in our criterion (OT). It is that the evidential or reporting use of theoretical concepts depends at least to some extent on the assumption that the theory T is true. In this sense theoretical concepts are relative to a theory and accordingly they can be said to be theory laden.

It should be obvious that "Rylean" concepts (in the sense of our Chapter 3) are to be classified as observational in the sense of criterion (OT). Thus action concepts, for example, are observational concepts. (We shall later usually denote by λ the set of observational predicates of a given theory T.)

No concepts for inner mental episodes are observational in our technical sense. Why not? Given the analogy theory, such concepts are introduced postulationally by means of the Jonesian theory, call it T. As will be argued later, in such a theory T the inner mental concepts are not definitionally reducible to λ-concepts. That is, they are semantically, epistemically, and ontologically open (to some extent, at least). It also follows that the evidential use of these concepts depends on T. Hence these concepts are to be classified as theoretical in the sense of (OT). We shall later make some additional comments on this argument for the theoreticity of inner mental concepts.

In this connection it may be emphasized that our characterization of theoreticity of concepts concentrates on the following three aspects:

(1) theory-ladenness
(2) explanatory power
(3) evidential use.

In the present context it is of interest to consider the dimensions (1) and (3) jointly. First, if a theoretical concept has no evidential uses, not even indirect ones (i.e., theory-laden ones), we are dealing with a pure "hypothetical" or theoretical construct in the strict instrumentalist sense. Such theoretical concepts may not have great explanatory power, but they may be of help in describing data more economically (see Tuomela (1973a), Chapter 6), and they may also be inductively effective (see Niiniluoto and Tuomela (1973)).

But let us concentrate here on cases where a theoretical concept has at least a few evidential uses. We may now examine how great a proportion *out of those uses* depend on theory T. (We also exclude cases with no theory-ladenness, and hence a discussion of our observational concepts.) Thus we are dealing with all the cases from some to all on dimensions (1) and (3). We may simplify the situation by using dichotomies to get the cross-classification shown in Figure 4.1.

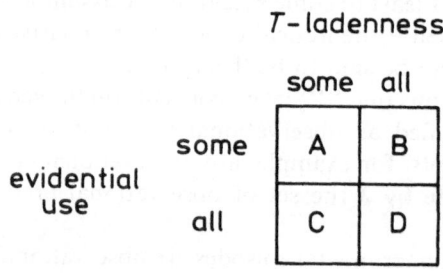

Figure 4.1

(Here 'some' means 'a small proportion'.) What do we get in our cells? In cells A and C we have concepts which are relatively close to observational concepts. I would like to call the elements of cell A *intervening variables*. They correspond roughly to the intervening variables of MacCorquodale and Meehl (1948) and also to the inner states of cybernetic systems (automata) and other comparable systems.

Observational concepts in the sense of (OT) would have a maximal number of evidential uses and they would have zero ladenness with respect to T (even if they may be highly laden with respect to some other theories). Now members of our cell C represent a peculiar form of theoretical concepts which are like observational concepts except for some amount of T-ladenness.

The elements of both cells B and D are theoretical concepts in a rather clear *realistic* sense. The elements of B can perhaps be taken to represent the typical or ordinary situation. The members of D represent a rather extreme case, but not at all an inconceivable case. It would accord very well with such extreme views as Feyerabend's (see, e.g., Feyerabend (1963)), according to which a good theory becomes its own "observation-language" so that a theoretician has so completely internalized his theory as to be able to use it fully evidentially, i.e., in direct commerce with reality.

When we move along the evidential use-dimension to zero from our cells A and B, we encounter instrumentalistic theoretical concepts. Those into which we get from B could perhaps be called *pure* instrumentalistic theoretical concepts.

Suffice it to say here of our dimension (2) that the greater explanatory power theoretical concepts have, the "better" they can be said to be. But, conversely, we may show that the introduction of theoretical concepts may increase both the deductive and the inductive explanatory power of a theory (cf. Tuomela (1973a)). We shall return to this question later.

I have gone through my most basic reasons for taking mental concepts to be theoretical. There will be one more argument to that effect, and it will be discussed in the next section. Roughly, it amounts to the argument that our common sense psychological concepts get their meanings *postulationally*. This argument, in fact, just says that our general meaning thesis (M), suitably qualified, indeed works in the case of psychology as well.

Except for a brief digression in Chapter 3 we have not discussed the reportive character of psychological concepts. We have been emphasizing the fact that agents have *epistemically privileged* noninferential access into their own mental life. In the present section this idea arose in connection with the evidential use of theoretical concepts: the evidential uses of theoretical psychological concepts are primarily, though not exclusively, first-person uses.

Privileged direct access and authority can be given even a greater importance. It may be argued that a certain form of privileged access is indeed "the mark of the mental" (cf. Rorty (1970)). According to Rorty it is a certain kind of *incorrigibility* that characterizes mental events and episodes (sensations, impressions, wantings, perceivings, and so on). Roughly speaking, this incorrigibility means that, in general, people in standard circumstances have the final say about what is the case about their mental life. What sensations, impressions and thoughts have in common in Sellars' account is merely their episodicity. On Rorty's account these mental episodes share the *epistemic* feature of incorrigibility. It should be immediately noted that this Rortian notion is different from incorrigibility-as-certainty. Strictly descriptive or "contentual" intrinsic features are out of the question both for Sellars and Rorty in the case of conceptual episodes, and, I think, correctly so. However, Sellars does think that there are such characteristic mental descriptive features in the case of nonconceptual episodes. Thus, for example, the impression of redness involves an intrinsic quality of red (cf., e.g., Sellars (1968), p. 22). This is central when discussing *functional* definitions and characterizations of mental episodes (see below). Sellars and Rorty both oppose "strict" functionalism and "topic neutral" translations (Smart, Armstrong), but on different grounds.

We are inclined to accept a position which is close to Rorty's. It seems that one cannot abolish the intuitive distinction (however it strictly is explicated) between mental and physical predicates by means of topic neutral translations (e.g., Ramsifications). The distinction must be preserved in any non-question-begging mind-body account.

How should one go about clarifying incorrigibility, which Rorty thinks is the mark of the mental? This is a very problematic question. Let us reproduce Rorty's latest analysis here (cf. Rorty (1970), p. 417 and (1974), p. 196):

(IR)   Subject S believes *incorrigibly* that p at t if and only if
    (i) S believes that p at t;
    (ii) there are no accepted procedures for resolving doubt about p, given that p fits into a pattern of sincere reports made by S, even though it fits into no more general theory;
    (iii) S's report that p counts as prima facie evidence for p.

One might ask what "accepted procedures" really are and require other relevant clarifications here. Still, this definition seems clear enough for our purposes as it stands. Given (I) one can try to characterize the notion of a mental (or inner psychological) event by (cf. Rorty (1970), p. 419):

(IM)   If there is some person who can have an incorrigible belief in some statement P which is a report on a particular event m, then, and only then, is m a mental event.

In other words, incorrigible believability is here taken to be both necessary and sufficient for something being mental. (IM) need not be taken as a strictly semantic analysis of mentality but still as something non-accidentally true of the mental or what in a certain culture is called mental.

In (IR) and (IM) a report is a description of a particular event or state of affairs. This excludes incorrigible a priori beliefs such as the belief that $5 + 7 = 12$. However, there is another problem related to "appears" statements: 'This pencil appears yellow to me'. As Rorty (1970) notices one can take statements like this either in a pre-Jonesian or in a Jonesian sense. In the pre-Jonesian sense the statement can simply be taken to express one's hesitation to endorse that the pencil is yellow. It need not then be regarded as a report of one's *state* of hesitation. In the Jonesian sense it could, however, be such a report. Rorty stipulates that 'report' is to be used "to mark off among incorrigibly believable statements those which are about *mental* states" (Rorty (1970), p. 419, my italics). Clearly, this use of 'report' makes (IM) circular. Therefore we have to drop the word 'mental' in this stipulation. Can (IM) then be accepted?

First we notice that (IM) does not apply as clearly (and as well) to beliefs, intentions, moods, etc., as it does to episodic thoughts and sensations. Beliefs and intentions (etc.) are dispositional (or, rather, mongrel categorical-dispositional). They are dispositions to act. Thus, a person's actions may give strong evidence as to what his intentions, etc., are. He may be mistaken in his own beliefs about them. The person's own authority does not always, in the first place at least, override others'.[1]

Secondly, what is mental is culture-dependent to the extent incorrigible reportability is. Incorrigible reportability again is culture-dependent to a great extent if we accept the analogy account and the Jones myth (cf. Chapter 3). Incorrigible reportability is something which is socially learned. Hence what is called mental may change from one culture to another. This is acceptable, even if people entertaining a Cartesian view of man (involving the Myth of the Given) may find it odd.

In the case of thoughts and sensations, we may perhaps accept that mental events can be incorrigibly reported, as (IM) specifies. Final authority can even be thought as something "given" to a person because of learned and internalized cultural conventions. What I find more problematic about (IM) is that incorrigible reportability entails mentality. If incorrigible reportability is something learned and something relative to culture, then it is of course possible that in some advanced culture people are taught to give incorrigible reports about such things as the functioning of their hypothalamus or their liver (and other — for us — nonmental entities). What this shows is that thesis (IM), if taken as a factual thesis, does not have a very high degree of lawlikeness. We thus have to stick to our own culture to avoid counterexamples of the present kind.

We shall regard (IM), at least roughly, acceptable (true) in the case of sensations and thoughts as long as we speak of our present culture. It has the same status as we have given to the Jones myth — something to be compared to contract theories in social philosophy. Perhaps one could also interpret (IM) in an explicitly *normative* way: we *ought to* think of mentality and incorrigibility as covarying as specified by (IM). In any case, given (IM), mentality becomes connected to something epistemic.

As everything epistemic belongs in part to the normative dimension it follows also from this way of thinking that mentality has an irreducible normative aspect (cf. our discussion in Chapter 3). (IM) is neither purely factual nor purely ontological, we can say. It does not give straight-forward descriptive information about mental events. In particular, it does not entail that mental events must have any nonphysical (nonmaterial) *descriptive* properties. The partly normative character of (IM) is just what makes it central as a criterion of mentality, I would say, and this notion of mentality is compatible with materialism. We can also put the matter by saying that (IM) represents a semantic norm about the use of the concept (or concepts) representing the mental event *m*, and this is independent of the truth of any reasonable form of materialism. (How far Rorty would accept the above interpretation is not clear to me, nor is it central here.)

A few remarks still have to be made about beliefs, intentions, etc., to which (IM) does not always apply. We noticed that they are dispositional. As believings and intendings (i.e., the states) become more particular and specific they become more and more like episodes (cf. my perceptual believing that I am holding a pen in my hand now). They thus become less and less dispositional as well. As Rorty has emphasized, they get closer and closer to having the incorrigibility proper mental episodes are claimed to enjoy. They can be said to be *nearly-incorrigible*. Given this, *incorrigibility* really seems to be a central mark of the mental. As we have emphasized earlier, *intentionality*, which, however, is reducible to metalinguistic categories, is another important mark of the mental, although it only applies to conceptual episodes and only gives a necessary condition for them.

A third important criterion of mentality is the *explanatory role* of mental episodes, as we have already emphasized and will emphasize again later. Mental episodes are introduced to explain people's behavior, we have claimed. However, it may seem that this criterion, though necessary, does not yet give a sufficient condition of mentality. However, if all that is distinctive about the mental belongs to the realm of the normative, as we have suggested, or if there is nothing more distinctive about the mental than about, say, the biological, then this criterion will be *descriptively* sufficient as well.

In any case, one necessary and sufficient criterion for us is (IM), qualified as above. (IM) gives an explication of the role and need for a cognizing *subject* in the philosophy of mind.

2. CONCEPTUAL FUNCTIONALISM AND POSTULATIONAL CONCEPT FORMATION

We have now presented our general reasons for construing mental concepts as theoretico-reportive concepts, and we have discussed the relevant philosophical issues at some length. In this and the next section we shall deepen our analysis in some directions. In this section we shall discuss a specific version of our meaning thesis (M), viz. a thesis of conceptual functionalism. This thesis will be illustrated by relating it to a specific example of postulational concept formation, which in addition clarifies how inner psychological concepts function as theoretical concepts.

According to conceptual functionalism the meanings of (inner) psychological concepts derive their meanings from overt intelligent behavior and thus from public culturally laden facts. Let us now consider Jones' mythical psychological theory, call it T, which is supposed to a) introduce inner episodes as ontologically "genuine" entities and to b) semantically characterize inner episodes in terms of overt behavior (and other public entities). Let the extralogical vocabulary of T consist of the set λ of predicate constants designating overt "Rylean" properties ('property' broadly understood) and of a set μ whose predicates are theoretico-reportive predicates standing for mental entities. Assuming the feasibility of the Frege–Davidson programme, T can be formalized either in first-order predicate logic or in a suitable second-order extension of it.

We can then claim in accordance with our meaning thesis (M) that the meanings of the predicates in μ are "implicitly defined" by the entire theory $T(\lambda \cup \mu)$, where the members of λ are antecedently understood (independently of the μ-predicates — thus assuming Sellars' first criterion of adequacy of Chapter 3 fulfilled). Inner psychological predicates are then taken to be semantically characterizable — in the manner of the "usual" kind of theoretical concepts — in terms of the lawlike connections into which they enter.

We can in fact adopt the following semantic *Thesis of Conceptual Functionalism* for a behaving system (e.g., organism, agent) to which psychological predicates apply:

(CF) Consider a theory $T(\lambda \cup \mu)$, describing a system S, such that the predicates in μ represent inner events, states, etc. of S and are theoretical with respect to the predicates in λ, which represent overt aspects of S. Then the meanings of all the predicates in μ are assumed to depend entirely upon their usage with the predicates in λ within $T(\lambda \cup \mu)$.

The principle (CF) amounts to saying that the meanings of the inner predicates μ are to be specified in terms of the input–output behavior of S. (Still, epistemic criteria such as incorrigible reportability may in addition have to be used to distinguish between mental and physical predicates; cf. Section 1.) Notice, that in (CF) 'overt aspects' is to be understood broadly enough so as to cover not only behavior but also various environmental factors such as inputs, stimuli, and suitable preconditions for behaving. (The members of μ are taken to be theoretical and those of λ observational in our sense (OT).)

The thesis (CF) corresponds to a principle of "semantic empiricism" that, e.g., Rozeboom (1963) has advocated. I have elsewhere (in Tuomela (1973a)) tried to argue against the adoption of this principle in the case of physical theories. However, for the reasons indicated earlier in this book (CF) seems, by and large, appropriate for psychological theories (and for social theories in general). I say 'by and large' because my claim strictly speaking applies only to the situation described in the Jones myth. When applied to our psychological concepts — as we have them now — (CF) may seem a little strict. However, that is presumably due to the existence of *epistemic* variance and slack (problems related to the use of the μ-predicates in epistemically imperfect cases), I would argue. Purified from epistemic problems the semantic situation seems correctly reconstructible by a suitable specification of (CF).

It should be emphasized here that, as we are later going to specify, (CF) is far from any old reductionist position (such as logical behaviorism) as a semantical doctrine. Furthermore, (CF) is not an ontological thesis and, in fact, inner events and states of S are usually to be regarded as ontologically autonomous, even if semantically they are not.

Below we shall illustrate (CF) and, more generally, postulational concept formation in a simple case. This is done by presenting a fragment of Jones' theory or, strictly speaking, what could have been such.

We are going to consider a psychological theory which is supposed to explicate the meaning of wanting (or at least a substantive part of it). We introduce a set of postulates T which postulates are such that any normal speaker using the term 'wants' must accept them. Still T will be a factual theory, which has contingent consequences. T thus has both a meaning specifying (or analytic) and a factual (or synthetic) component, we argue. We shall assume that our axioms will hold for every agent in a given culture. Whatever interindividual differences there may be with respect to meaning postulates or with respect to factual assumptions will be ignored here and in the sequel.

To begin, we consider the following meaning specifying set of postulates for wanting:

(A1) Agent A wants that p if and only if, given that A has not been expecting p, if A now realized that p is or will be the case, then A would immediately be disposed to feel satisfaction about this.

(A2) A wants that p if and only if, given that A has been expecting p, if A now realized that p would not be the case, then A would immediately be disposed to feel dissatisfaction about this.

(A3) If A does or would find daydreaming about p pleasant, and A believes that there is at least some considerable probability of p's occurring, then, under favorable conditions, A wants that p.

(A4) A wants that p if and only if for every action X which A has the ability and opportunity to perform, if A believes that his performing X is necessary for his bringing about p and that his doing X would have at least some considerable probability of bringing about p, then A is disposed to do X, given that A has no stronger competing wants.

(A5) If (i) it would be unpleasant to A to entertain the thought that p is not (or probably is not) the case, or the thought that p will not be (or probably will not be) the case, and (ii) A believes that there is at least some considerable probability of p's occurring, then, under favorable conditions, A wants that p.

(A6) If A wants that p, then (i) A is disposed to think or daydream about p at least occasionally, and (ii) in free conversation A tends to talk, at least

occasionally, about p and subjects that he believes are connected with p.
(A7) A wants that p if and only if, under favorable conditions, A has a
disposition to want-out-loud that p.
(A8) If A wants that p, then, if A is frustrated in his attempt to bring
about p, then A is disposed to develop latent aggression.

Our theory T = (A1) & (A2) & . . . & (A8) about wanting can be taken to
contain the semantically relevant information about (extrinsic and
intrinsic) wanting.[2] There would be much to be said to clarify them. We
shall here keep our comments to the absolute minimum needed for the
purposes of our present developments.

The phrases 'favorable conditions' and 'disposed to' obviously must be
analyzed so that they don't trivialize the propositions in which they occur.
(See Audi (1973a) for an analysis of the former and Chapter 5 of this
book for the latter.) Here we cannot discuss them or any other of the
over thirty extralogical concepts occurring in these axioms constitutive
of wanting, but have to rely on the reader's understanding of them
(against the background of current philosophical literature on the
topic).

Let us now consider the problem of which of the concepts in (A1)–(A8)
are Rylean concepts (belong to λ) and which are to be regarded as Jonesian
(the members of μ). (Note that we are not only analyzing wanting but also,
in part, several other mental concepts, i.e., all the predicates in μ, occurring
in (A1)–(A8).) I think that some of the extralogical concepts can be
classified relatively easily, and on a priori presystematic grounds. Thus all
the action concepts (e.g., action, doing, bringing about, performing) can
safely be (and must be!) taken as antecedently understood Rylean concepts. Similarly, e.g., free conversation, wanting-out-loud, and talking
belong to Rylean notions.

Clear cases of μ-concepts would be wanting, expecting, believing,
daydreaming, entertaining a thought, feeling satisfaction, latent aggression,
etc. I shall not discuss here how to resolve ambiguities in the case of
borderline cases such as frustration, for I am only trying to illustrate how
to think of (A1)–(A8) from the point of view of functionalist concept
formation. (To give a full conceptual analysis of our example would take
too much space.)

I have not above explicitly used Rylean thinking-out-loud concepts
other than wanting-out-loud in (A7). But obviously that terminology could
have been used elsewhere, too.

One central idea in the type of functionalism we are advocating is the ontological aspect of introducing "the stream of consciousness", i.e., inner mental events, states, and episodes. This is not quite explicit in (A1)–(A8). But, first, we could have written, e.g., "*A* wants that *p* if and only if *A* is in such a state that . . ." in (A1), (A2), (A4), and (A7) and used analogous phrases in other cases to indicate that wantings are states. Secondly, wanting involves basically two types of states (or events). There is first what may be called the duration state which lasts from *A*'s starting to want that *p* till his ceasing to want that *p*. Next, there is a singular state or event which is an inner actualizer of the want (and in most cases a behavior-activator). In our formalization (Ex) Wants $(A, p, x)(= A$ wants that $p)$ the variable x takes as its value such a state (or event). (See Chapters 5 and 6 for a more detailed account.)

One can formalize the theory (A1)–(A8) within the Frege–Davidson programme as indicated earlier and hence essentially within a second order extension of first-order predicate logic. I shall not go into the tedious details here.[3]

It can be argued that none of the axioms (A1)–(A8) alone is strictly analytic *ex vi terminorum*. They also seem to vary as to their "degree of analyticity". Perhaps (A4) is closest to being strictly analytic. Whether it really is strictly analytic depends on how its analysans-concepts are refined and disambiguated. In any case the other axioms would not become superfluous even if (A4) were regarded as strictly analytic. Thus there would obviously have to be strong analytic connections between the analysans of (A4) and the analysantia of several of the other axioms. We would thus still be left with an analyticity problem about wanting.

On the other hand, e.g., (A3) and (A8) seem to be to a less extent analytic than (A4), even if they too must be regarded as intimately connected with wanting.

The theory (A1)–(A8) taken as a totality is clearly synthetic. This can be argued for on the basis of an examination of the individual axioms (cf., e.g., the frustration-aggression hypothesis (A8)) or by showing that (A1)–(A8) in conjunction entail statements which clearly seem to have factual content. Let us consider a couple of such (informally) derived statements as examples:

(i) If, given that *A* has not been expecting *p*, his now realizing that it is or will be the case would immediately lead him to feel satisfaction about this, then A would tend to talk about p at least occasionally in free conversation.

(ii) If A does not want-out-loud that p (under certain standard conditions) then A either does not find daydreaming about p pleasant, or then he does not believe that there is at least some considerable probability of p's occurring.

(Also see Audi (1973a) for similar examples.) I shall not here press this point about synthetic content more because of our insufficient analysis of the extralogical psychological concepts in (A1)–(A8). Still, I hope the general message has been made acceptable: Psychological statements assumed to be constitutive of some concept or concepts and stated in normal "surface" natural language contain synthetic elements.

3. THEORETICAL ANALYTICITY AND MENTAL EPISODES

In the preceding section we argued that psychological theories seem to perform two functions at the same time: 1) they specify the meanings of mental predicates, and 2) they introduce inner entities and thereby require various factual connections to hold between the mental predicates (and entities) themselves and also between mental predicates (and entities) and overt behavioral predicates (and entities). Can these functions be clearly separated? Contrary to many doubters, we claim that in a sense they can. However, one can distinguish between the analytic and factual or synthetic components of a psychological theory only when the theory is stated so that its logical form is made explicit, and even then those components cannot be very "perspicuously" separated from each other. Thus, if we are given the logical forms of our axioms (A1)–(A8) of $T(\lambda \cup \mu)$ none of the axioms themselves will be purely analytic or purely synthetic. The analytic and the synthetic components are to be "ground out" differently.

I shall not here try to give a general philosophical defense of the possibility of distinguishing between analytic and synthetic statements in general on the level of deep structure. Let me just point out generally, first, that language is (ultimately) a product of human beings and their social practice. It thus has to contain some *conventional* elements (irrespective of whether the conventions and stipulations are fully deliberate or not). Secondly, school examples of analytic statements such as 'All vixens are female' can indeed be taken seriously!

Meaning postulates like 'All vixens are female' are rules. Rules can be approached both from a descriptive and a prescriptive point of view. The descriptive aspect of the above rule is, by and large, taken into account by the universal generalization '$(x)(V(x) \rightarrow F(x))$', where '$V$' represents 'vixen', '$F$' 'female', and '$\rightarrow$' material implication.

The prescriptive aspect of this rule is best given in terms of this semantic rule of inference: For every individual x, given that $V(x)$ one is entitled to infer that $F(x)$ (and one ought not to infer $\sim F(x)$); cf. Section 4 of Chapter 3. We could equivalently write this as: For every x, '$V(x) \; \square\!\!\rightarrow F(x)$' is true, where $\square\!\!\rightarrow$ represents a suitable conditional (e.g., Lewis' (1973a) variably strict conditional).

Let me now try to sketch how the descriptive analytic and synthetic components can be separated. This problem has been discussed rather extensively in recent literature. (For instance, Carnap, Wójcicki and Przełecki have done important formal work on this problem.) I have elsewhere (Tuomela (1973a)) presented my own views on the analytic-synthetic problem and will below summarize some of my relevant earlier developments and apply them to the present situation.

We consider Sellars' Jones myth and the "postulational" theory created by Jones. This theory, call it $T(\lambda \cup \mu)$, is meant to be the meaning postulate for the μ-predicates. As we have seen, even when $T(\lambda \cup \mu)$ is regarded as such a postulate it has some synthetic content. Furthermore, we have reason to think that $T(\lambda \cup \mu)$ was not meant to be *merely* a meaning postulate, for the μ-predicates were meant to be explanatory predicates. Thus we can assume that $T(\lambda \cup \mu)$ is a (Jonesian) psychological theory consisting of our common sense platitudes about people's minds and behavior.

Given such a first-order theory $T(\lambda \cup \mu)$, in which λ and μ are as earlier, and assuming (CF), we want to divide $T(\lambda \cup \mu)$ into an analytic component T_A (the set of meaning postulates for μ) and a synthetic component T_S.[4] (Let us first assume, for simplicity, that, counterfactually, there are no analytic connections within the language $\mathscr{L}(\lambda)$, i.e., the language having only members of λ among its extralogical predicates.)

Now we require that (i) T_S must not impose any conditions upon the way of interpreting the predicates in λ and that (ii) T_S and T must not differ in what they say about the part of reality describable by means of the λ-predicates.

Of T_A we require that (a) it must endow the predicates in λ with the same meaning as they have in virtue of the full theory T and that (b) it must be deprived of any observational content (i.e. content in terms of

$\mathscr{L}(\lambda)$). (See Przełecki and Wójcicki (1969) and Tuomela (1973a) for a discussion of these requirements.)

As is often suggested, the Ramsey sentence of a theory may be taken as its T_S-component, while its Carnap sentence qualifies as T_A. (If $\mu = \{P_1, \ldots, P_n\}$, the Ramsey sentence of $T(\lambda \cup \mu)$ is $T^R =_{df} (E\pi_1) \ldots (E\pi_n)$ $T(\lambda \cup \{\pi_1, \ldots, \pi_n\})$, and the Carnap sentence is $T^C =_{df} T^R \to T$.) More generally, one can explicate the above desiderata for T_A and T_S in model-theoretic terms and find a variety of T_A-components for a given theory T. (T_S will remain the Ramsey sentence in the finite case.) This is what Przełecki and Wójcicki (1969) have done in their rather comprehensive and exhaustive investigation to be commented on.

The Ramsey sentence of a theory does not numerically restrict the number of model-theoretic interpretations a μ-predicate can have for a fixed interpretation of the λ-predicates. "Numerical" Ramsey sentences with numerical quantifiers, i.e., statements of the form $(E\pi_1{}^{\leq k}) \ldots$ $(E\pi_n{}^{\leq k})T$, do that (cf. Tuomela (1973a), p. 82). The value $k = 1$ makes all the μ-predicates explicitly definable in terms of λ-predicates in T. Lewis (1970) and (1972) has suggested that theoretical terms should be definable by means of definite descriptions in a sense which practically amounts to this kind of explicit definability. Let us briefly consider his construction.

Lewis' materialist argument relies on the semantic hypothesis that names of mental states are like explicitly and functionally defined theoretical terms where, as we have done earlier in this chapter, one thinks of common-sense psychology as a term-introducing scientific theory. This common-sense "Jonesian" psychology asserts all or most of the platitudes of our common knowledge regarding the causal relations of mental states, stimuli and responses. (Our account adds intentional actions here as well.) Now Lewis assumes that any given mental state M is, by definition, identical with *the* occupant of some causal role R (see Lewis (1972)). On the other hand this occupant of the causal role R is postulated by a physiological theory to be identical with some neural state N. It trivially follows that mental state M is identical with neural state N.

Both of Lewis' premises for his identity argument can be doubted. As science will ultimately tell us whether the second premise is true, I will not comment on it here. Instead I will state some objections related to his first premise. First, while common sense does warrant some kind of functionalist account for mental terms, as argued earlier in this chapter, it does not warrant an explicit definition in general (also cf. Chapter 6).

There is no single property or state, in general, that our a priori semantical intuitions specify as the referent or significatum of a mental name.

Furthermore, as argued at length in Tuomela (1973a), there are a number of methodological reasons for employing *open* (not explicitly defined) predicates especially in growing science. However, Lewis' main underlying reason for defining theoretical terms (or, rather, states and properties) explicitly seems to be the conceptual one of giving them unique denotation (cf. Lewis (1970)). But I am quite willing to live with multiply denoting theoretical terms. I do not think our concept of denotation, as used in the case of scientific terms, can be taken to a priori exclude multiple denotation (cf. Rozeboom (1963)). Perhaps science will ultimately show us that all good scientific terms do have unique denotation, but that is a different story.

I conclude that there are no compelling a priori reasons to restrict psychology to introduce its theoretical states and properties by means of any kind of explicit definition (not at least a definition in terms of λ-predicates alone), and hence not in terms of definite descriptions nor numerical Ramsey-sentences with $k = 1$. It also follows, then, that the analytic component of a theory is to be construed differently.

The above main conclusion is not avoided by claiming that as Lewisian definitions define mental states in terms of overt entities *and other mental states* they are not too strict (as Rosenthal (1976) does claim). The fact is, however, that in the definientia no predicate constants belonging to μ appear. The definientia contain only λ-predicates and thus the definitions refer to other mental states only indirectly via their explicit definientia in terms of λ. In the general case this is too strict and reductionistic. (Rosenthal (1976) uses Lewis' approach to defend materialism. As he does it by means of theoretical reduction he does not really need explicit definitions, and our formalism would indeed do this job well (cf. Tuomela (1973a), Chapter VII).)

Let us next comment on Przełecki's and Wójcicki's (1969) work. As I have argued in Tuomela (1973a), there are some problematic aspects in Przełecki's and Wójcicki's (and, for that matter, Carnap's) model-theoretic constructions, too. What interests us especially with respect to psychological functionalism is that their conditions concern *only* the *expandable* models of T. (A Tarskian model $\mathcal{M}'$ is an expansion of a model $\mathcal{M}$ if they have the same domains but $\mathcal{M}'$ contains some relations over and above those contained in $\mathcal{M}$.) But obviously in the introduction of inner mental events and states the very point is to enlarge the domains of the models of the language $\mathscr{L}(\lambda)$ (and theories in $\mathscr{L}(\lambda)$). If we go model-theoretically from $\mathscr{L}(\lambda)$ to $\mathscr{L}(\lambda \cup \mu)$ merely in terms of expandable models (as we do in the case of the semantical approach of Carnap, Przełecki and Wójcicki), we cannot *guarantee* the introduction of any new entities in the ontology (cf. Tuomela (1973a), pp. 136–139).

Another difficulty with the above model-theoretic approach is, seemingly, that it is subject to a somewhat paradoxical sounding result: if there is some object o in the domain of a model of T which does not belong to the interpretation of any λ-predicate, then we can find another isomorphic model for T in which o is replaced by another object o', which can be a natural number, or a red herring, or whatever you like (see Tuomela (1973a), p. 138). (This result also applies to our proof-theoretic construal, to be given below.)

However, I think this is not really a great difficulty for conceptual functionalism of any sort. Mental occurrences just do not *semantically* have any contentual aspects, and this is what the paradoxical sounding result amounts to saying. On the other hand, I take this result to be disastrous if one's *intended* models as a consequence will have to include red herrings, and the like. This is what happens to Przełecki's (1969) construal, if it is understood to be an enterprise within standard predicate logic (then the intended domain cannot be fixed as Przełecki would like to have it; see Tuomela (1973a), and below).

We can also milden our first criticism against the model-theoretic approach by pointing out that we *here* apply our desiderata to a theory $T(\lambda \cup \mu)$ which already has an enriched ontology. Thus these desiderata do not have to require what is already the case. Furthermore, if inner events and states really are ontologically autonomous, it seems that we should not directly logically (or semantically) connect them to observable events and episodes. Inner events and episodes thus are characterized only on the basis of the μ-predicates and the interconnections between μ-predicates and λ-predicates. — I shall not therefore, on the basis of these considerations, rule out the model-theoretic approach.

A consideration more relevant in favor of a linguistic (or proof-theoretic) approach comes directly from Sellars' Jones-myth. We can understand and reconstruct it to state that Jones starts with a Rylean common sense theory $T'(\lambda)$ which is (finitely) axiomatizable solely in the λ-vocabulary. Then he introduces $T(\lambda \cup \mu)$ as a conservative extension (in the standard logical sense) of $T'(\lambda)$. Of course, we might also want to say that he goes to $T(\lambda \cup \mu)$ via T^R, which might sound more like a functionalist approach. But we know that in general one cannot always go via T^R (see e.g. Tuomela (1973a), p. 61). We must take into account the fact that *new* entities in the ontology will be needed in some cases: i.e. $T(\lambda \cup \mu)$ may have models which are not expansions of models of $T'(\lambda)$ – new entities in the ontology must be introduced. (Though, if $T'(\lambda)$ only happens to have finite models we can do it.)

In case our mythical Jones was a great psychological theorist as well, he may be thought, alternatively, to have introduced $T(\lambda \cup \mu)$ as a *non*-conservative extension of $T'(\lambda)$. Then $T(\lambda \cup \mu)$ would have deductive consequences $F(\lambda)$ solely in the language $\mathscr{L}(\lambda)$ which $T'(\lambda)$ doesn't have. Then there is still more reason to suppose that new entities may have to be introduced into the ontology than we have in the previous case.

A relevant point to note in this connection is that some of the predicates in μ may do some important "observational" work in the sense of being non-eliminable in the semantical Ramseyan sense. More exactly, supposing $\mu = \{P\}$, we say that the predicate P is semantically Ramsey-eliminable from $T(\lambda \cup \mu)$ if and only if every model of $T'(\lambda)$, the maximal recursive subtheory of $T(\lambda \cup \mu)$ in $\mathscr{L}(\lambda)$, is expandable into a model of $T(\lambda \cup \mu)$ without adding new entities into its domain. Thus if P is *not* semantically Ramsey-eliminable, it does do some work in ruling out some observational models ("states of affairs"). For then there will be some models of $T(\lambda \cup \mu)$ which are not obtainable by expansion (without adding new elements in the domain) from any model of $T'(\lambda)$ (but only from a model of $T^R(\lambda \cup \mu)$ in $\mathscr{L}(\lambda)$). This can, however, happen only if $T(\lambda \cup \mu)$ has infinite models. (See Tuomela (1973a) Chapter III for a discussion of semantically Ramsey-eliminable and non-eliminable concepts.)

Keeping to the Jones-myth one may now prefer to consider a proof-theoretic (linguistic) approach for grinding out T_A and T_S. The earlier desiderata for them are satisfied by requiring this (we can take T_S, T_A and T to be non-empty sets of sentences):

(1) T_S is a *synthetic* component of $T(\lambda \cup \mu)$ if and only if $\vdash T_S \equiv T(\lambda \cup \mu) \cap \mathscr{L}(\lambda)$.

(2) T_A is an *analytic* component of $T(\lambda \cup \mu)$ if and only if (a) T_A is proof-theoretically non-creative with respect to $\mathscr{L}(\lambda)$, and (b) $\vdash T \equiv T_A \cup (T(\lambda \cup \mu) \cap \mathscr{L}(\lambda))$.

(See Przełecki and Wójcicki (1969), and, for further comments, Tuomela (1973a).)

It is easy to show that e.g. the following central properties become true of T_A and T_S in the defined sense:

(3) $\vdash T \equiv T_A \& T_S.$

(4)   For every statement F in $\mathscr{L}(\lambda)$,
           $T \vdash F$ if and only if $T_S \vdash F$.

(5)   For every F in $\mathscr{L}(\lambda)$, if $T_A \vdash F$ then $\vdash F$.

We may now, following the suggestion of Tuomela (1973a), formulate a proof-theoretic notion corresponding to the Carnap sentence:

$$T^{C^*} =_{df} T'(\lambda) \rightarrow T(\lambda \cup \mu),$$

where $T'(\lambda)$ is T_S. $T'(\lambda)$ is assumed to be finitely axiomatizable according to the Jones-myth. (A maximal subtheory $T'(\lambda)$ in the language $\mathscr{L}(\lambda)$ is of course not finitely axiomatizable in the case of an *arbitrary* theory $T(\lambda \cup \mu)$.) Thus we can take $T_A = T^{C^*}$ (and, as said, $T_S = T'(\lambda)$). For this choice of T_A we get the interesting additional result that

(6)   If $T(\lambda \cup \mu)$ is a consistent theory with nontautological con-
           sequences in $\mathscr{L}(\lambda)$, then: if $T^{C^*} \vdash F(\mu)$ (F contains only μ-
           predicates), then $\vdash F(\mu)$.

This result, proved in Tuomela (1973a), shows that the purely theoretical axioms of T remain synthetic when $T_A = T^{C^*}$. (For other special properties of T^{C^*} see Tuomela (1973a), pp. 140–142).)

It should be noticed that the predicates in μ are interpreted predicate *constants* and not variables as a strict Ramsey-functionalist (e.g., Lewis (1972)) would like to have it. The fact that the μ-predicates are constants enables them to preserve the "mentality" (e.g., in the sense of incorrigible reportability) these predicates involve, in contrast to the Ramsey-functionalist's topic-neutral treatment.

Our results show that we can have psychological theories with strictly separable analytic and synthetic components. But this distinction is not very "perspicuous" even when the theory is considered in its deep-structure form. (See Tuomela (1973a), p. 136, for an example of individual analytic statements.) Thus, given that λ-predicates are fully conceptual and meaningful independently of the μ-predicates, we can construe psychological concepts without charges of circularity and aprioricity, and the like, often directed against holistic construals like ours. (Note here: in general, we may at least initially need for each 'Xing-out-loud' in λ an 'Xing' in μ; cf. Section 2.)

One aspect about our account still worth emphasizing is this. In the cases when $T(\lambda \cup \mu)$ introduces new elements (states, events, etc.) into the ontology, the formalism of course says nothing about the new elements which goes beyond how the λ-predicates characterize the μ-predicates in T. Therefore, it is still in principle open for a pre-Jonesian to argue that these new elements are some kind of public elements rather than inner elements, for our theory T does not, of course, alone determine this aspect of the factual or extraconceptual character, if any, of the new entities. In this sense our formal approach does not solve the deeper ontological puzzles involved in the dispute between pre-Jonesians and Jonesians (see also Subsection 4.3. below).

Conceptual functionalism does not solve any problems (semantical or other) which have to do with the "hardware" or stuff-aspects of the entities the theory is about. Thus if there are genuine problems about "qualia" (how to account for one's *felt* pain contra something like the concept of pain, etc.) they are to be solved differently. My suggestion is that they could be dealt with in terms of suitable restrictions on the intended or typical domains of the theory (cf. note 1 of Chapter 3). Suppose we are given a set $\mathbf{M_o}$ of typical or intended "observational" models (for $\mathscr{L}(\lambda)$). (Let us take that for granted here, cf. Przełecki (1969).) Then we can partially define a set of intended or typical models $\mathbf{M}$ for the full language $\mathscr{L}(\lambda \cup \mu)$ by:

(7) $\mathbf{M}$ is a set of intended models for $\mathscr{L}(\lambda \cup \mu)$ only if for all models $\mathscr{M} \in \mathbf{M}$, the restriction of $\mathscr{M}$ to $\mathscr{L}(\lambda)$ is a model-theoretic extension of some model $\mathscr{M}_o \in \mathbf{M_o}$ and T_A is true in $\mathscr{M}$.

(See Przełecki (1969) and Tuomela (1973a) for more details and for a fuller discussion.) This characterization cannot in my opinion be strengthened to give a sufficient condition as well (not at least in terms of our present kind of verbal and conceptual tools). It obviously would become subject to the earlier paradoxical "isomorphism result" if we tried to do it directly.

Let me point out here another related consequence of conceptual functionalism, which especially concerns the dispute about *reductive* and *eliminative materialism* in the mind-body literature. Let Jones' theory be $T(\lambda \cup \mu)$ as earlier. Suppose Jones' rival, call him Smith, introduces brain entities (events, states, processes) and concepts designating them by means of a theory $T^*(\lambda \cup \eta)$, where η = the set of brain predicates, and such that T^* and T have the same observational content (i.e., $T'(\lambda)$).

Now, given (CF) and assuming that T and T^* are isomorphic (i.e., given e.g. $\mu = \{P\}$ and $\eta = \{P^*\}$), it holds that $\vdash(E\pi)[T(\ \lambda \cup \{\pi\}) \equiv T^*(\lambda \cup \{\pi\})]$ then, and only then, T and T^* (and hence the predicates in μ resp. η) are equivalent in meaning. Thus, there are in this case no purely verbal or conceptual means for distinguishing between mental talk and brain talk. Whatever descriptive differences, if any, there possibly are here must be due to the intended models of T and T^*, and, it seems, they cannot be completely defined by verbal means. (I doubt that there are other ways either in this particular case. Who can make any "discriminating" sense of unextended, purely "ghostly", Cartesian mental individual events or states?)

How do we, finally, handle the case with analytic statements in $\mathscr{L}(\lambda)$? These are taken into account by relativizing our theory $T(\lambda \cup \mu)$ (as well as T_A and T_S) to the set, say $A(\lambda)$, of meaning postulates for λ-predicates. Thus, it seems to suffice to always assume $A(\lambda)$ in addition to the other premises (whether their set is empty or not) when defining T_A and T_S and when deriving theorems about them.

4. THE INDISPENSABILITY OF MENTAL EPISODES

4.1. We have frequently remarked that one of the most central reasons for the introduction of mental episodes in an analogy account is their *explanatory power*: they were introduced just to explain the verbal and non-verbal behavior of people. In our present context, we may accordingly concentrate on the explanatory power of the μ-predicates (and thus of the entities expressed by them) within Jones' theory $T(\lambda \cup \mu)$. We may ask not only whether they are indispensable within tasks of explanation but also, more generally, whether they are *methodologically indispensable* or, at least, *desirable* within $T(\lambda \cup \mu)$ (or its extensions) on the whole.

We recall that pre-Jonesian people only employ a Rylean language $\mathscr{L}(\lambda)$. We assume that this language has been enriched by semantical discourse so that pre-Jonesians are able to speak about abstract entities, classify verbal behavior, e.g., as sayings, and so on. Pre-Jonesians also employ a comprehensive Rylean behavior theory to explain behavior. As in the previous section, we call this theory $T'(\lambda)$.

How do pre-Jonesians (or neo-Wittgensteinians, to be more modern) then go about explaining behavior? To begin, we take it for granted that pre-Jonesians can distinguish between intelligent and non-intelligent non-verbal behavior and between meaningful (non-parroting) and parroting verbal behavior. (In our later terminology of Chapter 10, intelligent behavior roughly comprises the class of action tokens$_1$.)

How do pre-Jonesians explain non-verbal behavior? Non-intelligent behavior like reflexes (and reflex-like behavior) is presumably explained in something like Skinnerian fashion in terms of overt stimuli and the past conditioning history of the organism. As we are mostly interested in intelligent behavior in this book, we shall not consider this type of case deeper here.

Let us imagine that in our mythical Rylean society people were originally "chatter-boxes". All intelligent behavior was accompanied by some meaningful verbal behavior, some kind of verbal "antecedent" or commentary. (A good approximation to this situation is provided by many Hollywood movies!) A piece of intelligent behavior can now be explained (in a sense) by citing the accompanying verbal behavior, which of course must be understood in the "normal" way by means of the available semantical discourse. Thus suppose an agent says sincerely that, as it is too warm in the room, he is going to ventilate the room by opening the window. This semantical episode – a piece of verbal behavior – explains the agent's subsequent action of opening the window. It classifies and redescribes the behavior by stating the agent's intention "in" the behavior. We may construe our example as an explanation by means of a *practical syllogism* as long as intentions and beliefs in it are construed in a Rylean fashion (see Chapter 7 for more details). Used in this way, an explanation in terms of a practical syllogism gives a kind of hermeneutic understanding of the behavior without reference to any causal determinants of behavior. (There are also other ways that verbal behavior may accompany intelligent nonverbal behavior, but, in any case, the practical syllogism is the most interesting type as it covers, rather precisely, intentional action as we shall later see.)

In principle, pre-Jonesians can also explain intelligent nonverbal behavior *unaccompanied* by verbal behavior. Now they refer to the verbal behavior which in certain other similar circumstances *would have* accompanied this kind of nonverbal behavior. The model of the practical syllogism can be used here again to represent the most central type of explanation. Also a Skinnerian type of explanation might be tried here, but it is very doubtful if any intelligent behavior is or can be conditioned to external stimuli in the required fashion (even if I regard that as conceptually possible).

This brief survey gives the (possible) content of a pre-Jonesian psychology. Their explanatory theory $T'(\lambda)$ is understood to include all the general knowledge (including meaning postulates) needed for these types of explanation (cf. Chapters 8 and 9).

Can and will Jonesian theorizing somehow supersede pre-Jonesian theorizing? I take an answer to this question to be ultimately given by scientific research. Still, we can presently give a number of more or less heuristic reasons pointing towards an affirmative answer.

We recall that the main idea in Jones' introduction of mental episodes is the fact that they are (potential) *causes* of behavior. If such episodes really do exist and if they are referred to in explaining behavior then we have gained something on two dimensions of explanation. First, as in the pre-Jonesian case, we gain in the (semantical) *understanding* of intelligent behavior (cf. especially reason-explanation by means of the practical syllogism). Secondly, we get a *causal explanation* of the piece of behavior in question (cf. Chapter 8).

At least potentially Jonesian explanation may thus seem to clearly supersede pre-Jonesian theorizing. This holds for all of the above ex-plananda. In the case of non-intelligent behavior reference to inner episodes (which in this case may turn out to be purely physiological) may help to explain behavior much better than Skinnerian behavioristic theorizing does. In the case of intelligent behavior (accompanied or unaccompanied by verbal behavior) the situation is as stated in the pre-ceding paragraph.

There is still another, related, general way in which Jonesian theorizing seems to supersede pre-Jonesian theorizing. Pre-Jonesians cannot really (causally) explain the verbal behavior (and dispositions to verbal behavior) of people, it seems. They use (potential or actual) verbal behavior as an explanans of non-verbal behavior, but the only way they can explain verbal behavior would be in terms of some gross correlations (presumably statistical) to some overt physical-social environmental factors. Jonesians of course again have the possibility of referring to explanatory inner episodes. The explanation of dispositions to verbal behavior especially corresponds to the explanation of observational laws by means of Jones' theory (see next subsection and Tuomela (1973a), Chapter VII, especially pp. 200–201).

One central aspect in this connection, emphasized by Sellars (see Rosenthal–Sellars (1972)), is the explanation and the understanding of a report of a thought. Consider the agent's report that it is snowing outside (e.g., 'Ulkona pyryttää' in Finnish). This report presupposes that the agent is *aware* of the thought that it is snowing outside. The thought that it is snowing outside, when reported, *expresses* the thought about the thought that it is snowing outside. This relation of expression can be taken both in a semantical and (presumably) a causal sense. If we consider it in a semantical sense only, I think we can allow pre-Jonesians the semantical discourse necessary for taking into account such metathoughts. However, as we shall later in Chapter 6 suggest, the state or episode of thinking a meta-thought may be *causally* connected to the state of having the corresponding first-order thought. If this is admitted, then of course pre-Jonesian discourse is inadequate for handling the situation. (I assume here that causal connections always require two ontologically distinct entities.)

4.2. We may also approach the problem of the indispensability and desirability of theoretical concepts and entities from a slightly more technical angle to complement our discussion in the previous subsection. As Jonesian inner psychological concepts are theoretico-reportive concepts, general considerations pertaining to the indispensability of theoretical concepts in scientific theories can be applied here. I have dealt at length with these problems elsewhere (see, e.g., Tuomela (1973a) and Niiniluoto and Tuomela (1973)). Here I will only present a couple of clarifying definitions and point out some connections between the psychological case and the general case.

We start by formulating a characterization of causal indispensability for inner episodes, which is also a good explicate of their ontological indispensability. We again divide events and states into inner and external (or overt) with respect to a system S. Let '$C(a, b)$' be a two-place causal relation to be read 'singular event (or state) a is a direct deterministic cause of singular event (or state) b'. I have characterized this notion by means of an explanatory backing law account in Tuomela (1976a) and it will be discussed below in Chapter 9.

Now, using '$+$' for composition in the sense of the calculus of individuals, we define, for a system S:

(CI$_1$)   Inner event or state s is *causally indispensable* (with respect to external events or states) if and only if for some events or states a, b

(1) $C(a + s, b)$, and
(2) not $C(a + s', b)$ for any external event or state s'.

Analogously we can define a notion of causal indispensability for concepts (predicates). This notion is a version of the concept of "logical indispensability for deductive systematization" of Tuomela (1973a), p. 200. We define:

(CI₂)   A theoretical predicate $P \in \mu$ is *causally indispensable* (with respect to the observational language $\mathscr{L}(\lambda)$) if and only if there are descriptions D (in $\mathscr{L}(\lambda \cup \mu)$) and D' (in $\mathscr{L}(\lambda)$) such that

      (1) the causal conditional '$D \rhd\!\!\rightarrow D'$' is true; and
      (2) for all descriptions D'' in $\mathscr{L}(\lambda)$, the causal conditional '$D'' \rhd\!\!\rightarrow D'$' is false. Here D is assumed to essentially contain the predicate P.

The explanatory conditional '$\rhd\!\!\rightarrow$' has been defined (and discussed) in Tuomela (1976a), and it will also be discussed in Chapter 9. Here it suffices to say somewhat roughly that a causal statement '$D \rhd\!\!\rightarrow D'$' is true just in case there exists a causal law which, together with D, deductively explains D'.

In (CI₂) it is assumed that D, D', and D'' are descriptions of episodes (state-event complexes). For instance, we could have $D = F(c)$ & $P(f)$ (omitting reference to the agent). Here c would name an overt singular "stimulus"-event describable as an $F \in \mu$, and f would stand for a singular episode instantiating $P \in \mu$. Analogous remarks can be made about D' and D'', only they are exclusively about Rylean entities.

As I have analyzed '$C(-, -)$' and '$\rhd\!\!\rightarrow$' they are intimately connected with explanatory power. Thus an episode (event or state) causally indispensable in the sense (CI₁) automatically has great explanatory power, and similarly a predicate P satisfying (CI₂) will have considerable explanatory power. (Notice: our definition (OT) of theoretical predicates already guarantees that they have *some* explanatory power.)

In Tuomela (1973a) a broader notion of logical indispensability for deductive systematization was considered, too. In addition, a corresponding notion for inductive systematization was defined there. One could also define causal indispensability analogously for the inductive case (cf. Chapter 11). The general point I am trying to make here is this. Psychological theories and inner mental entities (and predicates designating them) can be treated so as to show that (and *how*) general considerations pertaining to concept and theory formation in the ordinary natural sciences apply to psychology as well. Thus, if one can show that theoretical explanatory entities are in general (or under such and such conditions) indispensable or desirable (in some suitable sense), such considerations are pertinent in the case of functionally introduced psychological concepts, too.

Now, concerning (CI_1) and (CI_2) one cannot of course a priori show that psychological entities and concepts will satisfy them or even weaker criteria of desirability. Thus, there are not any a priori reasons against the possibility that future psychology can or will show such examples to us. *Scientia mensura* – the scientific realist's motto applies here.

It is shown in Tuomela (1973a) that deductive explanation may require the use of theoretical concepts, and that theoretical concepts may be methodologically useful both in the cases of deductive and inductive explanation. Furthermore, scientific change, even evolutionary one, may require non-eliminable theoretical concepts (see especially Chapter VI of Tuomela (1973a)). As shown especially in Niiniluoto and Tuomela (1973), in certain cases theoretical concepts may in addition be even *logically* indispensable for *inductive* systematization.

One interesting case of methodologically indispensable theoretical concepts is related to their role within the deductive systematization and explanation of observational laws (cf. Hintikka and Tuomela (1970), and Tuomela (1973a), Chapter VI). Let us briefly discuss it here because of its intrinsic interest and because we can now interpret those results in a somewhat new light.

These results for the methodological indispensability of theoretical entities relate to the notion of quantificational *depth* (roughly, the number of individuals considered simultaneously in relation to each other $\approx$ the maximal number of layers of nested quantifiers in the theory).[5] This notion of depth concerns the individuals (logically speaking) one quantifies over in one's theory. In the present psychological case these individuals can most plausibly be taken as singular episodes (events, states, processes).

Consider a psychological theory $T(\lambda \cup \mu)$ which includes Jones' theory and which may also contain some additional factual axioms (e.g. a suitable fragment of Freud's psychoanalytic theory). This theory has a certain deductive observational content, which can be obtained by means of Craig's well known elimination theorem (see, e.g., Tuomela (1973a), Chapter II). This observational subtheory is, in general, not finitely axiomatizable.

Our theory $T(\lambda \cup \mu)$ has some depth d as it is formulated in its first-order distributive normal form in Hintikka's sense (see Hintikka (1965)).

Let us write $T^{(d)}(\lambda \cup \mu)$ for it. Now it can be shown that $T^{(d)}(\lambda \cup \mu)$ can be expanded into deeper distributive normal forms $T^{(d+i)}(\lambda \cup \mu)$, $i = 1, \ldots$, which are logically equivalent to $T^{(d)}(\lambda \cup \mu)$. We may now apply Craig's interpolation theorem here in a special way by defining a notion of "reduction" (r) to explicate what $T^{(d)}(\lambda \cup \mu)$ *directly says* about the members of λ. This reduct $T^{(d)}(\lambda)(= r(T^{(d)}(\lambda \cup \mu)))$ can indeed be rigorously characterized (see Tuomela (1973a)). What we then get can be diagrammatically represented by Figure 4.2.

$$T^{(c)} \leftrightarrow T^{(c+1)} \leftrightarrow \cdots \leftrightarrow T^{(d)} \leftrightarrow T^{(d+1)} \leftrightarrow \cdots \leftrightarrow T^{(d+e)} \leftrightarrow \cdots$$
$$\downarrow \qquad \downarrow \qquad\qquad\quad \downarrow \qquad \downarrow \qquad\qquad \downarrow$$
$$r(T^{(c)} \leftarrow r(T^{(c+1)} \leftarrow \cdots \leftarrow r(T^{(d)} \leftarrow r(T^{(d+1)} \leftarrow \cdots \leftarrow r(T^{(d+e)} \leftarrow \cdots$$

Figure 4.2

Here single arrows represent logical implication and double arrows logical equivalence. The upper row represents the various equivalent distributive normal forms of the theory $T(\lambda \cup \mu)$. We assume that $T^{(c)}$ represents the minimally deep distributive normal form of $T(\lambda \cup \mu)$. The lower row of this diagram represents the Craigian reduct (transcription) of T and it is thus given by the infinite sequence $r(T^{(c)})$, $r(T^{(c+1)})$, ...

What interests us especially is that each $r(T^{(c+i)})$ is entailed by, but does not entail $r(T^{(c+i+1)})$ (see Tuomela (1973a) for an exact clarification of the underlying logical situation). However, if the members of μ become explicitly defined in terms of the λ-predicates at some depth $d + e$, from that on we get equivalences also in the lower row, and the observational subtheory obviously becomes finitely axiomatizable.

We may also think of a methodological situation of theoretical generalization and start with some observational theory such as the pre-Jonesians' theory $T'(\lambda)$. We may assume that $T'(\lambda) = r(T^{(d)})$. Here, the depth d measures the number of interrelated *overt* episodes (events, states) we have to take into account *simultaneously* when making observations and when formulating observational generalizations. Now Jones' explanatory theory $T(\lambda \cup \mu)$, which we now freely allow to contain in addition some factual psychological axioms, can be taken as $T^{(d)}$. Of course the explanation of $r(T^{(d)})$ by means of $T^{(d)}$ involves much more than merely logical entailment, but that is not so central here. (Cf. Tuomela (1973a), Chapter VII, and especially pp. 183–204.)

What Jones introduces is inner episodes, but he does it via the μ-predicates (cf. the previous section). However, $T^{(d)}$ is logically equivalent to the deeper theory $T^{(d+e)}$, and in this sense (i.e., at least conceptually) he also comes to introduce new episodes. Although the logical situation itself says nothing about the nature of these new individuals, we may plausibly take them to represent inner mental episodes. (Note also: the depth of a theory is not directly related to the size of the domains of its models, except for the minimum size.)

That Jones' theory really increases our possibilities to deal with more complex event- and episode-complexes is seen when comparing the observational reducts $r(T^{(d)})$ and $r(T^{(d+e)})$. The introduction of $T^{(d)}$ over and above $r(T^{(d)})$ enables us to deal with those complex observational cases with which $r(T^{(d+e)})$ is concerned and with which $r(T^{(d)})$ is incapable of dealing. (Recall that $r(T^{(d)})$ is logically weaker than $r(T^{(d+e)})$. I have called this gain a *gain in deductive observational richness* due to the introduction of theoretical concepts (see Tuomela (1973a), Chapter VI, for further discussion).)

Another gain obtained through the introduction of new concepts is a *gain in economy*, obtainable as follows. The introduction of $T(\lambda \cup \mu)$ to explain $T'(\lambda)(= r(T^{(d)}))$ also gives a more economical description of the observational situation than obtainable by means of $T'(\lambda)$. For we may assume that $T(\lambda \cup \mu)$ can be formulated not only by means of $T^{(d)}$ but also by $T^{(c)}$, $c < d$. That is, we may move to the leftmost distributive normal form $T^{(c)}$ in the upper row of our diagram and obtain a gain in descriptive economy. This gain is due to the fact that, whereas $T'(\lambda)$ requires d nested quantifiers in its formulation, $T(\lambda \cup \mu)$ can be formulated by means of only $c(< d)$ nested quantifiers.

The reader is referred to Tuomela (1973a), Chapter VI, for a more detailed examination of these methodological gains and for scientific illustrations concerning their methodological importance. Let me only emphasize here that these results together with the various other reasons advanced for the noneliminability and/or the introduction of theoretical concepts convincingly show at least this: There are not only "such and such" *possible* cases but also documented *actual* cases in real science where theoretical concepts are methodologically indispensable (or at least desirable in the sense of not "weakening" the methodological situation). To what extent some particular psychological theories include such methodologically indispensable theoretical concepts representing inner episodes must be investigated *in casu*, and at least the issue cannot be resolved negatively on a priori grounds (see subsection 4.3. below).

Although our μ-predicates are meant to be *mental* predicates we have not yet in the present section made use of any properties that mental predicates have but physical ones lack (cf. Section 1, however). I would now like to emphasize here that some (though not most) of the senses in which theoretical concepts can be methodologically indispensable require that these concepts have at least a few direct reporting or evidential uses (cf., e.g., Tuomela (1973a), pp. 201, 220). But this is just a property mental predicates, as distinguished from physical ones, at least currently do have (based on some suitable epistemic feature – e.g., incorrigibility as discussed in Section 1). To be sure, direct reportability is a matter of language learning and social conventions – but I am here speaking of our present social practices, which we, moreover, have had for millennia.

To summarize, given something like our current social conventions for "mental talk", it is ultimately up to science to determine whether predicates called "mental" as well as their designata are needed, no matter how heuristically convincing our above arguments for the non-eliminability of μ-predicates are taken to be. But science also determines what properties these designata have and hence whether they are "mental" or "physical" (provided one can make some initial sense of what a mental *entity*, e.g., an episode, as opposed to a physical one is).

4.3. Does psychological research provide support for Jonesian theorizing as compared with pre-Jonesian or Rylean theorizing? Not surprisingly, very little can conclusively be said presently. This is, in part, because psychologists have not directed their interests towards the crucial questions in this dispute. A few things can be said, however.

We must distinguish between three different disputes. First, in psychology there is an old dispute between mediational and non-mediational theories. Are *mediating* constructs ("organism variables") in some sense necessary for psychological theorizing? Secondly, given that this is the case, are *representational* (or propositional) mediating constructs needed? Thirdly, given that representational constructs are indeed necessary, are *episodic* or episodic-dispositional (as contrasted with purely dispositional) representational constructs needed in psychological theorizing?

As we know, extreme behaviorists (e.g., Skinnerians) notoriously avoid mediating constructs of any kind (or this is the "official" story, at least).

On the other hand, for instance many neobehaviorists do accept mediating constructs as long as they are not mentalistic and representational (cf. Jenkins and Palermo (1964)). Other neobehaviorists (e.g., Osgood (1953)) and, e.g., cognitive psychologists again do accept representational mental concepts, although they have had some difficulty in seeing that this means construing them as explicitly *propositional* attitudes or episodes.

Neither the first nor the second of the above disputes is the one with which Jonesians and pre-Jonesians are primarily concerned. The dispute between them concerns whether representational mental entities can be episodic (and hence categorical) or at least mongrel episodic-dispositional and whether such episodes can have causal impact on behavior. Jonesians answer affirmatively and pre-Jonesians (neo-Wittgensteinians) negatively to this. It is important to see that this third question rather than the first two is at stake here.

Let us still try to clarify the dispute between pre-Jonesians and Jonesians. Let us consider wanting and the explicative axioms (A1)–(A8) of Section 2. Suppose now that a pre-Jonesian accepts these axioms more or less as they have been formulated. This means that he has to construe the predicates in μ in a Rylean fashion.

How does he do it? To answer this, we, for simplicity, concentrate on wanting and restrict ourselves to a consideration of (A6). According to (i) of (A6), if A wants that p then A is disposed to think or daydream about p at least occasionally. Although most of the other explicative axioms for wanting make it a disposition to act, (A6) makes it a disposition for having thoughts. Now havings of thoughts are inner representational episodes for Jonesians. (Cf. our discussion of this in Chapter 5.)

A pre-Jonesian naturally cannot accept this. His way of construing (A6) would be as follows. He thinks of wanting as a stable *second-order* disposition, viz. a (more or less) stable disposition to have *short-term* dispositions to want-out-loud that p, given suitable circumstances. In other words, instead of speaking of a disposition to have thoughts (thinking-episodes) he speaks about a disposition to have short-term dispositions to verbal behavior under such and such circumstances. Furthermore, here all the dispositions (both the first-order and second-order ones) should be instrumentalistically construed so as not to involve any kind of intrinsic bases (cf. our discussion in Chapter 5). Although we have concentrated on wanting here, other representational psychological constructs get their pre-Jonesian interpretation analogously.

Can we not rule out the pre-Jonesian construal merely on grounds of simplicity? Consider my driving a car. Occurrent perceptual environmental inputs produce (that part of) my occurrent behavior by which I try to cope with the demands of the situation. Now, is it not more natural to think of my perceptions as occurrent perceptual processes rather than as short-term dispositions? Note that these short-term dispositions would typically have a very short duration indeed, and there would be plenty of them. Why not rather postulate occurrent perceptual processes, which also carry with them a promissory note concerning their neurophysiological bases? This is not a strong argument, however. It is rather up to science to find out whether μ-predicates (in our Jones myth) are *essentially* Jonesian (and thus essentially beyond the resources of a pre-Jonesian). If these predicates truly designate something inner and (what is regarded as) mental then Jonesian theorizing has become vindicated. For instance, if psychophysical event-event laws, e.g. synthetic identities, are found, then we would know not only that some mental event-predicates truly designate something but also that what they designate in fact is something describable as neurophysiological. Also weaker forms of correlation would support the Jonesian view that mental predicates do designate at least something; and, more generally, the theoretical terms of any well corroborated (Jonesian) psychological theory can be judged prima facie to designate something.

To go to another limiting theoretical possibility, science might alternatively tell us that the Jonesian and pre-Jonesian languages do not essentially differ as to their ontologies (viz. with respect to the *kind* of entities they are about). Thus Jonesians and pre-Jonesians would in principle have *syntactically*, *semantically*, and *ontologically* the same object language and they would also pragmatically *use* their languages similarly. Jonesians would accordingly have to give up their metalanguage in case it explicitly makes mental terms refer to inner mental entities (differing thus from the pre-Jonesians' metalanguage). This is what the defeat of Jonesian theorizing would mean. All the other possibilities imply the defeat of pre-Jonesian theorizing. *Scientia mensura*.

What does present psychological theorizing and evidence then say about various disputes?

First, it seems fair to say that psychological evidence has shown us that "organism variables" are needed in psychological theorizing, and it also seems that representational mediating constructs are needed. Rather little evidence is, to my knowledge, available for deciding about the need for episode-representing constructs, however.

The evidence one may avail oneself of in answering the above queries is of two sorts, broadly speaking. First, there are psychological theories (or "models") which postulate theoretical entities of the disputed kind and which have received at least some indirect experimental support. Secondly, there are experimental results which give rather direct support for the existence of such theoretical entities.

Of course, there are lots of psychological theories employing organism variables. I will not discuss the need for organism variables in general at all here (cf. Tuomela (1973b) on neobehavioristic theories), nor will I consider the varieties of theories employing representational "mediating" constructs (cf., e.g., the "classical" personality theories, e.g. Freud's, Jung's, Murray's, Lewin's, and Rogers' theories).

Instead I would like to mention modern information processing theories, which explicitly postulate cognitive processes. Good examples of such theorizing are Newell and Simon (1972), Neisser (1967), and Lindsay and Norman (1972). Such information processing theories are typically relatively easily programmable and thus simulatable by computers. The postulated cognitive processes are represented in computer programs as successive changes from one internal (logical) state of the computer to another (cf. our discussion in Chapter 5). It seems to me that the era of information processing theories is only at its beginning, thus making it difficult to rationally judge the evidence for the postulated cognitive processes in that field. Therefore, we do not here get anything like a hard proof for the adequacy of Jonesian theorizing. It is important to notice, however, that information processing psychology presupposes Jonesian theorizing and thus representational mental episodes.

Another example of Jonesian theorizing that we may take up here is given by Osgood's representational mediation theory, which in a sense originates from Hull's theory (see e.g. Osgood (1953), (1968), (1970), (1971)). In his theory Osgood postulates "meaning responses" (r_M) which are occurrent representational mental *events* introduced to explain an agent's linguistic behavior. These r_M's are *componential* events which can be compared with linguistic phonemes (or, better, sememes) in this respect. The r_M's in Osgood's theory are also open and non-reducible theoretical constructs, as I have argued (Tuomela (1973c); also cf. Tuomela (1973b)).

Osgood's r_M's are supposed to account for the psychological or pragmatic meaning of both words and sentences (and thus, e.g., for the meaning-similarity, rather than physical similarity, of signs). As for sentence meanings, the theory is about as insufficient as any other theory in the field. In any case the r_M's are representational events which can be taken to explain not only linguistic behavior but, I think, also the immediate representational psychological determinants of linguistic behavior. For we may take a person's activated proattitudes and beliefs to be such immediate determinants and Osgood's r_M's can be regarded as explaining, in part at least, how a person forms his beliefs and intentions relevant to his "symbolic" behavior (cf. Osgood's use of his congruence theory to this effect in Osgood and Richards (1973); also cf. our remarks on intention generation in Chapters 6 and 12 below). Although I will not attempt to judge the closeness to truth of Osgood's theory, we can at least cite his r_M's as good examples of occurrent representational psychological events in current psychological theorizing.

Let us then briefly consider experimental psychological evidence. There is rather much evidence to show that mediating constructs and even representational constructs are needed. This evidence comes from such diverse fields as verbal learning and conditioning (see e.g. Bruner, Goodnow, and Austin (1956), Dulany (1968), Maltzman (1968)), various vicarious learning processes (cf. Bandura (1969)), symbolic self stimulation (cf. McGuigan and Schoonover (1973)). This list could easily be continued and lots of other equally demonstrative experiments could be cited. This is not the place for an exhaustive review of the results, especially when these results after all seem to be open to the pre-Jonesian interpretation (cf. our earlier remarks).

Let us here consider in detail one example which shows the importance for behavior of how stimuli are *perceived* (interpreted). The example is that reported in Dulany (1968). A brief summary of his experiments and results can be given as follows.

Dulany placed the experimental subjects in a hot chamber (110 degrees Fahrenheit) and presented them with a verbal learning task. On each of 100 trials subjects were to choose between two sample sentences. The subjects' choices were followed, depending on choice, by a blast of air which was either at chamber temperature, 40 degrees cooler, or 40 degrees warmer. This air blast was presented following the choice of the one sentence in a pair which contained the article 'a' prior to a certain word.

Some of the experimental subjects were told that the air blast signified a correct choice, some were told that it indicated an incorrect choice, while others were told that it had no relation to the correctness or incorrectness of the choice. Dulany's results indicate that the effects of the air blasts were substantially influenced by the label (viz. reinforcing, punishing, or neutral) placed on them. Thus, when subjects perceived a blast of hot air as punishing they reacted to it by reducing their choice of the punishing response. But when the subjects perceived this same stimulus as a reinforcer they increased their choice of that sentence.

The same was the case for cool air blasts. Although this stimulus should have been physically very reinforcing, it could be made punishing essentially by simply telling the subject to interpret it as signifying an incorrect choice.

Dulany's experiment demonstrates rather dramatically the need for representational psychological constructs in psychological theorizing. But, as indicated, this experiment does not necessarily require Jonesian perceiving episodes, even if Dulany himself speaks of "causal volitional mental processes" when discussing his experimental results. A pre-Jonesian interpretation in terms of second-order dispositions, as discussed earlier, still seems theoretically possible here.

Experiments which connect psychological states with physiological ones seem, at least in principle, to better support Jonesian theorizing than experiments only involving the agent's actual overt behavior. For instance, in brain potential experiments, in people intending to perform some actions such as moving their hand, or even uttering something, one can notice (what has been interpreted as) "intention waves" preceding the action (see e.g. Becker *et al.* (1973), Pribram (1969), McAdam and Whitaker (1971), and our discussion of their results in Chapter 6 below). These types of experiments may ultimately give us neat psychophysical laws connecting representational mental entities with neurophysiological events in a way which forces a Jonesian event-interpretation on these mental entities. Presently, however, we are still far from having reliable knowledge of that kind.

NOTES

[1] We may consider Freudian cases of subconscious intentions to demonstrate this. Notice, however, that the psychoanalyst's ultimate criterion is the subject's acceptance, and in this sense the subjects's epistemological authority, of the former's interpretation.
[2] These axioms are well known in the literature – either exactly in their above forms or in closely related forms (cf. especially Audi (1973a)).

³ As an example I give a (somewhat incomplete) formalization of (A1) as follows:

(A1*) $(A)(p)\{(Ey)W(A, p, y) \equiv (x)(z)(u)[(B(A, \sim p, x)\ \&\ T(x) < t_0\ \&$
$R(A, p, z)\ \&\ T(z) = t_0\ \&\ N(A, u)) \rightarrow (Ev)S(A, v)]\}$

In (A1*) 'W' represents wanting, 'B' expecting (or believing), and 'R' realizing; all of these being dispositional states. 'T' stands for the time operator, and its value 't_0' for 'now'. 'S' stands for the occurrent state of feeling of satisfaction, which is the manifestation of the disposition to feel satisfaction under some suitable, presumably complex, normal conditions $N(A, u)$.

Let me emphasize that (A1*) must be nomic and must thus have explanatory force. This epistemic feature may be taken into account as in Tuomela (1976a) (also see Ch. 5 and Ch. 9 below) in order to get a satisfactory analysis of the dispositional element in (A1*) and to "explain" the apparent extensionality of (A1*). But here we can be satisfied with the mere logical form (A1*).

⁴ For simplicity, we shall not below consider any theories, nor any version of (CF), involving *probabilistic* statements.

⁵ Formally, the *depth* $d(T)$ of a closed sentence T can recursively be defined as follows:

$d(T) = 0$, if T is an atomic statement or an identity; $d(T_1\ \&\ T_2) = d(T_1 \vee T_2) =$ the greater of the numbers $d(T_1)$ and $d(T_2)$; $d(Ex)T(x)) = d((x)T(x)) = d(T) + 1$, where in $T(x)$ x has been substituted for some constant in T.

CHAPTER 5

PSYCHOLOGICAL DISPOSITIONS

1. A REALIST ACCOUNT OF DISPOSITIONS

1.1. Using our earlier terminology, both pre-Jonesians and Jonesians can speak about the dispositions (to act) of an organism. However, the pre-Jonesians construe dispositions in an instrumentalistic (or Rylean) fashion whereas Jonesians do it "realistically". In the case of psychological dispositions like wanting, believing and intending, Jonesians thus introduce categorical episodic features to go with them.

We shall start by sketching a realist account of dispositions in general. The resulting construal is later shown to be applicable to psychological dispositions of the mentioned kind, too. It should be emphasized that in this section we argue for a realist construal of dispositions without direct recourse to the Jones myth. If wants, beliefs, etc. indeed can be regarded as dispositional we have a slightly different route to the same realist end to which the Jones myth has in effect already led us.

One aspect about dispositional concepts that, perhaps, has not been sufficiently emphasized in the extensive literature on the topic is the dynamic and evolving nature of these concepts. Roughly speaking, in the initial phases of theorizing and theory-formation dispositional concepts are introduced as some kind of pure intervening variables or summary concepts for certain observable stimulus-response type behavior of the system studied. In such initial phases dispositions (dispositional concepts) are semantically, epistemically and ontologically reducible, in some rather strong sense, to data concepts (via observational generalizations and theories). In more advanced stages of realist theorizing dispositional concepts achieve more autonomy in all of the above three dimensions. They thus get closer and closer to full blown realistically conceived theoretical concepts. However, I would argue that, for conceptual reasons, it is essential that they preserve their intimate connections to observable stimulus-response behavior of the system. This is what distinguishes them from some other theoretical concepts. Yet, as dispositional concepts can also be regarded as more or less full blown theoretical concepts, it is not surprising that many arguments for the indispensability and desirability of theoretical concepts apply to the case of dispositional concepts, especially to the so called basis-problem of dispositions (i.e., do dispositions always have intrinsic bases?).

What, then, is common to such widely differing dispositional concepts as being soluble, fragile, brittle, generous, irritable, as well as wanting and believing? One general common feature seems to be that if the object having such a dispositional property (call it D) is subject to some kind of stimulation, or if something is done to it, it will, as a causal effect of this, respond in some specific manner: if an x having the disposition D is F'ed (under suitable circumstances C) then x will G in response. Here, D could represent solubility (in water), F putting in water, G dissolving, and C "standard" circumstances (as we understand them in this kind of case).

But dispositions also have something else in common: they involve a *basis* property. This is essentially the core of the realist interpretation of dispositions. As this claim about the existence of a basis may seem very problematic we must take a closer look at the matter.

We shall make some points in defence of the following claim:

(B)   Dispositions essentially involve intrinsic basis properties.

In (B) 'intrinsic' means intrinsic to the object or system studied (in distinction to a property relating the system somehow directly to its environment). We think that this intrinsic-extrinsic characterization can be sufficiently clearly understood for our present purposes. Now we can argue as follows.

For every true contingent proposition there must be something (actual) in the world which makes it true. (Here truth can be taken in a broad sense including the assertability-sense so that conditionals, e.g., counterfactuals, will be included.) Similarly, for every false contingent proposition, there is something actual in the world which makes it false. Now consider an unmanifested disposition, e.g., solubility (in water). Suppose an object x is soluble at t but loses its solubility at t', $t' > t$. Now (given that it has not been put in water during the time period concerned) there must be something actual in the world to account for the change in the truth value of 'x is soluble' during that time period. There is no sense in an account that allows unexercised potentialities to vary independently of any corresponding categorical changes. (For a related argument along these lines see Armstrong (1969).)

If we are granted the above conclusion we still have to show that the actual change in the world must have been a change in x's intrinsic or non-relational properties. I argue for this on two dimensions, so to speak. First, there is a semantical or conceptual line of argument. We can read 'x is soluble' as 'x is *such that* if it is put in water it dissolves' (cf. Mackie (1973)). The phrase 'such that' indicates the presence of some property which accounts for x's dissolving when put in water. Similarly, the absence of this property, supposing there to be only one such property, accounts for the lack of such nomic dispositional behavior.

Furthermore, we may claim on semantical grounds that merely the observable stimulus-response behavior is never enough for ascribing a relevant dispositional property. Practically anything put in water gets wet, but this in itself is not sufficient for ascribing any "getting-wet" disposition to an object. Somehow, then, the intrinsic nature of the object or its properties are involved in dispositional ascriptions.

In many typical examples we seem to be able to classify the objects exhibiting a certain stimulus-response behavior by indicating their *nature*: x is sugar, and that is why it dissolves when put in water. Thing-predicates like 'sugar', 'salt', 'gold', etc. give the natural kind of the objects concerned. Note, however, that the 'that is why' does not (usually) serve to give a "proper" or at least a very good explanation of the stimulus-response behavior exhibited because there is an analytic (or almost-analytic) connection between being sugar and dissolving when put in water. If a lump of sugar would not dissolve when put in water in normal circumstances, we would hardly speak of its being sugar any more. The object would have changed its nature, we would say.

However, natural kind predicates like 'sugar' in any case carry with them the promissory note that something "really" explanatory lies behind the manifest surface. Namely, sugarness can be explained in micro-structural terms by saying that a sugar lump has such and such crystallic structure. This microstructural property (say) P (such that sugar *is* P) is what we normally would like to consider as the basis property of sugar's dissolving disposition.

This last remark leads us to the second, epistemological (and partly also ontological) type of argumentation for the necessary existence of basis properties. The idea is simply that a basis property (which is in some cases a state) causally accounts for and hence explains the G'ing of x when x is F'ed. Thus independently of how we semantically carve up the world and construe (and refine) our disposition concepts, there is something underlying in reality to support the conceptual structure.

Adequate semantics and semantic reconstruction must go hand in hand with what we know or expect to know about the world.

The claim to be considered here is specifically that the extra causal contribution provided by the disposition really is due to a basis rather than some relational feature of the object (which could be a suitable environmental or perhaps ontogenetic property). Not surprisingly, we are facing here an aspect of the general problem whether theoretical concepts (or, here rather, properties) are indispensable in some strong causal sense. I have elsewhere (Tuomela (1973a)) extensively discussed and argued for the indispensability or desirability of theoretical concepts and properties. The sense of causal indispensability needed here is just our (CI_2) of Chapter 4. It can be applied to our purposes in the obvious way as follows. We again distinguish between a set λ of "observational" predicates and a set μ of "theoretical" predicates representing properties intrinsic to the system studied. To λ belong all the observable stimulus-response predicates needed in dispositional talk, and μ of course includes candidates for basis predicates. Otherwise our earlier comments on (CI_2) apply here. We thus get:

(CI_2) A predicate $P \in \mu$ is *causally indispensable* with respect to a language $\mathscr{L}(\lambda)$ if and only if there are event-describing statements D in language $\mathscr{L}(\lambda \cup \mu)$ and D' in $\mathscr{L}(\lambda)$ such that

(a) the causal conditional '$D \vartriangleright\!\!\to D'$' is true; and
(b) the causal conditional '$D'' \vartriangleright\!\!\to D'$' is false for all D'' in $\mathscr{L}(\lambda)$. Here D is assumed to include P and F essentially, D' is assumed to similarly include G, and D'' to include F, where F and G are stimulus- and response-predicates, respectively.

As discussed in Chapter 4 and Tuomela (1973a) there are lots of both actually realized and conceivable cases in which theoretical concepts and properties are indispensable in senses closely related to our situation concerning causal indispensability. Most of my arguments there can be applied to argue for the existence of causally indispensable categorical bases in the above sense. Although the matter is ultimately to be decided *a posteriori* on scientific grounds, several convincing examples from science and everyday life lend support to the claim that (what we have called) dispositions always involve bases. It would indeed be very incredible if, say, two pieces of glass having exactly the same intrinsic nature would exhibit different overt "fragility-behavior". (Apart from possible cases of the existence of different bases, the converse statement seems justifiable, too.)

If what I have just said is acceptable, what is more natural than to linguistically construe and refine our pre-systematic disposition concepts in the realist fashion so as to involve essentially some basis or other. Then our conceptual system and the world are in desired harmony: It is a conceptual truth that there exists a basis for each disposition. But the truth of this latter thesis is not *a priori* but *a posteriori*. That is, if science ultimately finds out that each disposition has a basis, what thus is found out is a necessary truth.

Note that we have not spoken about *what* the basis is or will be in any given case. It is completely up to scientists to find out the specific states or properties qualifying as bases.

Armstrong (1969, 1973) has argued that bases of dispositions are always *states*. If states are something which an object can be in and not be in without this affecting the *nature* of the object, then it seems that Armstrong's claim is too strong. I think that (if we exclude baseless dispositions) we must admit thing-kind predicates as possible bases. These thing-kind predicates are even allowed to belong to the "manifest image" as contrasted with the "scientific image" (to use Sellars' well known but not the clearest distinction). Thus both of the identical predicates 'water' and 'conglomeration of a certain kind X of H_2O-molecules' might qualify as bases.

1.2. On the basis of our discussion we now proceed to a kind of general "definition" of dispositions which have been explicitly analyzed as dispositions to G if F'ed. We represent 'x_1 has the disposition D to G if F'ed, at time t_1' by '$D_{G;F}(x_1, t_1)$' or, for short, by '$D(x_1, t_1)$' and try to clarify this time-relativized "single-track" predicate in terms of the following (not purely semantic) analysis:

(*1*) $D(x_1, t_1)$ if and only if
$(E\beta)(E\gamma) [\beta(x_1, t_1) \& \gamma(x_1, t_1) \&$
$(x)(t)(\beta(x, t) \& \gamma(x, t) \to (F(x, t) \to G(x, t)))$ is a true causal law such that it, together with some true singular statements $\beta(x_j, t_i)$, $\gamma(x_j, t_i)$, and $F(x_j, t_i)$, causally deductively explains any true singular statement of the kind $G(x_j, t_i)$, i.e., the causal law backs the singular *causal* conditionals of the form

$\beta(x_j, t_i) \& \gamma(x_j, t_i) \& F(x_j, t_i) \mathrel{\rhd\!\!\rightarrow} G(x_j, t_i),$

for all $i, j.$]

There are many comments to be made concerning this second-order definition (or analysis). Let us first go through the notation. Let first B and C be arbitrary predicate *constants* replacing the *variables* β and γ in the above definition. The predicate B represents a basis property ('basis' broadly understood); C represents the "normal" circumstances required to hold before a dispositional ascription applies. F and G represent the overt "stimulus" and "response" as before. All of the predicates have been relativized to time. (We accept the definition $B(x) =_{df} (t_i)B(x, t_i)$ and allow locutions such as 'x is salt at t_i'.)

The basis idea in our analysis is to analyze a disposition by reference to a basis and to a "basis law" which *explains* the object's G'ing by reference to it being F'ed, assuming the presence of a basis and the obtaining of normal conditions. In particular, the law does its explanatory work at time t_1. The deductive explanation in question is technically analyzed in terms of the DE-model of Tuomela (1973a); also cf. Chapter 9 below.

Let it be remarked that the law could also be, e.g., a "bilateral" one, i.e., have '$\equiv$' instead of the second '$\rightarrow$'. That does not affect DE-explainability nor anything else said here.

It can easily be seen that our time-relative analysis avoids the "mutability" difficulty affecting Carnap's original analysis of dispositions using bilateral reduction sentences. According to that analysis any object becomes absolutely soluble if it is put in water and dissolves and any object put in water and failing to dissolve becomes absolutely insoluble. Thus it is not possible in Carnap's analysis that an object changes its solubility between two times t_1 and t_2.

In our analysis there are, so to speak, three elements of "necessity". First, any disposition must necessarily have a basis (in the sense explicated earlier). Secondly, there is an element of nomic-explanatory necessity in the requirement that F'ing cum the basis causally necessitates G'ing. This aspect of necessity is formally accounted for and explicated by our conditional $\rhd\!\!\!\rightarrow$. Thirdly, the basis is necessary for F'ing to necessitate G'ing. This follows from the definition of $\rhd\!\!\!\rightarrow$ (see Chapter 9 and Tuomela (1976a)).

The reader may be puzzled by the fact that the law in our analysans is formulated in terms of a material implication $\rightarrow$ rather than some stronger implication, even if the causal conditional $\rhd\!\!\!\rightarrow$ is used in the case of singular instances of the law. Let us first notice that we are not analyzing lawhood here. Instead we take for granted that such an analysis can be given and in such terms that a law could have the above form (thus not all true generalizations of the above form have to be laws or have a high degree of lawlikeness). In fact, I have elsewhere (see, e.g., Tuomela (1976a)) discussed the problem of lawhood and taken as a central necessary condition its explanatory power. Thus an essentially general non-probabilistic statement must qualify as a premise in a DE-explanation in order to be a law. That will also guarantee that a law supports counterfactual and subjunctive singular conditionals stated in terms of $\triangleright\!\!\rightarrow$. As clarified in Tuomela (1976a) this represents the *prescriptive* dimension of lawhood.

It should be noticed that as DE-explanation is not closed under logically equivalent transformations of the explanans, it follows that lawhood is not either closed under logical equivalence. This takes care of one commonly expressed difficulty for any analysis employing the material implication or related truthfunctional connectives.

Next, a law will have to be a true statement. This requirement has the epistemic counterpart that the law must be well supported, both empirically and theoretically (see Niiniluoto and Tuomela (1973), Chapter 11, for an analysis of this). As a consequence, the fact that $\sim(Ex)(Et)(B(x, t)$ & $C(x, t))$ formally entails the falsity of our analysans-law causes no harm as that statement will always be false for laws (cf. Pap (1958) for an advocate of this type of criticism). As our analysis is formally given in terms of ordinary predicate logic, the model theoretic semantics should be clear enough on the basis of our discussion, perhaps apart from the existential quantification over β and γ (see below and Tuomela (1973a), Chapter III).

Next we have to say something concerning what kind of predicates qualify as substituents for 'β' in our analysis. The basic requirement is that such a predicate B should be capable of occurring in laws. This is what we in any case must require of natural kind-predicates and of any "good" projected, etc., scientific predicates in general. The "goodness" of predicates (in that sense), lawhood, and explanation go together roughly in the manner sketched above (also see Niiniluoto and Tuomela (1973)).

We cannot here attempt to give a detailed philosophical analysis of what such good scientific predicates are. One more formal remark has to be made, however. It is that, on *a priori* grounds, B must be either an atomic (non-compound) predicate or a conjunction of atomic predicates or possibly, though I am hesitant here, a negation of an atomic predicate (or of a conjunction of negated atoms). (Note: I am here allowing only the conceptual possibility of negative predicates, not more.) Other truth-functional compounds are thus excluded and, to mention a special case, so are predicates defined by individual constants (see Pap (1958)).

Should we require that B be a theoretical predicate as contrasted with observational predicates (e.g., in the sense of Tuomela (1973a))? This suggestion would be in accordance with our realist analysis of dispositions were it not that we decided to possibly accept 'salt' and other observational thing-kind predicates instead of their structural counterparts (e.g., 'has such and such crystallic structure', etc.). In connection with this, it should, however, be kept in mind that our analysans-law containing B should not, of course, be (at least completely) analytic. This poses a clear restriction on basis predicates: if an observational thing-kind predicate would make the law analytic one should use a suitable (theoretical) structural counter-part-predicate instead.

Our analysis does not identify dispositions with their bases (contra, e.g., Armstrong (1973) and Rozeboom (1973)). Our bases will have contingent (and lawlike) relationship to the observable dispositional stimulus-response behavior, whereas our dispositions are conceptually linked to that (see our definition). I admit, however, that in the case of dispositions occurring in some developed theories the analytic connection to the observable stimulus-response behavior may be less clear. These cases would be borderline cases where dispositions can (almost) be identified with their bases, which are usually some theoretical properties. Then instead of the simple law in our definition we would be operating with an arbitrary theory — or, more exactly, with the Ramsey sentence of such a theory.

Can there be several bases for the same overt dispositional behavior? Can the same basis lead to different overt behavior in the case of different objects? Our analysis gives an affirmative answer to both of these questions. Concerning the second, surely individuals may differ with respect to some parameters to produce the situation in question (this is in fact typical in the case of personality traits). As to the first question, we are here deal-ing with one aspect of the general problem of causal and explanatory overdetermination, which I cannot discuss here (see Tuomela (1976a)).

There is one more central point to be made concerning the adequacy of our definition (1). We may ask whether it really captures exactly those concepts called dispositions.

One thing to be noticed here is that what scientists and philosophers call dispositions does not comprise a well defined set at all. Our remarks

about the dynamic and changing nature of scientific concepts are pertinent here: people's linguistic practices concerning what they call dispositions may conceivably change as a function of the methodological stage
of theory formation, without there having to be any changes in the
elements our (*1*) deals with. Let us briefly examine this apparent argument
against our analysis.

If our arguments for the existence of bases given in subsection **1.1.**
above are acceptable, our analysis seems to give at least a necessary condition for $D(x_1, t_1)$. But does it really give a sufficient condition as well,
or is it too wide? It is in answering this question that the methodological
stage of theorizing and people's linguistic practices are relevant. For (*1*)
in effect states that there are certain nomic connections between either
natural kinds or theoretical properties (bases) and some observational
properties such that these nomic connections are worth emphasizing by
calling the whole complex (i.e., a putative right hand side of (*1*)) a disposition. At some relatively early stages of theorizing this may be a
methodologically useful linguistic practice, whereas later, when much
more is known, that linguistic practice is dropped.

Thus, e.g., the natural kind water may be claimed to make true the
right hand side of our (*1*) by substituting something like 'a conglomeration
of kind X of H_2O molecules' for β, and by using 'having a density less than
one' for F, and 'floats' for G in it. But, the objection goes, nobody
would say that 'water' is a dispositional word. How do we answer this
criticism? First of all, I think that, contrary to the objection, 'water' *is*
a dispositional word in important respects. For instance, all objects with
density less than 1 will float when put in water. However, we must recall
here that our (*1*) only is concerned with dispositions explicitly analyzed
to have non-empty F's and G's. Water "as such" does not thus qualify.
We do not even have a good term for the present case with F and G
specified as above, although undoubtedly water under this buoyancy
aspect is dispositional. We might without harm let "water-as-buoyant"
be a disposition, and 'water' analyzed, for instance, in this way be a
dispositional word, as follows from (*1*). It is only that at the present stage
of science we know too much about water to make it interesting to
emphasize its observational dispositional features as contrasted with its
theoretical categorical features.

Should we now try somehow to relativize our (*1*) with respect to the
state of science (e.g., by requiring that the science or inquiry be "young"
in some sense) or should we go on insisting against linguistic practice

that most (?) scientific concepts are indeed dispositional concepts? (Note that, e.g., concepts representing events or states which *manifest* dispositions typically are not dispositional themselves. Dispositions can be clarified and understood only on the basis of an antecedent dispositional-categorical distinction, which may, of course, be problematic.)

One way out of our dilemma would be to emphasize the first horn, but — because of the difficulty in taking it explicitly and exactly into account — rather speak about scientists' linguistic practices within a given research paradigm. Let us propose the following additional condition to the "definiens" of our (*I*):

(L)   It is in accordance with the current linguistic practices of the scientists (of the scientific community under discussion) to call *D* a dispositional predicate.

This additional qualification (L) brings some circularity into (*I*), even if what *is* dispositional and what a group of people *call* dispositional are clearly different things.

Our additional condition is somewhat vague as it stands, but it is not easy to make it more precise. One suggestion here would be to say, somewhat normatively, that as soon as a *specific* basis property is found to replace the variable β scientists should stop calling the conglomerate a disposition.

In this book we may leave the acceptance of condition (L) optional and thus dependent on how much one feels himself bound by accepted linguistic conventions. Below we continue discussing (*I*) in its original form.

Our definition for $D(x_1, t_1)$ concerns only a singletrack disposition, i.e., a disposition which is manifested in one way only. A corresponding definition for multitrack (say, *n*-track) dispositions can be given in the obvious way. The only change is that we have *n* laws and corresponding number (at most) of C_i's, F_j's, and G_k's $(i, j, k = 1, \ldots, n)$. I shall not bother to write out the definition explicitly here. One thing to emphasize is that the same basis *B* will be responsible for the different *n* patterns of overt behavior. This is surely a great strength of a realist analysis of dispositions. It is very hard to see how a Rylean type of instrumentalist analysis (operating without bases) could possibly give a satisfactory account of multitrack dispositions.

A couple of extensions of our analysis can be mentioned here. First, time-independent dispositions corresponding to $D(x_1, t_1)$ can simply be defined by:

(2) $D(x_1)$ if and only if $(t_i)D(x_i, t_i)$ for the same basis B.

An analysis of *power* is obtained as follows. To say, for instance, that salt has the (passive) power to dissolve or that an agent has the (active) power to perform a (generated) action entails that there are suitable circumstances C and that there are suitable stimuli or means such that the overt response will come about as a consequence of these antecedents obtaining. Thus if we let $P_G(B^*)$ stand for 'thing-kind B^* has the power to bring about G' we can define $P_G(B^*)$ on the basis of our (*1*). The only essential change is that now we quantify over F and C (instead of B and C). There is no need to write out the definition explicitly here. It is easy to see that (and how) the resulting definition, as well as our definition (*1*), can be "probabilified" by requiring the explanatory law to be probabilistic and the explanation to be a suitable kind of inductive-probabilistic explanation. This gives an analysis of probabilistic dispositions and powers (see Chapter 11). This will obviously be needed at least if, after all, indeterminism is a true doctrine.

An interesting topic relevant to any disposition analysis is how to explain laws such as those occurring in our (*1*). In general we do it by a suitable scientific theory (for my own account see Tuomela (1973a)). If B is a predicate within the manifest image (e.g., water) then the law will be given a microexplanation which connects B with suitable micropredicates. Here we cannot, however, discuss this broad topic of reductive explanation (involving both thing-kind predicates and attributes) at all.

2. PROPOSITIONAL ATTITUDES AS DISPOSITIONS

2.1. Most psychological concepts related to man's inner world are dispositional or dispositional-categorical. This is true especially of concepts representing (in some way or other) conceptual states such as those corresponding to propositional attitudes. Below we shall briefly discuss the dispositional and categorical (including episodic) features of such propositional states as wanting, intending, believing, perceiving, and so on. There are also other interesting dispositional notions in psychology, such as personality traits (e.g., courageous, thrifty), but we will not discuss them here.

According to our conceptual functionalism, wantings, intendings, believings, etc., are conceptually characterized roughly as follows. They are functionally construed as dispositional states with a certain propositional structure. The conceptual or semantical introduction of these states is done causalistically in terms of the inputs (e.g., perceptions or something like that) and outputs (linguistic and non-linguistic intelligent behavior, including actions and reactions). In the case of propositional attitudes, it is central that they be introduced as potential causes of behavior.

Thus, to intend to do p means, roughly, being in a dispositional state with the structure p (or with the structure that p has) such that this state, given suitable internal and external circumstances, will cause whatever bodily behavior is required for the satisfaction of the intention (cf. Chapter 6).

After forming an intention an agent is in a state of intending which lasts till the intention has become satisfied or otherwise "dismissed". This I call the *"duration state"* of the intending. A realist argues that this is a categorical state of the agent. However, for the time being we do not really know anything about its descriptive intrinsic features. It is up to future neurophysiology and -psychology to give such descriptive non-functional information about it. Thus this categorical state is, for the time being, "abstract" in the extreme. But this abstractness is (presumably) due to lack of knowledge, not somehow an inherent feature of the state. (We say "presumably" as we want to leave a small logical gap here.)

Notice that Rylean instrumentalists could speak about duration states, too. However, they would not correlate any categorical inner state of the agent with these states.

It should be emphasized here, as will be seen later, that our conceptual functionalism also gives wantings, intendings, etc., event-like and episodic features in the sense that they can have inner (mental) *actualizations*.[1] As in the case of the Jones myth we may speak of wanting episodes which actualize wanting dispositions. We recall that a central part of the meaning of these wanting episodes derives from spontaneous and candid wantings-out-loud (cf. axiom (A7) of Chapter 4). Wantings, etc., thus have a "stream-of-consciousness" aspect. In suitable circumstances they have actualizations which, furthermore, *activate* the disposition in the sense of causing (or causally contributing to the bringing about of) behavior. In this sense wantings, etc., may even occur as event-causes of behavior, for these activating "manifestations" may be sufficiently event-like to be called events or episodes rather than states. Note that as activated wantings are *occurrent* episodes or events no perplexing ontological problems about their being (potential) causes need arise.

According to conceptual functionalism then, *there are* states and processes in each person such that these states and processes are related in certain specified ways to themselves and to perception, various need-states, action, and so on. (These represent the person's *mental* states and processes.) Now some of the relations and connections mentioned here must be *analytic* or true for conceptual reasons. What interests us specifically here is that wants, beliefs, etc., have analytic connections to behavior.

For our present purposes, we may interpret, for example, the explicate axioms (A4) and (A7) for wanting, discussed in Chapter 4, as analytic. Let us briefly consider (A4) here. As it stands it is neutral with respect to the realism-instrumentalism basis-controversy (or, what amounts to the same here, the controversy between Jonesians and pre-Jonesians). Now, we may reformulate (A4) in a more clearly realist fashion as follows:

(A4′)   Agent A wants that p if and only if A is in some state B such that, for every action X, which A has the ability and opportunity to perform, if A believes that his performing action X is necessary for his bringing about p and that his performing X would have at least some non-negligible probability of leading to an instantiation of p, then, given that A has no stronger competing wants and that normal conditions obtain, B (partially) causes whatever behavior is required for the performing of X.

In (A4′) the basis state B represents the duration state we discussed above. It is, roughly speaking, a state in virtue of which A exhibits such and such wanting-behavior.

Before attempting to give a more detailed characterization of the nature of wantings, etc., as dispositions, a few general remarks are in order. First, it is to be emphasized that wantings are *theoretical* concepts to a greater degree than, for instance, solubility and other standard examples. Psychological concepts such as wantings, believings, etc., are to be introduced by means of a psychological theory which shows their interrelations (even the simple (A4′) indicates this). Still, to want to do X is to be disposed to do X in normal circumstances. Secondly, as we saw in Chapter 4, wantings are *multitrack* dispositions: they can be overtly manifested in a number of different ways, depending on the person and on the internal and external circumstances (i.e., his beliefs and the physical opportunities available).

Thirdly, in the case of wantings, etc., the basis B will always represent a *state* (and never, e.g., a thing-kind). Fourthly, wants and beliefs are *stimulus-independent* to a great extent (cf. Armstrong (1973)), as contrasted with "ordinary" dispositions like solubility in water. At least singular beliefs are argued by Armstrong to be stimulus independent. But consider a man who believes that the earth is flat. I think that we must say that it is conceptually true that the man is then disposed, under normal circumstances, to answer that the earth is flat, when questioned about the form of the earth. Does this not show that even singular beliefs still are to some extent stimulus dependent? If so, wants and beliefs are stimulus-dependent, at least to some degree.

Fifthly, wants and beliefs involve *propositional* states. Thus, there is the inbuilt feature that their bases have a certain internal structure, which is not the case for arbitrary dispositions.

Sixthly, wants, etc., are *epistemically* special in that the person is in a privileged position to have knowledge about his own wants, etc. Hence, these psychological states are prima facie "reportable", as we have seen. However, it should be recalled that their reportability and near-incorrigibility is *learned*.

Of the above features at least the fourth, fifth, and sixth perhaps show that wantings and related psychological states are somewhat special among dispositions. Still they are not too different to be called at least a brand of dispositions.

2.2. Some of our claims above can be clarified, it seems, in terms of an oversimplified automata-theoretic analogy. In this hope we now make a brief digression into a "cybernetic" account of man.

We shall now assume that for our purposes an agent's psychology can be suitably represented by a Turing machine in the manner of the so called *Functional State Identity Theory* or FSIT for short (cf. Putnam (1960), Block and Fodor (1972)). FSIT represents one — rather strict — version of functionalism. According to it, for any organism that satisfies psychological predicates at all, there exists a unique best description such that each psychological state of the organism is *identical* with one of its (Turing) machine table states (i.e., logical states of that Turing machine) relative to that description.

The Turing machine that is assumed to represent the organism may very well be nondeterministic. The essential thing is, in any case, that each type of (occurrent) psychological state is contingently identical with a type of machine table state. As dispositional states obviously are not (merely) occurrent states, the above version of FSIT obviously needs a refinement to account for dispositional states (cf. Block and Fodor (1972), p. 168). Consider this: Every dispositional state of the organism is a possible occurrent state of the organism. That is, for every dispositional state of the organism, there exists a machine table state which is identical with an occurrent mental state (of the organism) and which actualizes the disposition.

It is not possible to describe FSIT more exactly here; the example to be given below will hopefully make it sufficiently clear. We will not, however, discuss the adequacy of FSIT in any detail here (cf. Block and Fodor (1972)).[2] Let me, however, mention one difficulty here. FSIT seems too strict basically for the reason that it does not account well for the *interaction* of psychological states. Each logical state of a Turing machine represents the total psychological state of the organism at a certain time according to FSIT. Thus, if, for example, the organism's perceiving at t is affected by his wants and mood at t, we would need an account of this interaction, but, as it stands, FSIT cannot provide this. We need a decomposition of the machine and hence the simultaneous consideration of all the submachines, it seems. Thus we have to consider man as simultaneously realizing a number of interacting parallel Turing machines. This and some other considerations indicate that, perhaps, FSIT must be replaced by a more general kind of functionalism (or by a totally different account) in order to get an adequate representation of man's psychology. As Block and Fodor (1972) and Fodor (1975) suggest, psychological states should perhaps be suitably identified with the *computational* states of the machine. A computational state of the machine is a state which is characterizable in terms of its inputs, outputs and machine table states. This kind of *computational state functionalism* (CSF) is a version of our general form of conceptual functionalism, obtained from the latter by restricting it to psychological theories realizable by Turing machines.

For the purposes of the present chapter we can, however, accept FSIT in the revised form stated earlier, although our main argument is in accordance with CSF as well. Let us now consider an example of a very simple Turing machine (more exactly: finite transducer) defined by means of the machine table given in Figure 5.1.

	f		g	
	O	I	O	I
Y_0	Y_2	Y_1	O^*	O^*
Y_1	Y_0	Y_1	O^*	O^*
Y_2	Y_2	Y_1	I^*	O^*

Figure 5.1

Here the functions $f: Y \times X \to Y$ (transition function) and $g: Y \times X \to Z$ (output function) define the Turing machine, where $X = \{O, I\}$ = the set of inputs, $Y = \{Y_0, Y_1, Y_2\}$ = the set of internal states, $Z = \{O^*, I^*\}$ = the set of outputs. We take the elements of X, Y, and Z to be *predicates* (or types). Thus O stands for the set of token zeros of the input type and I stands for the set of token ones of the input type. Asterisks in O^* and I^* mark output: e.g., '$O^*(a)$' reads 'a is a token zero of the output type'. The interpretation of the internal states Y_i will be given later. Let us note here that they can be regarded as occurrent or categorical states (even if nothing about their "hardware" aspects is known or even claimed to be knowable in principle and in general).

We may now think that to the above machine table corresponds a psychological theory T which can be regarded as a "unique best description" of the organism. The observational vocabulary λ of T contains O, I, O^*, and I^*. The theoretical vocabulary μ of T will include Y_0, Y_1, and Y_2. This theory is formalizable within predicate logic.[3]

As there is an exact correspondence between machine tables (or flow diagrams or program schemas) and certain (linguistic) theories (with some existentially quantified predicates), we can below also speak about specific machine-describing theories. Let us thus assume that, anticipating the invention of theory T, we postulate some states Y_0 and Y_1 to account for certain observable dispositional behavior (to be read from the g-part of our machine table). Thus, we accept as true the following statements entailed by T:

(1) $(x)(y)(Y_0(x) \to (O(y) \to (Eu)O^*(u)))$

(2) $(x)(y)(Y_0(x) \to (I(y) \to (Eu)O^*(u)))$

(3) $(x)(y)(Y_1(x) \to (O(y) \to (Eu)O^*(u)))$

(4) $(x)(y)(Y_1(x) \to (I(y) \to (Eu)O^*(u)))$.

If we really want the theory T to describe a machine computing a psychological program, we need some additional predicates and function symbols to be included in the set $\lambda \cup \mu$ (see next subsection). That is, however, inessential for our present purposes.

We immediatly see that Y_0 and Y_1 play exactly the same role in the subtheories $T_1 = (1) \& (2)$ and $T_2 = (3) \& (4)$. Y_0 and Y_1 correspond to and account for the same simple overt dispositional behavior (as seen from the g-part of the machine table). However, Y_0 and Y_1 are still observationally different. For instance, if instead of one single input event we consider sequences of two inputs, we get the output I^* for Y_0 and the output O^* for Y_1 given the input of two successive zeros.

Could we now represent the basis state of a want by either Y_0 or Y_1 in our theory? That seems possible, and below we shall choose Y_0 for this.

We claimed earlier that wanting, believing, etc., can be regarded as dispositional states of an agent. They can be represented in standard first-order logic by three-place predicates and by employing in them a special variable for singular states, events, or episodes (as the case may be) which actualize them. Thus we get in the case of wanting:

$$(Ex)W(A, p, x) = A \text{ wants that } p$$
$$= \text{there is some singular state (event, episode)}$$
$$x \text{ of wanting that } p \text{ in which state } A \text{ is}$$
$$= A \text{ is in some singular state } x \text{ of wanting}$$
$$\text{which has the structure } p.$$

Here 'p' stands for a statement in the "language of thought". (What amounts to roughly the same, 'p' might be taken as a Sellarsian dot quoted expression; e.g., $p = \cdot q\cdot$, when 'q' stands for a suitable surface statement.) Within our present framework I would rather say that W, at least formally, represents the basis of wanting and I would not completely identify wanting with its basis (cf. our earlier remarks). (This is really only a terminological issue here, for one can say that the predicate W has the required connections to observational behavior due to how W is introduced into the language.)

Now, we shall interpret Y_0 as the wanting state of the organism represented by the machine T^* realizing theory T. For simplicity, we shall just assume that this state has the propositional structure p (even if p should presumably be taken into play via an external memory) and do not discuss that problem further here (cf. our discussion in Chapters 6 and 9).

Then

$$(Ey)Y_0(y) = (Ey)Y_0(T^*, p, y) = y \text{ is a singular basis state of}$$
wanting with the structure p of the Turing
machine T^*
$$= \text{the Turing machine } T^* \text{ wants that } p.$$

We thus take the one-place predicate Y_0 to be shorthand for a three-place predicate Y_0 with fixed first and second arguments (their variation is a separate matter). Alternatively Y_1 could also be taken to stand for wanting that p (or some related proposition). Whether Y_0 or Y_1 gives a better representation naturally depends on the observational and other differences of these machine states. (Y_2 might perhaps be interpreted as an occurrent mental state or episode, e.g., a sensation; but we shall here exclusively concentrate on wanting and Y_0.)

How does our above suggestion fit together with FSIT? It does quite well. The statement $(Ey)Y_0(y)$ means that the machine will be at least once in the state Y_0. That is, given any model of the theory describing a Turing machine and containing the above statement, a singular state (or event, perhaps) satisfying Y_0 must be found in the domain of that model. In exactly this sense a dispositional state of the machine is a "possible occurrent state" of the machine.

In our simple machine-example we can thus make a distinction between *actualizations* and *manifestations* of a wanting disposition. States or episodes satisfying the wanting basis Y_0 will be inner actualizers, whereas overt event sequences realizing, e.g., I and O^*, given that Y_0 is realized, may be regarded as manifestations of the want (cf. the axioms of the theory T). In an instrumentalistic account of wanting and other propositional attitudes of course no such distinction can be made. How to characterize actualizations and manifestations in "real life" cases should be roughly understandable on the basis of our oversimplified account. Recall here our example of note 1. According to it, Nixon wanted for forty years to become president (the duration state lasted for forty years). We may think that this want was actualized (and activated) only occasionally (e.g., before and during presidential campaigns), and then it also led into overt manifesting behavior (giving speeches, etc.).

It is important to note that each singular state satisfying the basis predicate Y_0 is an actualizing state which also *activates* the disposition. Thus when a machine is in such a state it causally contributes to the bringing about of the output behavior.[4] If the general thesis CSF is acceptable, then every such actualizer will also be an activator in this sense. (Whether "idle" wants, as we think of them, really count as counterexamples against CSF is an interesting problem, but I do not have any good *a priori* reasons to offer for or against.)

Corresponding to our previous classification of dispositional states, we now get this distinction. First, the *duration* state of the want is exactly such a state in which the organism (or machine) is while computing a program in which '$(Ey)Y_0(y)$' has the truth value true. What was just said should be qualified by relativizing it to a specific program and a given input (or set of inputs) to get a more exact clarification of the meaning of a duration state. Perhaps the qualification coming closest to our ordinary way of speaking about wants would be obtained by specifying the program and requiring '$(Ey)Y_0(y)$' to be true for every input sequence. (Cf. Manna (1974) and below for an exact characterization of programs and their features.) Secondly, the *activating* states are obtained as clarified above (together with the additional qualifications just imposed on the duration states).

As we have claimed, the task of giving anything like a full account of psychological dispositions like wanting and believing would have to involve the full "framework of agency" and the complex connections involved there. Here, we have chosen rather the way of simplifying the situation considerably and thereby attempting to get hold of some interesting aspects of psychological dispositions.

Keeping to our oversimplified example and assuming the truth of FSIT (in fact, we would only need CSF), we now propose a summary definition for our example. It is formulated on the basis of our (*1*), the characterization (A4′) and our Turing machine example:

(*3*)   A *wants* that p (at t_1)
         if and only if
         $(E\beta)(E\gamma)[(Ey)\beta(A, p, y)$ & $T(y) = t_2$ &
         $(Eu)\gamma(A, q, u)$ & $T(u) = t_2$ & $t_2 \leq t_1$ &
         $(y)(u)(\beta(A, p, y)$ & $\gamma(A, q, u) \rightarrow (v)(Ew)(O(A, v) \rightarrow O^*(A, w)))$
         and
         $(y)(u)(\beta(A, p, y)$ & $\gamma(A, q, u) \rightarrow (v)(Ew)(I(A, v) \rightarrow O^*(A, w)))$
         are both true causal laws, each of which DE-explains any true
         singular statement of the form $O^*(A, w_i)$ when considered
         together with some true singular statements $\beta(A, p, y_j)$,
         $\gamma(A, q, u_k)$, and $O(A, v_m)$ (or, respectively, $I(A, v_n)$); i.e., each

law backs, respectively, the singular causal conditionals of the form

$$\beta(A, p, y_j) \ \& \ \gamma(A, q, u_k) \ \& \ O(A, v_m) \rhd\!\to O^*(A, w_l),$$
and
$$\beta(A, p, y_j) \ \& \ \gamma(A, q, u_k) \ \& \ I(A, v_n) \rhd\!\to O^*(A, w_l),$$

for every i, j, k, m, and n.]

In this definition the times t_i are to be construed as intervals rather than points; $t_2 \leq t_1$ means that interval t_1 includes interval t_2. t_1 represents the full duration of the want in question; hence the duration of the activating event y is to be included in that interval. The times of the other events are partly specified through the requirement of causality concerning the singular conditionals (cf. Tuomela (1976a) and Chapter 9 below on this).

The causal laws, using explicit event-ontology, in (3) are in fact just reformulations of the laws (1) and (2) for Y_0 in our example, which serves as our "intended model". We shall not here attempt to give a general analysis of wanting at all. Still it should be easy to see how to handle the situation in general on the basis of our example (cf. our explicative axioms (A1)–(A8) of Chapter 4).

Note that we have only taken the simple (i.e., single event-input) overt behavior into our definition. Whatever else observational behavior Y_0 accounts for is to be derived as factual hypotheses from the Turing machine theory T or whatever theory is used to explain the laws in (3).

A normal conditions predicate $C(A, q, u)$ (replacing $\gamma(A, q, u)$) is here assumed to contain whatever internal and external conditions are thought necessary. For instance, it may be used to exclude Y_1 in the Turing machine example and it may require the presence of a certain belief that q (recall our (A4′)). Not everything covered under this predicate has to be propositional and mention q.

Clearly our (3) is as close to (1) (or actually the multitrack version corresponding to it) as can be hoped. Thus, given that a version of conceptual functionalism (not necessarily FSIT) is acceptable, psychological dispositions like wanting and believing can be shown to be realistically construable *dispositional* theoretico-reportive concepts.

2.3. In modern psychology computer simulation techniques are often employed in theory building. A notable example is the pioneer work done by Simon, Newell and others within the area of human problem solving. The organism is considered as an information processing system. Accordingly, an information processing theory posits a set of inner processes or mechanisms that produce the behavior of the thinking agent. These inner processes are functionally introduced. Thus, at least CSF is presupposed by any work done in this area. The investigations reported in Newell and Simon (1972) indicate that even the stricter FSIT (in our revised form) could be retained. (However, these authors build their approach on the somewhat informally characterized IPS (information processing system); accurate and detailed "translations" to the FSIT framework have yet to be made.)

An information processing theory describes the time course of behavior, characterizing each new act (or other mental event or state) as a function of the immediately preceding state of the organism and of its environment.

The theory is formalized in terms of a computer program, which plays a role directly analogous to systems of differential equations in theories with continuous state spaces (e.g., classical physics). In information processing systems, though, any state is represented by a suitable collection of symbolic structures (in the language of thought) in a memory (cf. our discussion in Chapter 3).[5]

In spite of the wealth of interesting philosophical and methodological problems connected with work done within Artificial Intelligence and Cognitive Simulation we shall below and in the next subsection just take up some questions of a more formal kind. We wish to point out some connections between these areas of study (assuming they rely on CSF) and the general kind of conceptual functionalism for concept formation in psychology we have been discussing earlier in this and the previous chapter. Although what we are going to say is better related to theoretical psychological concepts in general rather than to dispositions specifically, our automata-theoretic framework in the previous subsection warrants the presentation of our observations in this connection.

To point out these connections I shall give a very concise description of some basic technical concepts and results. (A more detailed exposition is to be found in such standard works as Scott (1967) and Manna (1974). I mainly follow Manna's terminology.) We start by assuming that the notion of a *flow diagram* is familiar to the reader. A (*flowchart*) *schema S* is a finite flow diagram containing statements of the following five kinds: START, ASSIGNMENT, TEST, HALT, LOOP. A flowchart schema contains one START statement, and every ASSIGNMENT or TEST statement is on a path from the START statement to some HALT or LOOP statement. A flowchart schema S can be formalized in terms of first-order predicate logic with uninterpreted predicate and function constants and by leaving the domain for the individual variables uninterpreted.

A (*flowchart*) *program* P is a couple $\langle S, I \rangle$ where S is a schema and I is an interpretation function (from the predicates and functions into a fixed domain) in the standard model-theoretic sense.

Each program will contain an output predicate $\phi(\bar{x}, \bar{z})$ which defines the success or correctness of program-computation. Here, $\bar{x}$ is the individual variable for the total input and $\bar{z}$ for the total output. A program P is said to be *partially correct*, given a certain input value $\bar{x}$, exactly when $\phi(\bar{x}, \bar{z})$ is *true*, given that the program-computation halts. Now, this notion of partial correctness can be used to define all the other relevant notions in the program-semantics.

What interests us here especially is the following central theorem by Manna, which connects programs with second-order predicate logic: For every program P, input value $\bar{x}^*$, and output predicate $\phi(\bar{x}, \bar{z})$, P is partially correct with respect to ϕ, given the input value $\bar{x}^*$, if and only if a certain second-order formula $W_p(\bar{x}^*, \phi)$ is true (Manna (1974), p. 306). Given this connection one can define similar connections for program schemas, for the equivalence of programs and of schemas, and so on.

What is the open formula $W_p(\bar{x}, \phi)$, with $\bar{x}$ and ϕ as free variables, like? It has basically the form $(EQ_1) \ldots (EQ_n) M(\bar{x}, \phi)$ where the Q_i's represent predicate variables connected with suitable cut points of the program (so that every loop will contain at least one Q_i). In the matrix M, $\bar{x}$ and ϕ occur free. (In the place of ϕ a special halt predicate H may be used, too; cf. below.) In addition there will be some predicate constants p_j, which have not been existentially quantified over.

To make this a little more intuitive let us consider an example. Let our problem be the computation of the factorial function, i.e. $z =$ factorial (x). We can state the algorithm T_1 for it as follows:

$$T_1 = (x)\{(1 \cdot x \,! = x!) \,\& \, (y_1)(y_2)[y_2 \cdot y_1! = x! \to \text{IF } y_1 = 0$$
$$\text{THEN } y_2 = x! \text{ ELSE } y_1 y_2 \cdot (y_1 - 1) = x!]\}.$$

Here, obviously x, y_1, y_2 are individual variables; y_1 and y_2 coming to stand for internal states of a machine computing this algorithm. 'IF wff THEN wff ELSE wff' is a derived logical connective with the obvious meaning (see Manna (1974)). Everybody can now probably understand this algorithm and draw a flow diagram and a program for it (if not, see Manna (1974), p. 245).

A simple psychological theory might almost have the structure of our T_1 (or, better, something resembling T_1 but naturally containing more predicates, etc.). In any case, all the predicates of T_1 are interpreted, and that is how we presumably formulate psychological theories. But T_1 is not the formula we get when formalizing the factorial program. Let us drop the quantifier (x), replace '$y_2 \cdot y_1 = x!$' by '$Q(x, y_1, y_2)$' and '$z = x!$' by a halt predicate '$H(x, z)$', but keep '$y_1 = 0$' as it is. Then consider

$$T_2 = (EQ)\{Q(x, x, 1) \rightarrow$$
$$(y_1)(y_2)[Q(x, y_1, y_2) \rightarrow \text{IF } y_1 = 0 \text{ THEN } H(x, y_2)$$
$$\text{ELSE } Q(x, y_1 - 1, y_1 \cdot y_2)]\}.$$

T_2 is just the formula W_p that Manna's correspondence theorem gives us.

Corresponding to a flow diagram for the factorial program schema we would get a formula T_3 obtained from T_2 by replacing '1' and '$y_1 = 0$' by the uninterpreted constants 'a' and 'p', respectively. Obviously, T_3 is too abstract to correspond to a psychological theory.

T_2 resembles somewhat the Ramsey sentence of T_1 except that it contains x and H as free and that '$y_1 = 0$' is a theoretical predicate. More clearly, we might imagine that we had $\lambda = \{H\}$ and $\mu = \{Q, p\}$ if we were dealing with a psychological theory.

It seems, then, that T_1, apart from its simplicity, represents the logical form of a psychological or any scientific theory much better than T_2 or T_3. We easily see that $T_1 \vdash T_2$ and $T_1 \vdash T_3$. Furthermore $T_1 \vdash (Ep)(x)T_2$, where the predicate variable p replaces '$y_1 = 0$'. Now $(Ep)(x)T_2$, which equals $(Ep)(EQ)T_1$, is just the Ramsey sentence of T_1.

So we see that the connections between programs and psychological theories simulated by means of programs can be made almost as close as could be desired.

One thing to be emphasized is that T_1 as discussed above carries with it its intended model, the natural numbers. (Notice here that the truth of T_1 for the natural numbers is automatically guaranteed. The "fit" problem in simulation is then that between a factually given model and an intended one.) Often, scientific theories are not that closely connected with specific domains; one cannot fix domains of real objects as neatly as of mathematical objects (see Tuomela (1973a)). Thus a scientific theory will not normally refer to specific individuals like 1 and 0 as T_1 does, and hence we might want to have here a little more abstract formulation (cf. however, e.g., Kepler's laws). Such a theory would of course be at least as strong as T_1, and hence the conclusion just reached above remains true, and our results from Chapter 4 carry over here.

In general, we can say that the internal program variables y_k on our discussion will represent (singular) logical states of the machine and, via FSIT (or CSF), inner mental states. The functions (e.g., some Q_i's) and predicates (the p_j's and some Q_i's) represent inner mental "functions" (acts, etc.) and other mental properties like bases for propositional attitudes and other dispositions. (See Scott (1967) for appropriate definitions of a machine and for relations correlating machines with programs.)

On the basis of the correspondences set up above it is easy to define and clarify such issues as program equivalence and isomorphism, computability by a machine, and their psychological counterparts under FSIT or CSF.

Thus, for instance, two programs P and P' are most naturally defined to be *equivalent* (represent equivalent "psychologies"), if and only if they determine the same computation function in the case of any machine (see Scott (1967) for the notion of computation function). Thus, in this sense we can say that two programs are equivalent just in case they are computable by the same class of machines.

A more detailed and more informative discussion of the various connections between the theory of automata and computation and psychological functionalism, as well as of the adequacy of the Cognitive Simulation approach, must, however, be left for another occasion.

NOTES

[1] Two supplementary remarks may be made here. First, let us consider Richard Nixon's want to become the president. This want, we assume, lasted forty years. This wanting state we call the duration state. (Somewhat metaphorically, it is as if the propositional content that Richard Nixon is the president had been stored in the want-compartment of Nixon's long term "memory" for forty years.) This want became actualized and activated from time to time (e.g. before and during presidential campaigns). This we take to illustrate the occurrent episodic features related to that want. Although the distinction between the categorial wanting-state associated with the whole duration of the wanting and the episodic actualizing events connected with that want is undoubtedly *context-relative*, it still seems methodologically central.

Secondly, motivation psychologists sometimes distinguish between motivation as a *structure* and as a *process* or function. This corresponds to our above distinction between the long term categorical state and the episodic actualizations connected to wants and other motivating factors. For instance, we may mention as examples Atkinson's motive vs. incentive value, Cattell's erg vs. activation of an erg, Young's attitude vs. set, and Freud's Id, Ego, Superego vs. primary and secondary processes, to illustrate our distinction.

² Block and Fodor (1972) present a number of arguments against FSIT. Most of them are untenable, I think. I cannot here discuss them except for one remark. According to Block and Fodor (1972), p. 175, "on the assumption that there is a computational path from every state to every other, any two automata which have less than all of their states in common will have none of their states in common".

First we have here the problem of describing the organism at the right level of abstraction. Assuming that that is solvable, then, in view of the connections between formalized theories in predicate logic and program descriptions to be stated below, our results concerning the finding of T_A and T_S-components from Chapter 4 apply here. Thus we do not get into the paradoxical sounding situation described in the cited passage.

³ Each flow diagram schema (corresponding, e.g., to a machine table such as in our example) is in the general case formalizable within a fragment of second-order predicate logic. This result essentially says that a program schema (= flow diagram schema), when it leads to a terminating computation, is correct under a given model-theoretic interpretation if and only if a certain existential second-order formula is true for that interpretation (see next subsection and Manna (1974), esp. pp. 306, 312).

⁴ In this connection we can also say that the *coming about* of the state Y_0 is an event which causes the relevant output behavior.

⁵ Fodor (1975) analyzes mental processes as *computational processes* in accordance with CSF. The language of thought in Fodor's theory is to be compared and "identified" with the *machine language* of a computer, whose input-output language, connected by biconditionals to the machine language, correspondingly is to be compared with our "surface language" (cf. Fodor (1975), pp. 65–79). What seems to follow is that the language of thought in Fodor's system is something *innate* as opposed to learned and that his notion of mental episode is something innate, too. Such a strict view is what we have been strongly opposing in the last two chapters in the connection of the Myth of the Given. Although I find myself in agreement with much of what Fodor says in his book, on this point I clearly disagree, if Fodor indeed is committed to the above mentioned strict nativism.

CHAPTER 6

WANTING, INTENDING, AND WILLING

1. WANTING AND INTENDING

Propositional attitudes like wanting, believing, and intending are complex dispositional states. At least this is what we argued in the previous chapter. In this chapter we shall continue our discussion by an emphasis on the content of the concepts of wanting and, especially, intending. Intending is going to be a key notion in our causal action theory, to be developed in detail in the later chapters. Therefore most of this chapter will be devoted to intendings and willings (or tryings), which are active and executive intendings. We have already discussed the nature of wanting sufficiently for our purposes. Beliefs will not be analyzed in detail in this book.[1]

Propositional attitudes are so called simply because of their propositional content. Wants, beliefs, and intentions are treated below as the propositional states of wanting, believing, and intending. Thus, for instance, the statement "agent A wants to have an apple" becomes in our analysis (roughly) "A is in a dispositional state of wanting with the content that A is having an apple". We thus assume, as usual, that direct object constructions are translatable into descriptive propositional that-constructions.

Before going on to discuss intendings, a few preliminary remarks will be made. First, we speak of wants in an inclusive sense in this book (cf. our explicative axioms (A1)–(A8) of Chapter 4). This broad sense of wanting includes desires and duties (and obligations). Thus we may speak about *intrinsic* and *extrinsic* wants. Intrinsic wants are those whose object is wanted for its own sake, so to speak. This is the most common notion or "species" of want. An extrinsic want is something whose object is not intrinsically wanted but the realization of which is believed more or less necessary for the realization of something wanted (ultimately: wanted intrinsically). Thus an agent acting in order to fulfil his duty or obligation (or because of a certain norm) is often acting only on his *extrinsic* want to fulfil the duty, as he does not intrinsically want to act so, but the action will bring about something the agent wants intrinsically.

The general view about the relationships between wanting, intending, and acting that we accept is roughly as follows. A person's intentions (i.e., intendings) are *generated* in some sense or other by the person's wants (and some relevant beliefs). Both intrinsic wants (e.g., "primary" biological needs and drives as well as "secondary" psychological and social needs) and extrinsic wants (based on duties, obligations, norms, challenges, etc.), together with some relevant beliefs, can be said to generate (or contribute to the generation of) intentions. Or, to use different terminology, a person *forms his intentions* on the basis of either intrinsic or extrinsic wants. Intending involves wanting, but is not reducible to wanting (nor to wanting cum believing).

When a person has formed an intention, he is not yet usually embarking upon action. The intending must at least become *effective* before the execution of action begins. We shall below call effective intendings to *now* realize the intention *tryings* or *willings*. Such an effective intending causally results in action, provided the world suitably cooperates.

Conversely, actions are in the first place explainable by reference to effective intentions and in the second place by reference to the wants and beliefs or to the more general intentions which generated the (effective) intendings in question.

It may be remarked here that philosophers have paid too little attention to the problem of how a person's intentions are generated by his (standing or momentary) wants and beliefs. Psychologists have done rather much work in this area. For instance, various theories of motivation (e.g. Lewin, Freud) and the theories of cognitive balance (e.g. Heider, Festinger) may be mentioned in this connection. Most of the problems involved in want-intention generation are indeed factual psychological ones rather than philosophical. Still a philosopher can find lots of important conceptual problems here, too.

Psychologists working in this area have mostly neglected the conceptual investigation of mental concepts as well as of the varieties of action. Thus, to take a recent example, Atkinson and Birch (1970) have formulated an elegant theory of motivation, but they have not paid sufficient attention to the classification of behavior. They do not even really distinguish action from non-action, not to speak of such finer distinctions as intentional versus nonintentional or voluntary versus nonvoluntary action. Surely the motivational problems related to, say, reflexes must differ from those related to, e.g., habits or intentional actions. Furthermore, these authors do not either explicitly make clear distinctions between, e.g., the varieties of proattitudes (wants, intentions, hopes, wishes, etc.).

Although we are not going to study want-intention generation in any detail here, one general comment may be made. There are at least three important conceptual ideas involved in such generation. First, there is the Lewinian view that somehow competing wants are to be treated as forces so that the *resultant force* gives the agent's intention (or decision or "will").

Secondly, there is the "non-compensation" model according to which the momentarily strongest want (force) always wins and becomes the want on which the agent forms his intention to act.

Thirdly, there is the view currently popular in, e.g., decision theory that, in risk situations, agents calculate their expected utilities and intend to act so as to maximize them (cf. Bayesian theory). These three conceptions can be elaborated in various ways; their exact interconnections depend on such elaboration.[2]

After these preliminary remarks we can now go on to briefly discuss the notions of intending and trying, which are very central for our developments. We start with the notion of intending, which, as most psychological concepts, can be characterized from various points of view. One way to characterize intendings or havings of intentions is functionally via their "consequences". Intentions can in this sense be regarded as dispositions to act. One may distinguish between *previously* formed and *concurrent* intentions. Previously formed intentions (e.g. intention to do X) can of course exist (ontically) independently of action, even if the notion of the intention to do X and the notion of doing X are both grammatically and conceptually connected. Concurrent intentions can also exist as states separate from action, but sometimes ordinary ascriptions of concurrent intentions are merely additional "Rylean" ways of characterizing an action (cf. 'A did X', 'A did X with the intention Y'). We shall below only analyze those statements (and their uses) about intending which presuppose that in them the agent's intentions are realistically construable as his states of intending. (We thus do not deny that there may currently be important "Rylean" uses of statements about intentions.)

Another distinction we need is that between *complex* intentions and *simple* ones. Complex intentions are, for example, of the form 'A intends to bring about X by doing Y', whereas simple intentions have the standard form 'A intends to do Y'. We shall mostly be concerned with complex intentions, as they are the most important from the point of view of the teleological explanation of action.

Yet another distinction is between *absolute* and *conditional* intentions.

We shall concentrate here on absolute ones.[3] (For conditional intentions see, e.g., Kim (1976).)

Intentions may also be characterized in terms of their "psychological antecedents". Thus one may try to connect intentions to wants cum beliefs. It may well be that it is incorrect to think that there is only one pre-systematic notion of "intending to bring about X by doing Y". At least one has to admit that this notion, if there is only one such explicandum, is rather hazy and vague at its edges. No wonder that there are and have been many different approaches to this problem.

I do not know of any analysis which I could fully accept but a recent attempt by Audi goes some way toward an interesting and correct solution, though I think that, with respect to the notion of intending that I have in mind, it has some faults and shortages. Let us thus start by considering his proposal, which I here formulate by using different symbols (cf. Audi (1973b), p. 395):

(I)   For every agent A, every action X and Y, A intends to bring about X by doing Y if and only if

   (1) A believes that he will (or that he probably will) bring about X by doing Y; and
   (2) A wants, and has not temporarily forgotten that he wants, to bring about X by doing Y; and
   (3) either A has no equally strong or stronger incompatible want (or set of incompatible wants whose combined strength is at least as great), or, if A does have such a want or set of wants, he has temporarily forgotten that he wants the object(s) in question, or does not believe he wants the object(s), or has temporarily forgotten his belief that he cannot both realize the object(s) and bring about X by doing Y.

Can we take (I) as a semantically true statement about intending? This definition characterizes intention in terms of an (extrinsic or intrinsic) want to bring about X by doing Y and a means-belief plus some assumptions about "normal conditions".

Let us briefly consider Audi's conditions each in turn. First, condition (1) expresses the important fact that intending involves belief and hence something cognitive. However, I think (1) is somewhat too strong. In our earlier terminology it says that action Y generates (or probably generates) action X for agent A. But this requires too much. For instance, a good highjumper may intend to jump over 220 cm, even if he has previously succeeded only every third time in the average and, accordingly, cannot be rationally convinced he will succeed now. Another example would be provided by a mathematician who can be said to intend to solve a certain problem for which no recursive method of solution exists. We thus seem motivated to modify (1) slightly to require only that the agent believes that Y generates X *with some nonnegligible probability* such that $p(X \mid Y) > p(X)$, where p stands for a measure of an objective probability (cf. Chapter 11).

Condition (2) connects intending intimately with wanting: an intending "is" or involves a wanting. When wanting is considered in our broad sense, I can see no serious objections to (2). If one forms an intention on a basis of a certain intrinsic or extrinsic want, that want might perhaps change while the intention stays the same. Suppose I want to see a certain person at a party and therefore decide, and thus form the intention, to go to that party. However, suppose I then learn that the person in question will not be coming. Then my want to see that person is not any more the "basis" of my intention to go to the party. Still I may keep my decision and intention to go to the party. (I always consider myself obliged to keep my decisions, let's say, or I do not want to offend the host, etc.)

However, I think that in examples of this sort condition (2) does not seem to be violated, as there is a want of the kind required in (2), although not the same want as at the earlier moment of time. (2) says nothing about the underlying psychological "want-dynamics" and Audi's condition (3) in effect requires that the agent has one clearly strongest want in that situation. He has, by deliberation or by means of some non-intentional psychological process (e.g. subconscious process to achieve "cognitive balance"), come to a situation where one (single or combined) want supersedes all the others. I shall below regard (3) as an acceptable necessary condition.

Both condition (2) and (3) speak about temporarily forgotten wants (wants which the agent temporarily forgets or neglects to take into account even if he "has" them all the time). It might also be possible to temporarily forget one's beliefs (e.g., I have now temporarily forgotten whether I accept the axiom of determinateness or the axiom of choice of the competing addenda to set theory). Furthermore, ascriptions of intentions, beliefs, and wants in general include some amount not only of "global" rationality but also of "local" rationality on the part of the agent. Global rationality relates to the rationality requirements for the applicability of the conceptual "framework of agency" on the whole (cf. the intentions of an amoeba vs. of a man). Local rationality relates to temporary forgetting, emotional disturbances, and the like. Let us denote global rationality (whatever it strictly speaking contains) by rationality$_1$ and local rationality by rationality$_2$ (cf. Chapter 7). As our whole discourse presupposes that the agent is rational$_1$ we shall not write it out explicitly. But we do specifically require rationality$_2$.

One conceptual feature that Audi's conditions fail to take into account is that it is impossible for an agent to intend to do something unless he believes that he can, at least with some likelihood, do it. One can intend to do the impossible (e.g., a mathematically ignorant person may intend to square the circle), but one cannot intend what one *believes* to be impossible. We shall below include in our first condition the requirement that A believes that he at least with some nonnegligible probability can do X by doing Y. This formulation will be understood to contain the previous requirement that A believes that Y generates X with some nonnegligible probability.[4] What else 'can' requires of the agent's abilities, beliefs, etc. we shall not attempt to specify here.

Our analysis so far supports the following partial analysis of complex intending:

(*I*)   An agent *A intends to bring about X by doing Y* only if

    (1) A believes that he, at least with some nonnegligible probability, can bring about X by doing Y;

    (2) A wants to bring about X by doing Y;

    (3) A has no equally strong or stronger incompatible want (or set of incompatible wants whose combined strength is at least as great);

    (4) A is rational$_2$.

In this definition the conditions (1)–(4) are to be understood in the light of our above discussion. Note especially that Audi's condition (3) is to be regarded as contained in my (3) and (4). Also note that simple intending like "A intends to do X" can be given an analogous partial analysis by simplifying (*I*) in a rather obvious way.

(*I*) gives only four necessary conditions for intending. Could we not claim that they are jointly sufficient as well? To this my answer is no. First, intending starts with intention-formation, which involves activity. Secondly, intending guides behavior purposively in a sense wanting does

not. These are my basic reasons against the explicit definability of intending in terms of wanting and believing along the lines Audi has attempted.
It remains to spell out these reasons in some detail to make them look
plausible.

Let us first notice that there are basically three different routes to
intention-formation. First, an agent may form his intentions on the basis
of practical inference, and thus on the basis of more general intentions
(see Chapter 7 on intention transferral).

Secondly, he may, through rational deliberation, come to *decide* to
bring about X by doing Y or, which amounts to the same, decide to act *on*
his want to bring about X by doing Y. Roughly, deciding suffices for
intending, but is not necessary for it. (This could, of course, be spelled out
exactly in terms of another partial definition of intending, which makes
deciding sufficient for intending.)

Thirdly, it seems that an agent may "just form" a certain intention
without any deciding or any clear intention transferral occurring (cf.: I
suddenly form an intention to kiss my wife). What the psychological
processes, which led him to that, are, is left to psychologists to
determine.

In all these cases we can assume that there is some (psychological)
activity which leads to the agent's forming an intention, and intention
formation – an event or episode – itself is an activity of somehow (though
not necessarily through deliberation) settling in one's mind what to do.
In forming an intention one *commits oneself* to action, we can also say. It
seems that our above conditions (1)–(4) (and, for that matter, Audi's
conditions (1)–(3)) could be fulfilled *without* the agent's having formed an
intention to bring about X by doing Y.

First, we may consider examples like this. A deprived alcoholic is given a
glass of whisky which he quickly empties with a "reflex-like" movement.
(In our later terminology we accept this as an action token$_1$.) I think we
could imagine that our above conditions (1)–(4) are fulfilled but that,
nevertheless, the man did not really intend (and form an intention) to
consume the whisky by emptying the glass in front of him. No "movement
of the will" really took place here, we might want to say.

Admittedly borderline cases like this are hard to analyze and to use as
fully convincing counterexamples. Let me therefore go to the more
important aspect of intending which relates to their behavior-guiding role.
Suppose that an agent intends to kill a man by shooting him. This intention
generates the beliefs necessary for the execution of the intention. The

intention thus, so to speak, involves cognitive elements which to some extent specify the way the intention is carried out.

The agent's intention to shoot the man in a sense specifies (more or less roughly, though) the *intentional* action pattern for carrying out the intention. Thus, for instance, it would not be according to his plan (including the intention *cum* the beliefs generated by it) that his bullet misses the man by, say, three meters but reflects back from a stone wall and thereby kills the man. Or suppose our agent's bullet scratches the man's left ear, which scares the man so that he becomes mentally ill and dies fifteen years later. Or, alternatively, suppose the man to be shot stands behind a mirror so that a clever arrangement of other mirrors reflect his picture in the first mentioned mirror. Our agent shoots at this mirror and thereby kills the man behind it. But that killing is not intentional, for it did not take place according to the agent's intention. Intentions thus guide intentional behavior in the way mere wants do not. (See our later discussion of "intentionality paradoxes" in Chapter 9.)

We can also say that it was intentional of the agent that he formed the intention to shoot the man. Intending is in a sense the locus of purposive intentionality and it cannot be unintentional, for both forming an intention and acting because of (or on) an intention is intentional activity.

We may also say here that our agent's intending to kill the man by shooting him conceptually entails that he then intends to intend to kill the man, by shooting him. Second-order uses of 'intending' are uncommon, to be sure, and they do not have a clear presystematic meaning over and above involving that intending is intentional. But if we understand second-order intending roughly in that way, we can say, I think, that the state of intending (at time t) to intend (at t) to do X (at t') conceptually only amounts to the state of intending (at t) to do X (at t'). (In any event this is the case for active intending, i.e., intending actualized and effective at t.) Analogously with the strong notion of knowledge, which similarly collapses the hierarchy of knowing, we here have a strong iteration-collapsing notion of intending. Corresponding to the "KK-thesis" for knowing, according to which knowing that one knows that p equals knowing that p, we have here the "II-thesis", provided the symbol "I" could really be taken to represent both second-order and first-order intending (note: their *contents* (objects) are somewhat different in kind).

Going back to our conditions of intending we can now argue that intending guides behavior in the way mere wants do not. Wants *simpliciter* do not, for instance, generate beliefs so as to specify, more or less concretely, what kind of external events really qualify as *realizing* events among all events that technically satisfy the want (cf. our discussion in Chapter 9), as our above examples can be taken to show. A central difference between mere wantings and intendings is that the latter have much more (and much clearer) cognitive content.

It seems that there is no non-circular way to amend our (nor Audi's) conditions to make them sufficient for intending. For instance, we should add to Audi's condition (1) that the agent believes he will realize his want *according to his intention* (or *as he intends*, or something related) to get closer to a working analysis of intending. Needless to say, that kind of specification would make the analysis directly circular. Although wants can be said to be dispositions to intend, intendings are not reducible to wants cum beliefs. We shall below still return to discuss the difference between mere wants and intentions in terms of second-order propositional attitudes.

So far we have not discussed the logical "entailment" properties of intendings and other propositional attitudes. As we do not regard these properties very important for the purposes of this book not much will be said about them either. Most of our remarks will be made in Section 3 in the context of formalizing statements about propositional attitudes.

Here we may briefly comment on *conflicting* intentions. The point is that we will accept the following principle as a conceptual truth about intendings: If A intends to do X then A cannot, for conceptual reasons, at the same time intend to do X' if he considers that X and X' require, respectively, actions Y and Y' to be performed such that Y and Y' are considered by A to be nomically or logically incompatible. Notice that because of the double reference to 'considers' it is not the objective but the subjective conflict of X and X' as well as of Y and Y' that is concerned here. If one or both of the conflicts are objective the principle ceases to be a conceptual truth.

2. Trying

Above we did not explicitly involve time considerations into our treatment of intending. We now have to do it, as we are going to discuss the notion of *effective intending* (to do something *now*). Roughly, this kind of an *effective intending* is an executive inner episode which (from a dynamic point of view) corresponds to some theorists' notion of *willing* or *trying*. 'Willing' is perhaps not as good a term because of its relations to bad philosophical ideas. 'Trying' again is bad because here it does not have its most common use, e.g., it should *not* be taken as synonymous with 'attempting', which rather clearly merely expresses an "overt" notion.

To start, we consider a simple example related to practical inference (and expressible in terms of a practical syllogism, cf. Chapter 7).

Let us consider a practical syllogism concerned with my intention to record a piece of music. This piece of music will be on the radio five minutes from now, and when it begins I have to push down the recording button of my tape recorder. So I keep an eye on my watch while doing something else: In five minutes I will push the button, in four minutes . . ., in one minute . . ., *now* I will do it. I am engaged in a process of intention transferral or specification. Given the original intending and the requisite beliefs, a process of intention transferral may presumably be regarded as a *causal* one. (However, we shall not make this an a priori requirement — *scientia mensura*.)

In the present example I thus seem to act in accordance with the practical syllogism:

(PS)   From now (= *t*) on I intend to record the piece of music which will be on the radio five minutes from now.

From now on I consider that in order to record that piece of music I have to push down the recording button at *t* + 5. Therefore, I now intend to push down the recording button at *t* + 5.

Here I have a plan for acting as given by the "main" intention (in the first premise) and as spelled out by (PS). What remains for me is to abide by the intention expressed by the conclusion of (PS), and action will come about at *t* + 5, *conceptually* without any further "willing". (I shall not here complicate the issue by considerations of measuring the time, etc.) The essential point here is the zero moment, viz. the intending to push the button *now* (= time *t* + 5 as estimated by me). Then my intending becomes *effective* in the sense of leading immediately to my execution of the pushing action by means of my bodily behavior. It thus becomes my intentional *trying to bring about by my bodily behavior whatever is required to satisfy the intention.* In this case I set my body in motion in the way that my hand reaches the button and pushes it down. I am thus exercising my causal power, for the bodily behavior involved in the pushing is caused by my trying (willing).

What was just said sketches briefly the typical relation between intending and the intentional action realizing it. Many things must be said in clarification, however. In fact, a great part of the latter half of this book can be regarded as such an elucidation. But a few remarks have to be said immediately about our notion of trying (or willing).

We assume that trying is to be conceptually characterized according to the functionalist analogy theory. We recall that to specify the conceptual content of a mental episode is analogous to saying what a relevant linguistic expression means. Here for instance, most "non-predictive" uses of 'I will now do X' qualify as they express intendings *now*. Here I shall not investigate the variety of linguistic expressions which can play this role (obviously at least 'I intend to do X' is another such expression). In any case, a willing-episode has now been introduced in the manner of the Jones myth. Instead of using the word 'willing' I below prefer the *technically* used term 'trying' in the case of intentional action.

Consider the simple example of a man who tries to raise his arm. In one case he tries (both in the "inner" and the "overt" sense) and succeeds. In another case, where his arm, unbeknownst to him, has become paralyzed, he tries but does not succeed. In both cases he tried, however. We may always take a "sceptical" standpoint and use the location 'A tried to do X and succeeded' regarding any successful action. We construe this trying as a mental episode, which conceptually is, however, very much like an overt action. A trying event (or episode) clearly, according to our example, is an event separate from overt action. By the "sceptical on-looker" method one can argue that *every* bodily action involves a trying event although nothing like "making an effort" needs to be involved (cf. O'Shaughnessy (1973) and McCann (1975)).

Trying is *intentional* as it is based on the agent's intending. The agent has direct *self-knowledge* of it (as he has of his other mental episodes and of his bodily and basic actions). He can also reflect on his tryings and thus have corresponding "metathoughts" (cf. our discussion in the next section). Next, trying is under the agent's *control*: he could have refrained from trying. It is up to the agent whether he tries or not.

A trying is unlike an overt action (as we think of the latter) at least in these senses: 1) It is an inner event. 2) It is not a conceptual feature of trying that it is caused by another trying (or anything like that). 3) A trying event has no result (in the sense overt actions have). The first of these features requires no special comments here (cf. Chapter 3), but 2) and 3) do need some remarks.

A trying event is an intentional inner event (or episode) which, the world suitably cooperating, causes the overt action-event realizing the purpose involved in the trying. Thus A's trying to raise his arm is, roughly, an event which, under normal circumstances causes A's arm raising (under the non-functional description 'A's arm rises'). Later in this book a singular trying event will be denoted by t, and the singular overt action by u_o. We will call the sequence $\langle t, u_o \rangle$ the *full* action, i.e. $u = \langle t, u_o \rangle$, provided t purposively caused u_o, i.e. $C^*(t, u_o)$ in our later terminology.

This kind of functional characterization gives the essential conceptual content of trying. What else conceptual clarification can be given comes through the fact that a trying is an *effective intending now*. As intending is conceptually connected to wanting, believing, deciding, etc., trying also is (cf. especially the connection expressed by the practical syllogism). But what interests us here is that the notion of trying does not *conceptually* presuppose that it is caused by anything, whereas the notion of overt action (in our causal account) does. In particular it is not caused by another trying. Whatever sense can be made of 'trying to try' it cannot amount to more than trying. Even if one can today intend to try to raise his arm tomorrow, one's trying (intending *now*) to try (intend *now*) to raise his arm only amounts to trying to raise his arm. No infinite regress is thus involved here either (cf. our related discussion in Chapter 3).

The action of raising one's arm conceptually involves the *result* of the arm going up. A trying to raise one's arm involves no such corresponding inner result (the whole trying event could hardly be taken as its own result).

One central aspect about trying which has to be especially emphasized is the following. A trying is a *conceptual* (representational) event — a trying to do something, say X. A trying event begins the *execution* of the action X. This execution is concerned with whatever the agent is required to do *by his bodily behavior* in order to exemplify X. There are two important comments to be made here.

First, there is the problem of how many tryings are really needed for a given action or sequence of actions. Consider our earlier recording example. Could I not have *tried* (effectively intended) directly to record the music rather than having to perform the practical inference and end up with a more specific intending? Or perhaps my intending to push the button is not sufficiently specific and I have to perform some more intention transferrals, perhaps to end up with something like "flexing my arm muscles in such and such a way". Or consider my intentionally playing a melody on a piano. Do I need one trying or 16 or perhaps 1151 tryings — one for each separate note?

Our general answer to these problems, raised, e.g., by Ryle, is this. Conceptually one may of course perform as many intention transferrals and other inferences as one pleases. The real question is *how* (in terms of which units) a person has learned to act. Where a child needs several actions and tryings an adult may only need one (cf. tying one's shoe laces or the above piano playing example). The question is obviously what a person's "automatized" behavioral "subroutines" are *de facto* (not merely as a matter of conceptual possibility). Notice, furthermore, that the multiplicity problem, which was raised for tryings, of course concerns overt actions as well. Our general answer thus goes both for tryings and overt intentional actions. All this will of course be reflected in how the agent forms his intentions and plans for acting. He must essentially rely on his available "subroutines" when performing, and planning to perform, complicated actions.

We should notice that trying to do X means trying to do X *by one's bodily behavior*. Trying is an intentional executive event, which makes one's body work in a planned way. It causes *in a standard way* all the bodily behavior involved in an overt action. *How* it does all this is up to neurophysiologists and neuropsychologists to find out. Presently we know very little of the neurophysiological aspects of trying. However, some recent neurophysiological findings seem to support my view that the introduction of trying events to causally explain intentional action in the manner of the analogy theory is justified.

In an experiment subjects were asked to intentionally push a rod in a tube at irregular intervals (see Becker, Iwase, Jürgens, and Kornhuber (1973)). Simultaneously EEG recordings were made from several positions of the head, including the mid-vertex position C_z, which is central for limb motion. The results show that there exists a very clear readiness potential about 0.5–0.8 seconds before the pushing movement (depending on the rapidity of the movement). This indicates that the full action of pushing the rod starts a little before the overt pushing movement. If the pushing movement were blocked we would be left with this specific readiness potential together with its dramatic reduction right at the cerebral "beginning" of the movement. It is not yet quite clear whether our indicator (the readiness potential) has gotten hold of a specific (necessary and sufficient) cause of arm movement. In any case this readiness potential and its reduction seems *necessary* for voluntary arm movement. (Also see Pribram (1969) where "intention waves" are discussed.)

Analogous results have also been obtained in the case of speech behavior. A corresponding inner episode seems to precede voluntary verbal behavior, too (cf., e.g., McAdam and Whitaker (1971), Low, Wada, and Fox (1973)).[5]

It is now tempting to suggest that in the described experiment we are dealing with an *intending* to push the rod (a suitable readiness potential), which becomes a *trying* to push the rod. But still one should of course not overinterpret these experimental results. They clearly indicate the presence of certain activity related fairly specifically to various kinds of overt action, but we are very far from having anything like a specific neurophysiological account of the extraconceptual nature of intending and trying.

Our technical concept of trying was created analogically on the basis of overt intentional actions, especially intentional "willings-out-loud" or "intendings-*now*-out-loud". Analogously we may introduce corresponding explanatory antecedents for, e.g., non-intentional actions. Thus, for example, spontaneous action-like behavior which is not preceded by conscious decision or anything like that (e.g., my scratching my ear while writing this) could perhaps be regarded as caused by some inner conceptual episode along the lines of Jonesian theorizing (cf. Chapters 3, 4, and 10).

In chapter 10 we shall introduce two broad notions of action called the class of action tokens$_1$ and the class of action tokens$_2$. We cannot here explain what these classes comprise except for saying this. The class of action tokens$_2$ includes non-intentional action tokens which, however, could have been intentional under suitable circumstances. Among action tokens$_1$ we, however, have behaviors which, though in some respects like intentional actions, do not even share this kind of "potential intentionality" feature (e.g., some of Sellarsian thinking-out-loud behaviors belong here).

Corresponding to these two classes of actions, we may introduce the notions of volition$_1$ and volition$_2$ (for lack of better phrases). Volitions$_1$ are causal antecedents of action tokens$_1$ and volitions$_2$ of action tokens$_2$. Both volitions$_1$ and volitions$_2$ are conceptual episodes and hence "thoughts". However, in general, volitions$_1$ and volitions$_2$ need not be intentional nor anything one can decide to do, although in a very broad sense they are acts. For instance, my inferring-out-loud-that-so-and-so is an action token$_1$ (and is not an action token$_2$), and it is preceded by a volition$_1$. This volition is not a mental action in the sense as, e.g., a trying is, although it can be part of a larger intentional mental episode involving a trying to do something related to, and including, inferring (e.g., deliberating on the premises of the task of inference and deciding to draw some auxiliary constructions on a sheet of paper).

In the case of volitions$_1$ and volitions$_2$ as well as trying it is up to scientists to tell us more about them and thus help to make finer classifications and refinements. Scientific research and conceptual clarification must go hand in hand. Science will also ultimately have to tell us whether our talk about tryings, volitions$_1$ and volitions$_2$ will remain as genuinely Jonesian talk or whether these (or some of these) notions only can be regarded as Rylean dispositions to will-out-loud, and so on. Put still better, science will in the last analysis tell us precisely which uses of psychological terms and statements are Jonesian and which are pre-Jonesian. For the time being, I am only trying to make plausible the view that there are Jonesian uses of some key parts of psychological language.

Above, we have used as our technical concepts philosophically loaded and mocked concepts such as trying, volition and willing. I think, however, that the "ordinary" criticisms against volition do not have any bite against our account. (See, e.g., Melden (1961), Chapter 5 for the standard criticisms against the classical Humean type of account.) Let me once more briefly point out my answers to the three most important criticisms.

First, there is the behaviorist criticism against the existence of any proper mental episodes. To this our answer is simply that given by the analogy theory. As said, it is up to science to ultimately answer this and thus to show which mental episodes exist and which do not. In the meanwhile, at least the fact that people do have authoritative direct (learned) self-knowledge of their mental episodes is good prima facie evidence for their existence.

Secondly, there is the criticism which says that, conceptually, a vicious regress is involved in any account of volitions. Various forms of this general argument were discussed and rebutted in Chapter 3. In the present chapter we, in addition, argued against the argument by reference to the "noniterative" nature of tryings and volitions (e.g., trying to try only amounts to trying).

Thirdly, there is the argument according to which one cannot have both a conceptual and a causal connection between volitions and action nor between tryings and action. Were we to accept this premise, the argument would go on as follows. Suppose first the connection is contingent and causal. Then "wild" causal chains (both "internal" and "external" wayward chains) are to be allowed (cf. Davidson (1973) for these notions). But that would be intolerable, we believe (see our discussion in Chapter 9).

If volitions were conceptually but noncausally connected to action then we would be pre-Jonesians (or neo-Wittgensteinians) and could not have volitions as causal explanatory entities. That would of course be an unfortunate situation for a Jonesian and a causal theorist.

But it is quite possible to have a conceptual connection between tryings and volitions and action while the former are causes of the latter. For we analyze causality as singular event-causation here. For instance, a trying t is ontically independent of the overt action u_o it causes. Still there is a conceptual connection between the concept of trying to do X (expressed by a predicate 'Tr') and the concept of an action X. Thus even if 'Tr' and 'X' are grammatically and conceptually connected and '$Tr(t)$' and '$X(u_o)$' are true, we can have $C^*(t, u_o)$, i.e., singular causation (cf. Chapter 9).

The causation in question is purposive, i.e., it preserves purpose, as will later be seen. This is, roughly, due to the fact that people have learned and internalized language in a certain pattern governed way. Statements like 'I will do X now' have been conditioned from early childhood to go with my doing X. (This is terribly oversimplified and crude when capsulized in one sentence, but that is the basic idea; recall our discussion in Chapter 3.) This rule obeying behavior, which in principle (but only in principle, it seems) can be described in naturalistic (and non-intentional) terms, will, roughly speaking, take care of the problem of wild causal chains. We shall later in Chapter 9 present a technical elaboration of this in terms of our notion of purposive causation (to be distinguished from "mere" causation).

3. A FORMALIZATION OF FIRST-ORDER AND SECOND-ORDER PROPOSITIONAL ATTITUDES

3.1. Although the present chapter is mainly about conative propositional attitudes we shall, below, also occasionally discuss other propositional attitudes (especially beliefs). We are going to sketch out how to formalize statements about propositional attitudes, somewhat in the manner statements about actions were formalized in Chapter 2. We shall discuss both first-order attitudes and second-order ones or meta-attitudes.

Let us start by considering an example. Suppose, to continue our example from Chapter 2, that Jones wants to butter the piece of toast (this Mr. Jones is presumably nonidentical with the mythical genius Jones).

This, we translate roughly as follows: Jones wants the statement 'Jones butters the toast' to be true. Technically this suggestion in our analysis amounts to:

(2) $(Ey)\ W$ (Jones, (Ex) Butters (Jones, the toast, x), y).

Here 'W' is a three place predicate such that its first argument refers to the agent (or to the subject of wanting). The second argument gives the object of the wanting (a statement). The third argument contains the bound variable y ranging over singular events (including states) and hence singular "wantings". Notice that the operator W is opaque only with respect to its second argument: names of agents and wanting-states occur quantificationally transparently.

It should be noticed that (2) only says that *there is* a wanting-event or state of Jones. That is, there is at least one want-actualization. What this actualization is otherwise like we are not told in (2). Thus, we are not told whether this actualization (or those actualizations, if there are several) is overt or covert, active or passive, and so on. However, we are going to accept the realist interpretation of wanting sketched in Chapter 5, according to which these actualizations can be inner events, episodes or states. We may thus, by way of analogy, think of these as tokens in a storage or memory of a computer. That an agent wants something means that there is a token of an appropriate want-statement in the memory. It may be activated and taken into the processing of information at a suitable phase of the computer program.

We may assume that Jones' want to butter the toast in the bathroom was actualized episodically right before he in fact did butter his toast. Let us call this particular wanting a. Then it is true that

(3) W(Jones, (Ex) Butters (Jones, the toast, x), a).

Analogously with the case of wanting, we may formalize believing and intending, using the operators B and I, respectively. Thus, generalizing and using our earlier terminology, we get the following formalizations for action-wants, -intentions, and for action -beliefs:

(4) $(Ey)W(A, (Ex)(\text{Mod-Verbs}(B, o, x)\ \&\ T(o^*, x)), y)$
            ($= A$ wants that B Mod-Verbs o such that T is satisfied);

(5) $(Ey)I(A, (Ex)(\text{Mod-Verbs}(A, o, x)\ \&\ T(o^*, x)), y)$
            ($= A$ intends to Mod-Verb such that T is satisfied);

(6) $(Ey)B(A, q, y)(= A$ believes that q).

In (4) and (5) the object (or content) of the conative attitude has been made explicit. As earlier I have used a three place action verb. The symbols 'o' and 'o^*' stand for some suitable objects and 'T' for a complex predicate as clarified in Chapter 2, section 3. An interesting difference between wanting and intending is that an agent A cannot intend another agent B's actions, whereas A can want B to do something. This is indicated in our formalism.

Many other kinds of propositions besides action-propositions may of course be the objects of one's wants and intentions, although we are not so much interested in them here. In (6) I have let the object of the agent's belief be any proposition q, though we shall below mostly be interested in beliefs concerning action generation (cf. the practical syllogism).

The statement which is the object statement of a propositional attitude must be a statement in the "language of thought", as discussed in Chapter 3. We have been claiming that when the statement is about actions then this statement has the logical form displayed above (and in Chapter 2). If we are on the right track, presumably these logical forms might be taken to coincide with Sellars' dot-quoted statements. Thus, given a statement q in our language of thought it should abstract the *representational role* from many surface statements (in English as well as in other natural languages), i.e. $q = {\cdot}p_1{\cdot}$, $q = {\cdot}p_2{\cdot}$, etc. for a number of surface statements $p_1, p_2, \ldots .$[6]

The language of thought must be finitary and recursive in character as discussed in Chapters 2 and 3. This is a prerequisite for any language learning and of course for theory building in psychology as well. As we have said, it should be at least in principle possible that a psychologist employ the conceptual framework developed in this book when building his theories. (In any case, psychological theories should be reconstructible so as to fit our framework.)

A psychologist theoretician must then assume that propositional attitudes can be grouped together so as to form a system recursively generatable from a finite basis set. He would then start with a finite number of "object statements" $p_1, \ldots, p_n$. The rest of the object statements would be suitable logical constructions (e.g., truth functional combinations) out of these basic object statements.

Then we could classify our basic wants accordingly and call them $W_1, \ldots, W_n$ on the basis of their objects, where each W_i has the structure $W(A, p_i, x)$. Perhaps our psychologist could decide that he needs, in his theorizing, only a small finite number of wants including the basic ones: $W_1, \ldots, W_n, W_{n+1}, \ldots, W_r$. Completely analogous remarks and assumptions can be made in the case of other propositional attitudes. Let me point out here that while the construction of this type of system might be possible in principle, to carry it out in practice seems extremely difficult when attempted on a comprehensive basis. Furthermore, it might be possible that agents can create totally new propositional concepts which do not belong to any known recursive system envisioned above. To the extent that this is possible, scientific psychology is limited in principle. But this problem cannot be solved a priori.

Another puzzle about propositional attitudes is the question of how much is true about them on purely conceptual grounds. We may not, for instance, want to allow for flat and obvious inconsistencies (e.g. beliefs in mutually incompatible propositions), although people's actual belief systems often may be inconsistent in some less transparent sense. Here some standards of *rationality* and *coherence* have to be imposed.

Philosophers have in general concentrated on analyzing completely or almost completely rational agents when discussing propositional attitudes. Our strategy in this book is almost the opposite. We want to speak about the beliefs and wants, etc., of any language using animals, including children and morons. Thus we wish to minimize the rationality assumptions to be made on a priori grounds, and we rather wish to emphasize the finitude of human mind and the accompanying failure of the logical closure principles concerning the contents of propositional attitudes.

In any case, some formal devices for expressing rationality and coherence assumptions are needed. As the most interesting way of accounting for this is in terms of second-order propositional attitudes, we shall postpone our discussion until the next subsection.

One topic we may briefly comment on in discussing the formalization of statements about propositional attitudes is *negative* attitudes, i.e. negative wants, beliefs, etc. To illustrate this problem area we consider an example about a negative belief:

(7)   Jones does not believe that the earth is flat.

How do we formalize (7)? Believing is a propositional state expressible by means of a three place predicate 'B'. Its first argument names the agent (J). Its second argument gives the object of the belief. Let us formalize 'the earth is flat' by '$F(e)$' with 'F' standing for 'is flat' and 'e' naming the earth. Then we may consider the following three ways of formalizing (7):

(8) $(Ex)B(J, \sim F(e), x)$

(9) $\sim (Ex)B(J, F(e), x)$

(10) $(Ex) \sim B(J, F(e), x)$.

Of these, (8) reads 'there is a belief state x of Jones having the content that the earth is not flat'. Formula (9) reads 'there is no state x of Jones such that x is a believing that the earth is flat'. Finally, (10) reads 'there is a state x of Jones such that x is a not-believing that the earth is flat'.

(8) and (9) are, as such, non-equivalent and they both clearly explicate a sense of negative believing. We shall in a later context briefly discuss the semantics of propositional attitudes, which should also clarify the differences between (8) and (9).

(10) is compatible with (8), and it amounts to saying that not every singular state x of Jones is a believing that the earth is flat. Obviously, (10) does not represent negative believing, unless perhaps when some strong additional restrictions on x are imposed. Furthermore, if one could interpret '$\sim B$' as expressing disbelieving, (10) could be an explicate of (7). (I do not see how the ordinary "operator-account", such as Hintikka's, could plausibly account for this explicate of (7).)

One may give (7) the so called relational interpretation ("Jones does not believe of the earth that it is flat") and quantify existentially into the belief context by replacing the name e by a variable y and quantifying over it. Analogous explicates for negative believing are obtained also for this more complicated situation. (Our further comments on quantifying into opaque contexts will be deferred to Chapter 9.)

3.2. It seems that in terms of second-order propositional attitudes one can say interesting things about several difficult philosophical problems. First, one can clarify the *rationality* and *coherence* assumptions embedded in the notions belonging to the framework of agency (e.g., wanting, intending, believing, etc.). Secondly, one can analyze *reflective thinking* and "metathinking". Thirdly, the notion of *will* and the related notions of *akrasia* and freedom of the will can be elucidated in these terms. (The connections to our previous discussion of the notions of intending and trying (willing) can thus also be seen.) Finally, and this relates to all of the three mentioned features, one can fruitfully discuss the notion of *personhood* in terms of second-order propositional attitudes. We shall below mostly discuss the first and third of these issues, although we will make some remarks on the two other issues as well.

If we do not think of man as a completely or almost completely rational being (whatever exactly these notions mean) our framework must be able to accommodate the whole range of rationality, so to speak. We cannot a priori make our notions of intending, believing, etc. so strong that no real human beings ever can be said to intend, believe, etc. We must certainly be able to speak of the beliefs and intentions of, e.g., children. How should we proceed here?

Let us consider an example related to the concept of believing, which philosophers sometimes give much a priori logical content.

First, we, as earlier, distinguish between the state of believing and the content of believing (expressed by the second argument of our belief predicate). It might, for instance, be the case that

$$(11) \qquad (Ex)[B(A, p, x) \ \& \ B(A, q, x)]$$

is true. There is, thus, a singular state x such that x is (or instantiates) both A's belief that p and his belief that q. Obviously (11) entails, by means of pure logic,

$$(12) \qquad (Ex)B(A, p, x).$$

This "conjunctive" type of entailment corresponds exactly to the type of entailment discussed in the context of action statements in Chapter 2. It is obvious how this example can be generalized, and thus we need not devote space to it here.

Let us next consider entailments by virtue of belief contents. Consider these two statements:

$$(13) \qquad (Ex)B(A, p \ \& \ q, x)$$

$$(14) \qquad (Ey)B(A, p, y).$$

Does (13) entail (14)? Can we say on a priori grounds that if A believes in some conjunction (e.g., $p \ \& \ q$) that he cannot fail to believe in each conjunct (e.g., p) as well. I think that in the general case we must answer in the negative. If the conjunction $p \ \& \ q$ is complicated enough, a less clever agent might believe in it but not in p. Notice that we do not have to characterize the concept of believing in terms of this kind of logical properties of belief contents at all but rather use an explicative set analogous to that constituted by (A1)–(A8) for wanting (cf. Chapter 4).

Still it might be the case that (*13*) entails (*14*) for suitable types of belief contents. Perhaps we should say that a person who sincerely says that he believes, e.g., that this ball is red and that this ball is round but that he does not believe that this ball is red does not understand what it means to believe (that something is the case).

If we now accept that (*13*) entails (*14*) how do we characterize this in logical terms? One possibility is just to specify in terms of metalinguistic semantic rules of inference that any statement of the type (*14*) cannot be false if a related statement of the type (*13*) is true. One may imagine that one would get a *finite* number of such inference schemas for believing, and similarly for all the other propositional attitudes. Thus the finitary character of the language of thought (here taken to be a first-order language or an extension of such) would have been preserved. However, this solution has a somewhat *ad hoc* character and it is not philosophically very illuminating.

A second strategy would be to give conditions for the singular states (or events or episodes) x and y in terms of inclusion, etc. by means of, e.g., the calculus of individuals. Thus we might try this: (*13*) entails (*14*) if and only if y is a *relevant* part of x. The trouble here is that it is not sufficient to require that y be merely included in x. Intuitively, y must correspond to the "p-aspect" of x rather than to its "q-aspect".

If x and y are taken as structured events, then it might not be very difficult to define such a notion of relevant inclusion. But in order to be interesting such a semantics should be connected to the neurophysiological or, more generally, physical properties of x and y, and about that we do not really know anything informative yet.

A somewhat different strategy to account for entailments between the contents of propositional attitudes is via second-order attitudes. Let us consider the following statement:

$$(15) \qquad (Ez)B(A, (x)(B(A, p \;\&\; q, x) \rightarrow (Ey)B(A, p, y)), z).$$

This statement says, roughly, that there is some second-order belief-state of A with the content $(x)(B(A, p \;\&\; q, x) \rightarrow (Ey)B(A, p, y))$. This content says that whenever A is in a singular state of believing that $p \;\&\; q$, he cannot fail (as a matter of nomic fact) to be in some singular state y instantiating his belief that p.

We may now ask whether it is the case that (*15*) is true if and only if (*13*) entails (*14*). As I understand second-order believing this is not quite the case.

It seems plausible to accept that if (*13*) entails (*14*) then (*15*) must be true. In saying this I take the entailment of (*14*) by (*13*) to be something nomic (to some degree at least) and thus non-accidental, of course. When an agent has fully *internalized* a semantical or logical principle (i) he should be capable of reflecting on it and (ii) he should also be disposed to act on this principle. We shall not here discuss (ii), which otherwise is central for our theory of action (cf. acting on or because of a reason, discussed in Chapters 8 and 9).

In so far as (i) is acceptable we must require the truth of (*15*), given that (*13*) entails (*14*) as a nomic matter of fact. (Here (*15*) is still more plausibly true if at least the second-order belief predicate '*B*' is taken in the broad sense of 'thinks that'.) In the case of mature language users at least we then accept that if (*13*) entails (*14*), then (*15*) cannot be false. We may argue in addition that direct self-knowledge about the fact that (*13*) entails (*14*) presupposes that (*15*) is true.

However, if (*15*) is true, then (*13*) need not be taken to entail (*14*) even in the case of mature agents. Condition (*15*) is compatible with inco-herence and "errors" in believing (cf. the case of complicated p's and q's). One can be regarded as a rational person and still make errors and mistakes such as believing in p & q but failing to believe in p.

It would seem now that the second-order criterion (*15*) would be a good rationality criterion for believing. It is a descriptive criterion which, however, involves a norm. The norm it involves is obviously that if one believes that p & q then, in epistemically standard conditions, one *ought to* believe that p and at least ought not to believe that $\sim p$. The content of the second-order belief in (*15*) gives a descriptive characterization of this norm (cf. our remarks in Section 3 of Chapter 4).

I have made my point in terms of a trivial example. A more general account can obviously be given in terms of our above discussion. To see how conative attitudes like intending could be handled consider an analogous example:

$$(16) \qquad (Ez)B(A, (x)(I(A, p \& q, x) \rightarrow (Ey)I(A, p, y)), z).$$

This condition says roughly that (at least a mature) agent A does believe (or think) that he cannot intend (to do something) p & q without intending (to do) p. Condition (*16*) would seem to be a plausible candidate for a conceptual truth about intending (at least for "everyday" p's and q's). As (*15*) also (*16*) is compatible with slips of thought and similar mistakes in intending.

"Metathinking" as expressed by (*15*), (*16*), and other related statements, seems to characterize, in part, what the concept of *person* is. A person is capable of 1) reflective thinking, and of 2) internalizing norms while at the same time 3) occasionally making errors in thinking. These three features of personhood find a nice expression in our above analysis. As pointed out in Chapter 4, it is hard to see how a pre-Jonesian (or neo-Wittgensteinian) could handle metathinking in an interesting way, at least if the singular state (or episode) z is thought to have causal impact on the states x and y (see our formulas (*15*) and (*16*)).

Can anything be said about the relationship between second-order states or episodes, such as z, and first-order states or episodes, such as x and y? In general, we can only say what our conceptual functionalism permits us to say—as long as no neurophysiological, etc., characterizations are available. This means that presently nothing very interesting can be said. Even under a Jonesian interpretation these states can have almost any factual intrinsic features (and even fail to have any factual intrinsic features).

Consider next briefly our "*II*-thesis" of the previous section. According to it, intending (at t) to do X amounts to intending (at t) to intend (at t) to do X. This intending in this situation "involves itself". (Cf. especially the notion of trying obtained for $t = $ now.) Accepting this, we consider the following example:

(*17*) $(Ez)I'(A, (Ey)I'(A, (Ex) \text{ Opens } (A, \text{ the window}, x), y), z)$.

Formula (*17*) is supposed to formalize that A intends (at t) to intend (at t) to open the window. (I have assumed unabashedly that the same predicate 'I' can be used in the second-order case equally as in the first-order case.) Let us now assume these intending states or episodes in fact obtain or take place. We call the first-order intending a and the second-order one b, and get:

(*18*) $I'(A, I'(A, (Ex) \text{ Opens } (A, \text{ the window}, x), a), b)$.

If the first-order intending involves itself then b must be a part of a, i.e. $b < a$ in the symbols of the calculus of individuals. But intending to intend also involves first-order intending as its part, if what I have said about the *II*-thesis is accepted. Thus we also have $a < b$. Hence $a = b$ (in the sense of the calculus of individuals). This entails that whatever is descriptively predicated of a is true of b, and vice versa. Thus (*17*) is equivalent to

(*19*) $I'(A, (Ex) \text{ Opens } (A, \text{ the window}, x), a)$.

Our final topic is the notion of *will* as elucidated in terms of second-order propositional attitudes (cf. Frankfurt (1971)). Let us start by discussing a simplified example. We consider the case of a "moral alcoholic". Being an alcoholic he obviously has a strong want to drink, and this want is stronger than his want to stay sober. (This holds at least during his drinking periods.) Symbolically, we may put this as follows:

(*20*) $(Ex)(Ey)(W(A, d, x) \ \& \ W(A, s, y) \ \& \ S_w(x, y))$.

Here *d* is a statement expressing that *A* is (or will be) drunk and *s* is a corresponding statement expressing that *A* stays sober. 'S_w' reads 'stronger as to the intensity of the want-aspect', and thus it in fact comes to express a comparative relation of wanting (viz. *A*'s wanting *d* more than *s*).

That our alcoholic is "moral" we take to mean that he prefers wanting to be sober to wanting to be drunk. This preference is a second-order one. We may write this out in terms of second-order wants as follows:

(*21*) $(Eu)(Ev)(W(A, (Ey)W(A, s, y), u) \ \& \ W(A, (Ex)W(A, d, x), v) \ \&$
$S_w(u, v))$.

Perhaps we could even have this:

(*22*) $(Eu) \sim (Ev)(W(A, (Ey)W(A, s, y), u) \ \&$
$W(A, (Ex)W(A, d, x), v))$.

Formula (*22*) says that our moral alcoholic even lacks the second-order want to want to be drunk.

Earlier in this chapter we made a division of wants into *extrinsic* and *intrinsic* ones. Our present example and other similar ones indicate that extrinsic wants (wants related to the fulfilment of duties and obligations, etc.) at least in many cases are, so to speak, taken into account primarily as second-order wants, and accordingly they (at least potentially) involve "self-reflection". Building a child's character just involves teaching it extrinsic wants ("a conscience" or "superego"). In the case of small children this takes place primarily in terms of reward-punishment conditioning, to be sure. But more full-fledged human agents must also have been taught to be in principle capable of reflecting on their wants in each behavior situation in which they act intentionally according to a norm, etc.

What I just said hardly makes a sharp distinction between intrinsic and extrinsic wants, but at least it indicates some kind of a difference between biological intrinsic wants (e.g., those related to hunger or sex) and so-called secondary, social extrinsic wants (social duties, obligations, etc.).

Let us consider our statements (20) and (22). One could say that the moral alcoholic's "passion" is stronger than his will, assuming that he goes on with his drinking. His passion would be given by his first-order wants (cf. (20)), his will is reflected in his second-order wants (cf. (22)). Can we make this any more precise?

Consider the following second-order wants:

(23) $(Ex)W(A, (Ey)W(A, s, y), x)$

(24) $(Eu)W(A, (Ev)W(A, d, v), u)$.

I would say that both (23) and (24) express the agent's *potential will*. What is his actual will, then? Could we take the stronger of the second-order wants expressed by (23) and (24), respectively, to be his will? That is a possible option. It is roughly in that sense that our moral alcoholic's passion is stronger than his will.

But there is another aspect involved in the notion of will, and this, not surprisingly, is related to willing (or effective intendings). We might take the agent's will, then, to be his second-order intending based on either (23) or (24). Thus, as discussed earlier in this chapter, we may think that the agent forms an intention to act on his want to be sober. If our earlier argument for the *II*-thesis is acceptable we see that this makes the first-order and second-order intendings (named by a and b) identical in the following statements:

(25) $I(A, s, a)$

(26) $I(A, I(A, s, b), a)$.

(See our earlier argument to this effect.) The alcoholic's intending to be sober will then at the right moment turn into an effective intending *now*, i.e. a trying, provided he is able to keep his intention. Given that there are no external factors inhibiting his sober-staying behavior, he will indeed stay sober as long as he has the intention expressed by (25). His *strength of will* is reflected simply in how long he is able to stick to that intention. Conversely, *akrasia* or *weakness of the will* is reflected in that the agent is not even able to form the intention (25), but remains "undecided" with his conflicting second-order and first-order wants.

Freedom of the will in our analysis means that the agent has the ability or power to make either one of the second-order wants expressed by (23) and (*24*) his effective will. That is, if the agent is able to form an intention to act according to both of them as he chooses (assuming (*23*) and (*24*) to exhaust the alternatives), we say that his will is free in this context. One problematic point here is what is involved in such an ability to form intentions, and it seems that a philosopher cannot say much of interest here. He can analyze more carefully what the relevant concept of mental power involves, to be sure. (We shall not here go into a more detailed discussion of it.) Anyhow, the most interesting questions in this context relate to the psychological, sociological, and, perhaps, neurophysiological determinants of one's ability to form intentions, and these are scientists' rather than philosophers' problems.

What we have said about the freedom of the will is compatible with determinism. Roughly, it is possible that intention formation on the basis of one's first- and second-order wants (which are non-actions) is a deterministic process, though intention formation is intentional mental activity. The question obviously turns on the question whether our requirement that the agent could have (effectively) intended otherwise, had he chosen, is compatible with determinism, and we claim it is. The concept of a condition preventing the intending (or trying, for that matter) to do *X* can be argued to be different from the concept of the condition under which the intending or trying to do *X* is physically impossible. This is the crux of the "proof" that freedom of the will is compatible with determinism. (Sellars' (1966) and (1975) discussion and "proof" along these lines seems to me acceptable once it is carved into the conceptual framework above.)

NOTES

[1] Audi (1972) analyzes beliefs as theoretical concepts analogously with how we analyzed wanting in Chapter 4. I find Audi's approach acceptable and illuminating (apart from some details). Technically this kind of approach can be made precise by means of the formalization of propositional attitudes we have discussed in the previous chapters and will be discussing in detail later in the present chapter.

[2] The mentioned recent theory of motivation by Atkinson and Birch (1970) assumes that an organism acts according to the dominant or strongest resultant action tendency. Thus Atkinson's and Birch's theory is in this sense Lewinian. The basic idea from which they derive their Lewinian conclusion is that the rate of change in the strength of a tendency (*T*) at any moment in time (*t*) is equal to the instigating force (*F*) minus the consummatory force (*C*), i.e. mathematically, $dT/dt = F - C$.

Atkinson's and Birch's theory is also compatible with the Bayesian expected utility maximization view, but it is incompatible with, e.g., Festinger's theory of cognitive dissonance, which is a balance theory (cf. Atkinson and Birch (1970), Chapter 6 and pp. 329–330). Notice that their theory also in principle applies to cognitive processes and hence, for instance, to the problems of intention generation.

In this connection the useful survey of intention generation by Fishbein and Ajzen (1975) should be mentioned. Contrary to most psychologists, they use a conceptual framework fairly similar to ours. Thus they consider intentions to be the closest determinants of actions and discuss both the theoretical and experimental aspects of intention-determination due to proattitudes and beliefs.

3   Conditional intentions (intendings) have conditional statements as their objects, as opposed to absolute or categorical intentions. Thus A's intention that p if q would be a conditional intention. In the context of conditional intentions one may ask whether there are conceptually valid "detachment" principles for them. A good candidate is the following: If A intends that p if q and if A believes that q then A intends that p (cf. Kim (1976)). I doubt whether even this innocent looking principle is acceptable for all A, p, and q without assuming A to be rational in a strong sense. For in the case of suitably complex (e.g., mathematical) p and q, A might satisfy the antecedent conditions of this principle and yet fail to connect them and thus fail to absolutely intend that p.

4   The requirement that one can intend only such things of which one believes that he can with some nonnegligible probability bring them about may still need another qualification read into it. Suppose an agent believes he cannot ride a bicycle. Still he may intend to ride bicycle next summer, say, for he believes that he can *learn* the skill before that. Thus an agent can intend to do things which he believes he now cannot do but intends to learn to do and believes he can learn to do (cf. von Wright (1972)).

5   In their pioneering study McAdam and Whitaker (1971) recorded slow potential shifts over Broca's area up to 1 second before the moment when a normal subject intentionally uttered a polysyllabic word. No corresponding negative slow potentials could be recorded when subjects produced simple oral gestures such as spitting.

Let me here take the opportunity to thank Prof. Risto Näätänen for helpful discussions concerning neurophysiological and neuropsychological matters such as those above.

6   Our claim that propositional attitudes have as their objects propositions (in the discussed deep structure sense) is not completely unproblematic. For one thing, it depends on the treatment of indexicals in the deep structures.

Consider the statement

(*)   John intends to go to the theatre.

We may translate (*) non-technically into

(**)   John intends true that he goes to the theatre,

and technically into

(***) $(Ey)\ I\,(\text{John},\ (Ex)\ \text{Goes}\,(\text{John},\ x)\ \&\ \text{To}\,(\text{the theatre},\ x),\ y)$.

Now (***) uses 'John' for the 'he' of (**). It is thus assumed in our treatment that a suitable name or description be found for each such indexical as the above 'he' occurring in the surface form. What I am aiming at here seems to accord very well with Lakoff's (1975) analysis of indexicals, according to which they do not directly occur in logical forms at all. The present problem thus in no way undermines the idea that the objects (contents) of propositional attitudes are suitable deep structure propositions, i.e. logical forms.

For our present purposes we may assume that Sellarsian dot-quotes play the role of an operator that converts a sentence into its logical form, viz. '·p·' represents the logical form of the sentence 'p'. In other words, we may assume here and later in this book that the dot-quote operator has the same properties Harman (1972) gives to his operator '≠', which we discussed in Chapter 2.

CHAPTER 7

CONDUCT PLAN AND PRACTICAL SYLLOGISM

1. CONDUCT PLAN

1.1. Although this book is supposed to be mainly about human actions and their explanation, we have so far said rather little that is systematic and detailed about this topic. In this and the next chapter, we will start our more comprehensive treatment of action theory and in so doing will rely on the philosophical groundwork built in the previous chapters of this book.

In Chapter 2 we already discussed actions in a preliminary way; that discussion will have to serve our purposes in this and the next two chapters. A systematic treatment of various types of actions and related concepts will have to be deferred until Chapter 10. This is because our "intentionalist" causal theory of action requires lots of groundwork before it can be fully and explicitly stated and discussed. This is perhaps somewhat unfortunate from the standpoint of the reader, but it cannot be helped.

A brief preview of what is to come in the next two chapters goes as follows. In our view (intentional) actions are caused by effective intendings (called tryings or willings). As we have seen, however, intendings are very complex mental states. Among other things, we claim, they are central parts of the agent's broader plans for acting. Such plans—to be called conduct plans—will also serve to *explain* the agent's actions. This is because the agent's conduct plans may become activated so that they *purposively cause* actions (i.e., the agent's *having* a conduct plan may serve as a cause). In the previous chapters we discussed the nature of mental states and episodes and defended a kind of functionalist analogy view. Our notion of a conduct plan is, of course, to be conceptually construed along those lines. In the present chapter we will be mostly interested in some special details of conduct plans.

One particularly important type of conduct plan becomes defined through the practical syllogism. We shall devote a large part of the present chapter to a discussion of some forms of the practical syllogism (and thus of practical inference, too). Conduct plans and practical syllogisms are intimately connected to the teleological or "intentionalistic" explanation of actions. We shall argue that the connection is due to the fact that conduct plans and what is expressed by the premises of practical syllogisms purposively cause actions. This will require a separate discussion of causation and purposive causation.

When discussing relationships between actions, the notion of *action generation* seems to us to be of special importance. We will need this notion both when discussing conduct plans and, still more importantly, when defining various types of action in subsequent chapters. Let us start our developments in this chapter by a treatment of action generation.

In Chapter 2 we discussed Davidson's switch flipping example. We said that for a fine grain theorist all the six action descriptions of the diagram represent different actions. Still these actions are related in the sense that each of the actions except the first one, is generated by another one. This generational mechanism seems to create tree-like structures, which is what we shall now try to clarify.

Goldman (1970) has presented an elaborate account of action generation. Although my theoretical approach is somewhat different from his I basically agree with his analysis of the underlying situation — "the data". Goldman distinguishes four different categories of generation whereas I only speak of two, viz. a) *factual* and b) *conceptual* nomic generation. Goldman's categories are 1) causal generation, 2) conventional generation, 3) simple generation, and 4) augmentation generation. The first three of Goldman's notions can, roughly, be taken to explicate locutions of the form '*A* does *a* by doing *b*'.

A typical case of causal generation is this: *A*'s flipping the switch causally generates *A*'s turning on the light. In cases of causal generation the defining characteristic in Goldman's account is that there is a causal relation between *A*'s flipping the switch and the event of the light going on. Thus, an action token *a* of agent *A* causally generates action token *b* of *A* only if (1) *a* causes a certain event *e*, and (2) *b* consists in *A*'s causing *e* (Goldman (1970), p. 23). What I call factual nomic generation is slightly different from Goldman's causal generation. I analyze the above example as follows. We have two events: flipping the switch and the light going on. It is assumed that the first of these events is a cause of the second. (Goldman would say, rather, that the action of flipping the switch is a cause of the light going on.) Thus, in my account of causal action generation, the *results* of the two actions in question are in a cause-effect relationship. Other examples of causal generation, e.g., *A*'s arm moving causally generating his opening the window, are analyzed similarly. In addition to causal relationships, my notion of factual nomic generation also may involve other, non-causal, nomic factual relationships.

According to Goldman, A's action of extending his arm out the car window conventionally generates his signaling for a turn. More generally, action token a of agent A is said to *conventionally* generate action token b of agent A only if the performance of a in circumstances C (possibly null), together with a rule R saying that a done in C counts as b, guarantees the performance of b (Goldman (1970), p. 26). (Goldman only gives this necessary condition.) In our example the circumstances C include the agent's driving his car on the road, and so on. The rule R says that extending one's arm out the car window while driving on public roads counts as signaling. Another example of conventional generation would be this: A's moving his queen to the king-knight-seven conventionally generates (in chess) A's checkmating his opponent. This example is to be analyzed analogously.

A's coming home after midnight may in suitable circumstances generate A's breaking his promise. Examples of this sort are for Goldman cases of *simple* generation. Goldman argues that while conventional generation might be schematized as 'a and R and C jointly entail b' simple generation can be schematized as 'a and C jointly entail b'.(See Goldman (1970), pp. 26–28.) I would like to claim, however, that even cases of simple generation in a sense involve general rules. In our example C includes the promise that A will be back by midnight. The rule R is a meaning postulate for the concept of promising to the effect that speech acts called promising logically involve their fulfillment. Thus, I see no fundamental difference between conventional and simple generations of at least the above kind, though perhaps a detailed botanization of such rules R, e.g., in terms of their specific culture-dependence, might bring out some differences worth being emphasized.

Other examples of simple generation perhaps should be mentioned here, too, for this category is a broad one. Thus, we consider the following generational claims (presumably acceptable by Goldman to belong in this category):

(i) A's jumping 2 meters simply generated A's outjumping B;
(ii) A's asserting that p (while disbelieving that p) simply generated A's lying that p;
(iii) A's killing B simply generated A's revenging himself;
(iv) A's breaking a precious vase simply generated A's doing a bad thing.

In example (i) the generation involves a semantic rule relating outjumping and jumping. Examples (ii) and (iii) are concerned with A's mental states (belief and motive). From a semantic point of view we may think that they, respectively, deal with the concepts of lying and revenge and the respective semantic rules.

Example (iv) deals with an evaluation and is in this sense special. But (iv) may be taken to be true or acceptable in part because of cultural conventions (concerning what is good and bad). I think (iv) can be treated on a par with the other examples placed in this category (even if Goldman does not mention examples of this sort).

Goldman's last type of generation is *augmentation*. It is represented by examples like this: A's extending his arm generated A's extending his arm out the window in those cases when A extends his arm out the window; A's saying "hello" (in certain circumstances), generated his saying "hello" loudly. Intuitively, augmentation generation is based on the fact that actions are always performed in some manner or other; and this extra information may be taken into account in redescribing the action. Furthermore, contrary to augmentation-generation, in all of the other species of generation a generated action is formed by making use of some additional fact not implicit in the generating action.

Notice that an agent's extending his arm out the car window may both generate and be generated by his extending his arm, according to what I have said (contrary to Goldman). However, this only holds for some tokens of these action types.

While causal, conventional, and simple generation, at least roughly, can be characterized in terms of the preposition 'by' ('A did X *by* doing Y'), this preposition is not applicable to augmentation generation. (A did not extend his arm out the car window *by* extending his arm.)

We have seen that Goldman's conventional, simple, and augmentation generation concern a wide variety of generational relationships. However, they all deal only with "mere" redescription or reconceptualization of actions (cf. our discussion in Chapter 2). They do not involve factual nomic producing (between action-results) in the sense causal generation does. Thus even if, e.g., my jumping counts as outjumping John only given certain circumstantial facts, this case is essentially different from my trigger pulling causally generating my killing John.

I will accordingly below lump together conventional, simple and augmentation generation and call them cases of *conceptual* generation. Causal and other nomic generation will be called *factual* generation. We will assume that, so understood, factual and conceptual generation are mutually exclusive and jointly exhaustive.

Both the exclusiveness and the exhaustiveness assumption may be questioned, perhaps. The exclusiveness assumption is obviously connected to the broader philosophical issues concerning what is conceptual and what is factual. We shall not here worry about that, especially as the exclusiveness assumption is not very central for our developments. On the contrary, the exhaustiveness assumption is central for us. Nothing we have said so far, nor what Goldman's (1970) rich treatment contains, amounts to a very conclusive proof of this assumption. There are some considerations that still can be brought to bear on the issue here. Let us consider them.

There are relationships between actions which are not generational in the sense discussed above. These are exemplified by a) *subsequent* actions, b) *co-temporal* (or, more generally, *independent*) actions, and c) actions relating to each other in a *part-whole* relationship. Of these Goldman (1970) discusses subsequent and co-temporal actions as well as temporal parts of wholes.

We may say in linguistic terms that an action token *b* of *A* is subsequent to action token *a* (of *A*) just in case it is correct to say that *A* did *a and then* did *b* (cf. Goldman (1970), p. 21). For instance, when having lunch I may first eat my salad and then my beef. The idea here is to try to capture temporally distinct and consequent but (conceptually and factually) independent action pairs. It is easy to see that this linguistic criterion, irrespective of how good it otherwise proves to be, tells us at least that our cases of action generation discussed earlier do not become subsequent actions (e.g., *A* did not flip the switch *and then* turn on the light).

Subsequent actions are independent, we said. But there are also other types of conceptually (or semantically) and factually independent actions. Goldman (1970) considers *co-temporal* actions as another type of case. For instance, an agent may wiggle his ears *while also* writing a letter at the same time. Two co-temporal acts in this "while also" sense fail to be in any generational relationship to each other, and this can be checked against our earlier examples concerning action generation. We shall later be much concerned with co-temporal actions in the context of our treatment of complex actions. For in many cases complex actions in part consist of such co-temporal actions. For instance, when flushing a nail I do the hammering with one hand while holding the nail with the other. The action of flushing a nail comes about due to actions of these two kinds. Another example would be an agent's shooting the basketball at the basket while jumping in the air. The co-temporal actions of jumping and shooting jointly generate the agent's scoring.

There are lots of other types of actions which are independent in the sense of not being in a generational relationship. For instance, if I build a house I must obviously do lots of things which are generationally independent in this sense but which are neither co-temporal nor subsequent to each other (e.g., I must lay the ground and put on the roof). This kind of independent actions (which nevertheless contribute to the end result, e.g., the house becoming erected) will be incorporated in our later discussion of complex actions of which building a house is an example.

Two actions which are in a generational relationship, in our sense, must be contiguous in time; they cannot be subsequent actions. We must also exclude the possibility that one of the actions is a temporal part of the other. By this we mean the following. Suppose that I want to play a certain melody. Then I must play the note c followed by a, followed by f, and so on. The whole melody consists of the playing of this sequence of notes but it is of course not generated by any single element in this sequence. (Similar remarks hold for analogous spatial and spatio-temporal wholes.) It is quite another matter how to conceive of the playing of the whole melody as a complex action which is somehow generated or put together from the playings of the single notes. Our later treatment of complex actions is applicable here, too. Indeed, we will claim that our analysis serves to clarify the structure of everything we do or may do.

Let us now go back to action generation and make a few further remarks. First, it is to be noticed that we have not yet formally characterized our notions of factual and conceptual generation. Let us denote by the two place predicate '$R_1(a, b)$' the locution, 'event token a factually and directly generated (generates) event token b' and by '$R_2(a, b)$' the analogous locution, 'event token a conceptually generated (generates) event token b'. The only formal properties that we here attribute to these relations are rather trivial. On the basis of our earlier discussion we can say that both R_1 and R_2 are irreflexive. R_1 is asymmetric (cf. causal asymmetry), but of R_2 we can only say that it is not symmetric (recall the earlier arm extending example!). As R_i, $i = 1, 2$, is concerned with *direct* generation we do not want to make it transitive.

We shall later in Chapter 9 return to give a more formal characterization of our relations of event generation after discussing causality. Our definition of R_i will proceed on analogy with our backing law treatment of singular event causation. Still it will be useful and instructive already here to examine Goldman's (1970) final definition, which brings up some aspects that we have not discussed so far. I use Goldman's symbols and framework in presenting the definition (see Goldman (1970), p. 43):

> "Act-token A level-generates act-token A' if and only if
> (1) A and A' are distinct act-tokens of the same agent that are not on the same level;
> (2) neither A nor A' is subsequent to the other; neither A nor A' is a temporal part of the other; and A and A' are not co-temporal;
> (3) (a) there is a set of conditions C^* such that the conjunction of A and C^* entails A', but neither A nor C^* alone entails A';
>   (b) if the agent had not done A, then he would not have done A';
>   (c) if C^* had not obtained, then even though (agent) S did A, he would not have done A'."

In Goldman's above definition 'level-generation' represents (roughly) the same thing as my 'generation'. 'A' and 'A'' represent *statements* describing acts [my actions] as well as the acts themselves, and 'C^*' stands for statements describing conditions as well as the conditions themselves. (I think Goldman's semantic apparatus is somewhat confusing and, considered together with the above definition, subject to the paradox that every singular action generates every other action, as I have pointed out in Tuomela (1974b); but let's waive that objection here.) Let us now comment on Goldman's conditions one by one.

Condition (1) employs the notion of same-levelness. It can be clarified by means of an example. Suppose that I move this table. Suppose, too, that this table is the heaviest object in this room. Now for Goldman my moving this table and my moving the heaviest object in this room are *distinct* action tokens. (For us again here one action is described in two ways; cf. Chapter 2.) Thus, since he clearly cannot say that one of them generates the other one, he says that they are on the same level. Otherwise I find the above condition (1) acceptable.

On the basis of our earlier discussion, condition (2) gives an acceptable necessary condition for action generation.

Condition (3), however, seems too strong. (3) (a) states that A and C^* are both necessary for generation. It seems that A must be necessary in something like this (or a stronger) sense. (This will be guaranteed by our later characterization.)

The necessity of C^* is problematic, however. I think there could be cases of causal generation, for instance, where A without any extra circumstantial conditions could be said to nomically entail A'. Of course, much depends on what Goldman means by 'entails' here. If it means simply entailment on the basis of the logical axioms then I agree that at least some general backing law or meaning postulate in addition to the singular statement A will be needed. As a general point I would, however, like to emphasize that it is a scientific and a posteriori matter what such a C^* will be. This remark is especially pertinent in the case of factual generation.

How about (3) (b) and (3) (c)? I think that although these counter-factuals should hold true in the general case, there are some exceptions. For there may be cases of generational overdetermination in which A' is generated by two independent actions A and A^* so that (3) (b) is falsified. Thus, if I kill you by shooting you, your death might have come about at the same time by the poison I gave you. What is more, if my shooting (A) would have been only slightly different (the path of the bullet being different by a microscopic amount, say), this alternative shooting (A^*) would certainly equally well have led to your death and thus falsified (3) (b).

Goldman seems to think that (3) (c) is needed to ensure the proper direction of action generation. For example, let C^* represent the rule that extending one's arm out the car window while driving counts as signaling, plus the circumstance that the agent was driving when he extended his arm out the car window. Then if C^* had not obtained, then even though the agent extended his arm out the car window, he would not have signaled for a turn. C^* must here be taken to represent the minimal set of conditions necessary for A generating A', says Goldman. (Otherwise (3) (c) is obviously false.) But, first, there is no guarantee that there is such a *unique* minimal set at all. If not, (3) (c) fails to hold. Secondly, the same type of cases of overdetermination which undermined (3) (b) also serve to disqualify (3) (c). This is because associated with each of the overdetermining actions A^* there will in general be some circumstances described by some suitable $C^{**}(\neq C^*)$.

Our backing law analysis of action generation, which will be stated in more exact terms after our discussion of singular causality in Chapter 9, does not accept condition (3). Still, being a backing law account, it will respect the nomicity indicated by (3).

1.2. As we have construed mental structures and episodes in this book, they will have *explanatory power* with respect to the agent's behavior. That they are explanatory entities is not their defining feature, rather it is the primary methodological reason for their introduction. Ultimately and ideally explanatory power is thought to guarantee the existence of such postulated mental entities (cf. Chapter 1).

For the purposes of action theory, the basic and primary explanatory entities in explaining actions, are the agent's activated wants and intentions (and other conative attitudes) plus the relevant beliefs (and other doxastic attitudes). We will accordingly think that an agent's actions are to be explained in the first place by reference to his (conscious or subconscious) plans for acting — his *conduct plans*, as we shall call them. Such plans (or the agent's havings of such plans, rather) are thus basically *explanatory* entities. However, as wants, intentions (and other conative attitudes) and doxastic attitudes have conceptual connections to actions (cf. our earlier discussion), conduct plans will also have some conceptual connections to them.

Before going on to a more detailed discussion of conduct plans, there is one more thing to be emphasized. Conduct plans are mongrel dispositional-categorical entities. They, as any entities, can of course be described in various ways. However, it is central here that we describe these action plans, such as practical syllogisms, only in ways which are at least potentially *acceptable* to the agent (even if these descriptions need not be formulated exactly as the agent has formulated or would formulate them).

Let us start our detailed discussion by reference to a simple example. We consider Figure 2.1 of Chapter 2 and assume that it represents (a part of) the content of an agent's plan for acting. We assume that the agent's basic intended goal is to illuminate the room. He thinks (or believes or considers) that his flipping the switch (in those circumstances) will generate his turning on the light, and the latter will amount to his illuminating the room.

Let us now slightly modify the further details of the example from how it was described in Chapter 2. We assume that our agent A believes that his flipping the switch will generate his hurting his finger with probablity q, that his turning on the light will generate his exploding a bomb with probability r, and that his illuminating the room will generate his alerting a prowler with probability s. (We have not yet discussed probabilistic action generation; see Chapter 12 for an exact definition of it.)

Figure 2.1 thus comes to represent the content of A's intentions and beliefs. The single arrows in the tree represent factual generation (as believed to be the case by A) and the double arrow represents conceptual generation (as believed by A). The nodes of the tree represent A's actions. As we assumed, some of these nodes, furthermore, represent his *intended* goals.

To make the example more interesting, we might also assume that with each node of the tree there is associated a conative attitude. Thus, we might assume that A does not *want* to hurt his finger, explode the bomb, or alert the prowler.

To see what A's conduct plan now amounts to in our example, consider the following descriptive statements, which summarize what we have just said:

(1) A intends to flip the switch.
(2) A intends to turn on the light.
(3) A intends to illuminate the room.
(4) A believes that his flipping the switch will factually generate his turning on the light.
(5) A believes that his turning on the light will conceptually generate his illuminating the room.
(6) A believes that his flipping the switch will factually generate his hurting his finger with probability q.
(7) A believes that his turning on the light will factually generate his exploding the bomb with probability r.
(8) A believes that his illuminating the room will factually generate his alerting the prowler with probability s.

We could go on and add the three statements about A's wants referred to above, but to keep the example simple we will leave them out.

We will now use our earlier way of formalizing statements involving propositional attitudes and propose the following (partial) formal translations for (1)–(8):

(1') $(Ez)I(A, (Ex)$ Flips $(A,$ the switch, $x), z)$
(2') $(Ez)I(A, (Ex)$ Turns-on $(A,$ the light, $x), z)$
(3') $(Ez)I(A, (Ex)$ Illuminates $(A,$ the room, $x), z)$
(4') $(Ez)B(A, (x)(Ey)$ (Flips $(A,$ the switch, $x) \rightarrow$
                        Turns-on $(A,$ the light, $y)), z)$
(5') $(Ez)B(A, (x)(Ey)$ (Turns-on $(A,$ the light, $x) \rightarrow$
                        Illuminates $(A,$ the room, $y)), z)$

(6') $(Ez)B(A, \text{p (Hurts } (A\text{'s finger}, y)/\text{Flips } (A, \text{the switch}, x)) = q, z)$
(7') $(Ez)B(A, \text{p (Explodes } (A, \text{the bomb}, y)/$
$\qquad$ Turns-on $(A, \text{the light}, x)) = r, z)$
(8') $(Ez)B(A, \text{p (Alerts } (A, \text{the prowler}, y)/$
$\qquad$ Illuminates $(A, \text{the room}, x)) = s, z)$.

Statements $(1')$–$(8')$ can be taken to be in the "language of thought" if our earlier tentative assumptions concerning the identification of logical form, syntactic-semantic deep structures, and the thought structures are accepted. $(1')$–$(8')$ are simplified translations in that, for instance, time considerations and references to relevant initial conditions, etc., have been left out. Yet they seem to be sufficiently realistic for our purposes.

The intention predicate I and the belief predicate B are to be taken as in Chapter 6. Thus, they are both three place predicates relating the agent to a statement (in its logical form) and to a singular state. In the contents (i.e., second arguments) of the propositional attitudes we use action predicates which should be self-explanatory. The arrow '→' symbolizes a nomic implication and also in part (generic) action generation (in (5'), though, the arrow is to be interpreted as a conceptual implication). If our arguments in Chapter 5 (and elsewhere) concerning the form of laws are acceptable, → can simply be taken as a *material* implication (possibly accompanied with a "pragmatic and epistemic commentary"). We will later assume in our formal analysis of action generation that each "deterministic" singular factual generation claim is "backed" by a suitable general law (often a "derived" or "applied" law), presumably of the kind occurring in the above belief contents (e.g., (4')). Conceptual singular generation is handled analogously.

Statements $(6')$–$(8')$ concern beliefs about probabilistic generation. An exact treatment of probabilistic generalizations will be given in Chapter 11 and of singular probabilistic action generation in Chapter 12, as stated previously.

After these preliminaries we may finally turn to conduct plans. In general, an agent's conduct plan as a plan must be taken to involve some goal or goals the agent *intends* to achieve. Typically, although not always, there is one goal, which could be taken to be the agent's *principal* intended goal.

A conduct plan must also somehow relevantly involve the agent's means-end *belief* or beliefs (means-end beliefs broadly understood) concerning the agent's achieving his goals. When I say that a conduct plan must involve intendings and believings, I mean that they, when causally effective, must involve mental states which play (at least roughly) the *role* intendings and beliefs do according to our earlier analyses. Thus, they may in some cases be *called* wantings, desirings, decidings, etc. or thinkings, considerings, knowings, etc. as well.

Let us emphasize here that conduct plans need not (though they may) involve rational *deliberation*. They need not either be fully *conscious* to the agent. If we are to believe Freud, they may even be fully repressed. As to the standard objections against "intellectualist" construals of mental entities, of which our approach perhaps is a brand, see our discussion in Chapters 3 and 4. If our replies there are acceptable, our concept of a conduct plan or any related concept is not such a prototype of philosophical confusion nor such excessive conceptual garbage as has often been claimed.

Let us now go back to consider our example. What is the agent's conduct plan there? It is simply what the conjunction $(1')$ & . . . & $(8')$ describes. In other words, any model (in the Tarskian sense) satisfying $(1')$ & . . . & $(8')$ is a candidate for being a conduct plan. But we just cannot accept arbitrary such models: the models qualifying as conduct plans must in a certain sense be intended models (e.g., number theoretic structures are trivially disqualified).

Several remarks are needed to qualify this brief answer. First, we assume (although that is not absolutely necessary) that the singular states (and episodes) of intending and believing that the variables z stand for can be Jonesian mental states and episodes (or state-cum-episode complexes, if needed).

Secondly, as we have assumed the contents of the propositional attitudes in $(1')$ & . . . & $(8')$ to be stated in the language of thought, an intended model for $(1')$ & . . . & $(8')$ will be a complex of mental states-cum-episodes, which have the propositional structures described (or, better, named) by the second arguments of the propositional attitude predicates in $(1')$ & . . . & $(8')$. For convenience of exposition, we shall not now discuss in detail the intended semantics of statements such as $(1')$ & . . . & $(8')$ but leave it until Chapter 9.

We shall below accordingly call conjunctions of the kind $(1')$ & . . . & $(8')$ *conduct plan descriptions. Conduct plans* are intended Tarskian models of conduct plan descriptions. (When no harm is to be expected we may occasionally, for simplicity, call conduct plan descriptions conduct plans, too.) The notion of an intended model will be clarified later, we said. How about conduct plan descriptions? Can they be given an exact definition? Let us consider this matter briefly.

Conduct plan descriptions obviously rely on the notion of a conduct plan, and hence we seem to move in a circle. Although we shall leave the technical semantics of conduct plan descriptions aside here, a few things can be said in clarification. We note first that conduct plans involve *representational* mental states and episodes, viz. mental states with certain propositional contents. These contents form the tree-like (or graph-like, rather) structures of conduct plans. Recalling Figure 2.1, it is essentially the belief contents of the conjunctive statement *(1′)* & . . . & *(8′)* that give it a tree-like structure. The nodes again are associated with conative attitudes, at least some of which are intendings (or play the role of intendings).

Let us consider a finite structure $S = \langle D, P_1, \ldots, P_k, R_1, R_2 \rangle$. Here D is a set of action tokens characterized by the action types or generic actions $P_1, \ldots, P_k$ (which are the extensions of action predicates in D). R_1 and R_2 may here be taken to represent respectively singular factual and conceptual (strict or probabilistic) generation. When D consists of six action tokens and $P_1, \ldots, P_6$ represent the generic actions of our example, S can be taken to represent the tree structure of Figure 2.1. This gives the *content* of the agent's conduct plan in our example.

All conduct plans do not, however, have such a simple tree-like structure. Trees must start from one nongenerated action. But, as will be seen, there are complex actions which, so to speak, start from several nongenerated actions (cf. my flushing a nail or my singing a song while accompanying myself). We must thus allow for *merging trees*. In fact, one can think of many kinds of graph-like structures that actions can have. A tree structure is, however, a basic form and it represents simple basic (or nongenerated) actions plus what they generate. Perhaps we should call conduct plans of this simple type *action plans*. Out of such action plans one can build compound trees (representing compound basic actions plus what they generate) as well as more complex directed graphs (representing arbitrary complex actions).

The idea of representing arbitrary complex actions by means of directed graphs is naturally very attractive. In a certain sense we will just do that later. However, from a mathematical point of view such a representation is not, after all, very rewarding. This is because no interesting formal results seem to be obtainable for the kind of small finite graphs we are dealing with here.

I have not found any informative and interesting way of exactly defining conduct plans in graph-theoretic terms (such as figured above). Next, an exact explicit definition in terms of intended goals and relevant means-end beliefs of the plan might be tried. I have not been able to find such a characterization either (essentially for the reasons I will use in criticizing simple forms of the practical syllogism in Chapter 8).

The basic idea in the notion of a conduct plan is that it is a *dynamic explanatory* mental entity. 'Explanatory' means that it is a potential prima facie explainer of action. 'Dynamic' here means that it tends to causally produce behavior and that it itself has process-like and episodic aspects. Thus, the best we can say about conduct plans in general is just that they are functionally introduced dynamic explanatory mental state-cum-episode complexes which in a relevant way involve the agent's intentions and means-end beliefs.

There are some additional remarks that can be made about our conduct plans. One is that, being in part process-like entities, conduct plans might be taken to involve information about the process itself (and not only of its propositional content). Thus, we might think that we include statements about how the agent's wanting to do X factually generated his intending to do X (assumed to be a state separate from the wanting). If we think of a conduct plan as a psychologist's explanatory device, this suggestion sounds quite plausible. An objection might, however, be raised to the effect that here the agent's wanting may at best be used to *explain* the agent's intending, which belongs to the conduct plan. Thus the wanting cannot belong to the conduct plan. But even if one accepted the premise of this objection one would not have to accept its conclusion, for the conduct plan containing both the wanting and the intending might still be taken to satisfy our functional characterization of a conduct plan. We shall not below exclude the possibility that a conduct plan may be of this broader kind. After all, if the conduct plan has process-like aspects and if the process is dynamic (e.g., causal) one has to carve it into suitable explanatory pieces. We cannot guarantee a priori that these pieces will be instantaneous snap-shots, as it were.

Another problem with conduct plans is how much logical or conceptual content we really should give them. Consider the conjunction $(1')$ & . . . & $(8')$ and assume it describes the agent's conduct plan. Should we include all of the logical (and conceptual) consequences of $(1')$ & . . . & $(8')$ in the "conduct plan consequences"? As we have emphasized (cf. Chapter 6), a real agent's mind is, in an important sense, only finite. Very few logical transformation properties can thus be accepted as valid, as seen in Chapter 6.

The same negative result holds for the "practical" (as contrasted to "theoretical") consequences of $(1')$ & ... & $(8')$. An example of such practical consequences is exhibited by the practical syllogism. If the conduct plan description consists of (or contains) the premises of a practical syllogism, what about the ensuing action-conclusion? If intentions are characterized as in Chapter 6, we seem forced to accept such practical conclusions as true (or assertable). But this will hold only for a special narrow type of practical syllogism, as will soon be seen.

Except for a detailed treatment of the semantics of conduct plans in Chapter 9, we have now finished our general discussion of conduct plans. However, we shall have much more to say about a specific type of conduct plan, viz. the practical syllogism. The next two sections and parts of the next chapter will accordingly be devoted to practical syllogisms and their significance for the description and explanation of actions.

2. PRACTICAL SYLLOGISM

2.1. We shall below study a form of *practical inference* sometimes called the *practical syllogism*. The practical syllogism basically consists of two premises and a conclusion. The first premise of this pattern of thought is a statement about an agent's intention to achieve a certain end. The second premise is a statement about what he believes to be required of him to do in order to achieve this aim. The conclusion is a statement which, roughly, says that the agent does or proceeds to do what is required of him in the second premise. (Sometimes a third premise concerning so-called "normal conditions" is added to the syllogism.)

The above is an outline of the content of the practical syllogism, which represents a pattern of practical inference. It represents *practical* inference in contrast to *theoretical* because it is concerned with the agent's aims, beliefs, and doings. Theoretical reasoning again can be said to be concerned with propositions and thus with the contents of propositional attitudes. However, if the premises and conclusions of the practical syllogism are regarded as propositions (or statements, something true or false), the distinction between theoretical and practical inference patterns begins to vanish. We shall return to this topic.

The practical syllogism has a long history from Aristotle, through Kant and Hegel, to such present day writers as Anscombe and von Wright.

We shall not here discuss this history, however. Rather we shall regard the practical syllogism (or at least its premises) as a special kind of conduct plan with an important role in the theory of action. The practical syllogism can be employed for the purposes of *understanding* behavior, as well as for *explaining* and *predicting* actions. As the whole topic is a broad one, we shall in the present section concentrate on a discussion of the content and basic nature of the practical syllogism, while its most central methodological uses, such as those related to teleological explanation, will be left to the next chapter.

There is much literature on the practical syllogism and we will not attempt to give a fully systematic and detailed treatment of this topic (see, e.g., von Wright (1971), (1972), and Manninen and Tuomela (1976)). We shall below briefly outline the basic features of this pattern of thought. Only topics where I have something new to contribute will be discussed at greater length. The version of the practical syllogism to be discussed is basically that elaborated by von Wright in his *Explanation and Understanding* (1971) and other works (see especially von Wright (1972) and (1976a, b)). However, it should be said right at the beginning that I see the nature and methodological role of the practical syllogism through the spectacles of a causal theorist, and thus my philosophical views are bound to differ somewhat from von Wright's intentionalist views on this topic.

To begin, we consider the following schema of practical inference of von Wright (1971), p. 96:

(9)   (P1) A intends to bring about a certain end E.

   (P2) A considers that unless he does action X he cannot bring about E.

   (C) A sets himself to do X.

This schema represents the basic form of the practical syllogism. Its first premise speaks about A's *intention* to do something which has as its consequence the fact that E is realized or brought about. We could also have said, alternatively and more specifically, that A intends to do Y such that the result of the action Y is the realization of the event (or state of affairs) E. Instead of 'intends' one could say 'is aiming' or 'pursues an end' or, sometimes, 'wants'. However, it is central that whatever words are used, (P1) must be concerned with A's intention (viz. the words must be used so that they specify A's intention to achieve E). What we said about intentions (viz. intendings) must be taken to hold true of the intentions (P1) speaks about.

Premise (P2) speaks about A's *means-end belief* in the broad sense that his doing X is factually (or conceptually) necessary for his achieving E (or alternatively, his doing Y). Again, phrases like 'believes', 'thinks', etc., could alternatively be used, but they have to play the role of describing A's means-end belief.

The conclusion (C) describes A's beginning to do X. (C) is assumed to "logically follow" from (P1) and (P2) in a sense to be discussed later. (We could have used 'therefore' in front of (C), too.) One could also say 'embarks on doing', 'proceeds to do', and sometimes simply 'does'. In any case, (C) is meant to entail that behavior has been initiated. We shall later return to what this can be taken to involve.

As said, (9) is only a *schema* for practical inference. The symbols 'A', 'E', and 'X' play the role of variables (or placeholders, if you like). To obtain a specific practical inference we have to substitute for them respectively, a name or definite description of an agent, a goal or result (or goal-action, when using the action terminology), and an action. (Here, actions are obviously to be taken as action *types* or generic actions.)

Understood in this fashion, an exemplification of the schema (9) consists of two *descriptive* premise statements and a *descriptive* conclusion statement. The premises and the conclusion can be regarded as having definite truth values. They are either true or false. This is in accordance with what we said about statements about propositional attitudes in Chapter 6. Whether the conclusion also should be taken as a *descriptive* rather than *normative* statement is a more problematic issue. We shall comment on it later. Another problematic issue is whether the *third person* version of the practical syllogism represented by (9) can be adopted to take care of the *first person* case as well.

Let us briefly consider the first-person counterpart of the practical syllogism (9). We may write it as:

(9′)   (P1) I intend to bring about a certain end E.
        (P2) I consider that unless I do action X I cannot bring about E.
        ———————————————————
        (C) I will do X.

The conclusion (C) uses the word 'will' because my doing X (or my setting myself to do X) normally takes place in the future rather than right now. (C) speaks about my doing X rather than my setting myself to do X. This stronger version seems to me to apply to all typical cases. Only when I am very uncertain about my success in doing X do I use the weaker phrase.

(C) is not merely or primarily my prediction about my future action. It represents my *commitment* to doing *X*. Furthermore, it can be regarded as my *declaration* of this intention (or commitment). As such this declaration is neither true nor false. Still my *making* this declaration is something true or false. Or, if I do not actually make this declaration, we can still say that I *have an intention* to do *X* (I am in a state of intending to do *X*). That again is something which has a truth value. This is what our third person version (*9*) is implicitly concerned with. We could write the conclusion of (*9*) as '*A* intends to do *X*' and get a valid case of intention transferral.

Given that there is this kind of a close correspondence between the first person and the third person cases, we shall from here on be exclusively concerned with the third person case. (For a "pre-Jonesian" discussion related to my above "Jonesian" argument see von Wright (1972).)

To illustrate the basic form of the practical syllogism we consider the following instance of it:

(*10*)   (P1) *A* intends to illuminate the room.
            (P2) *A* considers that unless he flips the switch he cannot
            illuminate the room.
            _____
            (C) *A* flips the switch.

Here we assume that *A* successfully flips the switch, and hence we speak about *A*'s flipping rather than his setting himself to flip. Using our formalisms of Chapters 2 and 6 (and Section 1 of this chapter) we now consider two formalizations of (*10*):

(*11*)   (P1') (*Ey*)*I*(*A*, (*Ex*) Illuminates (*A*, the room, *x*), *y*)
            (P2') (*Ez*)*B*(*A*, (*x*)(*Ey*) (Illuminates (*A*, the room, *x*) →
                  Flips (*A*, the switch, *y*)), *z*)
            _____
            (C') (*Eu*) Flips (*A*, the switch, *u*)

(*12*)   (P1'') *I*(*A*, (*Ex*) Illuminates (*A*, the room, *x*), *a*)
            ·(P2'') *B*(*A*, (*x*)(*Ey*) (Illuminates (*A*, the room, *x*) →
                  Flips (*A*, the switch, *y*)), *b*)
            _____
            (C'') Flips (*A*, the switch, *c*).

Of these practical syllogisms (*11*) is in a sense concerned with the generic action of flipping the switch and (*12*) with a singular exemplification *c* of it. In (*12*) we have assumed that the intending *I* and the believing *B* have been exemplified respectively by some singular states or events (or state-event complexes) a and b.

It should be emphasized here that a practical syllogism can be understood either in a pre-Jonesian or in a Jonesian fashion (to use our terminology of Chapters 3 and 4). Most of what will be said below applies to both the pre-Jonesian and the Jonesian interpretation. However, the central idea in the Jonesian interpretation is that intending-believing state-event complexes can be "ontologically real" and that they serve as (a certain kind of) causes of action. This is of course not the case for a pre-Jonesian interpretation (such as von Wright's).

If the logical form of a practical syllogism such as (10) would be fully given either by (11) or by (12) obviously it would not be logically valid. For neither in (11) nor in (12) does the conclusion follow from the premises by means of the rules of inference of (first-order) predicate logic. Either our formal translation is completely on the wrong track or then there are some premises missing. We shall adopt the latter alternative and think that (10) and (11) are enthymematic and that accordingly some extra premises (or perhaps extra rules of inference) are missing. Supplying them will make (10) and (11) logically conclusive, we suggest. (12) is concerned with something very specific viz. concrete singular states and/ or events. If we succeed in making the inference (11) logically conclusive by means of a generalization, for instance, perhaps a suitable instantiation will suffice to render (12) logically valid, too.

Let us thus turn to discuss the logical validity of practical inferences of the basic form (9). Suppose that the premises of a practical syllogism of the form (9) are true. Can the conclusion fail to be true? Our answer to this question can only be given after a lengthy and many-sided discussion, to which we now turn.

First, we notice that we shall below understand 'intending' in the rather strong sense explicated in Chapter 6. Thus, for instance, if A intends to bring about E by doing X it is assumed that A believes that he at least with some non-negligible probability can bring about the result E by doing X. With a slightly different twist, the underlying idea can be formulated by saying that A believes that there exist some actions (perhaps unknown to him at the time) the doing of which, together with doing X, somehow enables him to bring about E (and to do Y we assume). Thus, we notice again that intending involves cognitive elements.

If A intends to do Y by doing X, he must also have resolved his want-conflicts, as we assumed in Chapter 6. Thus, even if A would have arrived at his intention through some kind of rational deliberation (which he need not) we see that the practical syllogism is not really concerned with such deliberation and decision making. It deals, at best, with the end result of such a process. Thus, it does not explicitly take into account the agent's *other* competing goals (intentions, wants, desires, etc.) than the one mentioned in premise (P1). Nor does it explicitly consider more than one means-action for achieving the goal. These restrictions are dealt with by implicitly assuming that 1) A then had no other goal E' which he preferred to E and 2) there was no other action X', also necessary for E, such that A preferred X' (or, rather, X' together with its various consequences) to X (with its consequences).

The practical syllogism then, explicitly, at most deals with only a small final fragment of rational deliberation. In being so restricted it of course fails to take notice of the existence of the variety of (rational) decision-making principles. That is, it fails to recognize that there are, in general, several different but intuitively equally rational principles for combining goals and information to yield action (or a decision to act). For instance, A might slightly prefer E to E' but also strongly prefer X' to X. Whether he would do X or X' depends on his decision-principle, and what that is is a contingent matter not to be decided merely on conceptual grounds. (Compare here the situation with decision-making under ignorance and under risk within statistical decision-making, where Milnor's paradox and other comparable puzzles show the existence of incompatible but in a sense equally rational decision-principles.)

The practical syllogism is also very restricted in another related respect. It is that, in contrast to conduct plans in general, it is not explicitly concerned with the agent's wants, wishes, and other conative attitudes. Furthermore, conduct plans in general may be concerned with several parallel actions and chains of actions, which, furthermore, need not be considered strictly necessary for the attainment of the agent's goals. In addition, the practical syllogism neither specifies how the agent's wants, desires, obligations, etc. generate his intentions nor how they serve as the agent's basis in his intention formation.

The above remarks should be taken not so much as a criticism of the practical syllogism but rather as delineation of its scope. Basically, its somewhat limited scope is due to the fact that it is regarded as logically binding and to the fact that it deals with the agent's *intending*, which is conceptually a very strong notion. On the other hand a successful theory of action needs such a strong notion of intending, as will later be seen.

As to the second premise concerned with the agent's means-end belief, I will here only make two remarks. First, in order to make the practical syllogism valid, it should be read as supporting the following counter-factual:

> A believes that should he fail to do X he would not be able to bring about E.

This means that A must consider X necessary for E in a rather strong sense.

Secondly, in some contexts the means-belief may be explicitly relativized to a situation: A considers that unless he does X in situation S he cannot bring about E. If we use this situationally relativized belief premise, we must also add the following corresponding belief premise to the practical syllogism: A believes that he is in situation S. Below, we shall for simplicity not usually make explicit any such situational assumptions.

Let this suffice for the meanings of the notions of intending and believing in the practical syllogism. We still need several qualifying conditions before we even get close to a logically valid practical syllogism. As the qualifications to be mentioned below have been rather extensively discussed in the literature I will be brief in my remarks.

Assume that the premises of the practical syllogism (10) are true. If A's hands are paralyzed A might not be physically *able* to flip the switch. Furthermore, even if A were physically able to flip the switch he might mistakenly flip the fan on instead, not knowing which switch is for the light. The agent must be required to *know how* the action mentioned in the conclusion of the practical syllogism is to be performed. This entails that he must have an *active correct belief* about which switch is for the light.

We might include the above know-how-requirement in a broad requirement that the agent be able (psychologically, socially, and physically) to perform the required action. This means, roughly, that whenever there is an *opportunity* to perform the action, then if the agent tries to do it he will succeed (at least with a high probability). (See our definition of ability in Chapter 9; also see the discussion of the ability and knowledge requirements in von Wright (1971) and (1972).)

Does the conclusion of (10) then logically follow from its premises? It still does not. The agent might be externally *prevented* from flipping the switch. For instance, another agent might physically prevent him from flipping the switch just when he is about to do it. The switch might at the critical moment become broken, furthermore, and thus the world would not cooperate with the agent's attempt, even if there was a physical opportunity for successfully performing the action right before the agent set himself (or attempted) to do it.

There is still one more general kind of qualification to be read into the practical syllogism before it has any plausibility of being logically conclusive. This is a kind of rationality requirement, which basically contains the idea that the agent's intention, which perhaps was formed a long time ago, as well as his belief, are still in *effect* at the time of acting (assuming that the intention or belief contains an idea bout this time). Thus we assume this (cf. Chapter 6):

(a) A has not forgotten his intention or belief;
(b) A has not changed his intention or belief;
(c) A is not emotionally or physically disturbed at the moment of his acting;
(d) A has not forgotten (and is not mistaken) about the time.

Following the terminology of Tuomela (1976c) I will say that these amount to the requirement that A be rational$_2$. In that terminology rationality $_1^i$ means the rationality "type" of A as a rational decision maker. For instance, A could be a minimaxer, a maximaxer, a Hurwiczian optimist, etc., with the values of i ranging over such strategy types. But as the practical syllogism does not deal with rational decision making (except at most the end result of it) we do not need this concept of rationality here.

We can now summarize our brief comments on the qualifications to be read into the practical syllogism before it can possibly be regarded as logically binding. Lacking a better terminology I will be using the phrase 'normal conditions' for the present purposes. These *normal conditions* assumptions then are as follows:

(i) A is able to do the required action X;
(ii)   there is an opportunity for A to do X (i.e., the specific facts and circumstances of his situation allow him to do the action);
(iii) A is rational$_2$;
(iv) A is not prevented from doing X.

We shall still make some additional clarifying remarks on these normal conditions later.

Let us finally emphasize that as the practical syllogism represents our rational reconstruction of the agent's practical inference the agent does not necessarily have to think that the normal conditions hold in a given situation. Although the agent obviously must be required to have the concepts used in the practical syllogism, we may at best require that he upon reflection (or by psychoanalysis or something related) comes to think that the normal conditions hold true.

We are now in a position to present our final version of the (descriptive) practical syllogism. Let us assume that the agent's intention specifies the time of the end's realization and that the belief states a time limit for the doing of X. We get (cf. von Wright (1971) p. 107):

(*13*)   (P1) A intends to bring about a certain end E at t.
       (P2) A considers that unless he does action X no later than at time t' ($t' < t$), he cannot bring about E at t.
       (P3) Normal conditions obtain.
       (C) A sets himself to do X no later than when he thinks time t' has arrived.

I regard this schema as conceptually conclusive and (in a relative) sense "logically valid". It represents a conceptual truth about the notion of intending *within the framework of agency*, viz. relative to a certain *conception of man*. This conception is a kind of *broad factual* "theory" of man. I would like to argue that it is not true of other animals, and certainly not of, e.g., plants (cf. our remarks in Chapter 3).

What the above claim about the conceptual conclusiveness of (*13*) involves requires some further discussion and argumentation, however. There are three different issues which I will take up immediately. The first has to do with the time gap between the agent's intending and his action. Secondly, some of our normal conditions (especially the notion of preventing) must be analyzed further. Thirdly, the role of causality in the connection of this schema must be laid bare.

In Chapter 6 we claimed that an agent's intending (to do something Y) ultimately develops into his intentional trying (willing) to do by his bodily behavior whatever action Y requires. That is, when the time for doing a required action X is *now*, A's intending develops into his trying (willing) to do X. Let us thus connect this idea with the schema (*13*) by taking $t = t' =$ now. Obviously this 'now' must be taken to have some duration. No action is infinitesimally short. Rather, this 'now' lasts fractions of seconds, seconds, hours, or even months. What is more, not only does the *overt* action have some such duration but also the inner trying-event preceding and causing the overt action takes some time, it seems (see our remarks in Chapter 6).

The phrase 'sets himself to do X now' in the conclusion of (13) under our present interpretation entails that A has initiated the doing of X. This phrase represents in part an "overt aspect" of the inner trying-event. Setting oneself to do is not a complete pre-Jonesian counterpart to our Jonesian notion of trying, it seems. (I presume, however, that 'sets himself to do' cannot be applied unless some overt bodily movement is taking place.) If so, my notion of trying, which temporally seems to start earlier, is "somewhat more" than a Jonesian counterpart to setting oneself to do something. (It is not very central here how the semantics of 'sets himself to do' is made precise in this respect.)

What is A's trying to do X then? We said in Chapter 6 that it can be functionally characterized as an event *causing* whatever bodily behavior is needed for A's doing X. When A thus, in our sense, comes to try (to will) to do X, *nothing more is required on his part for his doing X.* The rest is up to the world. Let us now understand the notion of *preventing* in a broad sense so as to suitably comprise everything that is "up to the world" (viz., nothing bizarre or outlandish is allowed to occur between A's setting himself to do X and his actually doing X). We have then bridged the gap between *A's setting himself to do X now* and *A's doing X now* intentionally. This, together with our earlier analysis of trying, serves to indicate that 1) there is indeed a *conceptual connection* between the premises of a practical syllogism and its conclusion. At the same time 2) it is a *conceptual* truth that there is a *causal* connection between the episode which instantiates the premises and the episode instantiating its conclusion in the now-case. (More exactly, the premise-instantiating episode, non-functionally described, is a purposive cause of the conclusion-instantiating episode, non-functionally described; cf. Chapter 8.)

In most methodologically interesting applications of the practical syllogism there is, however, a time gap between the intending and the action. That is, the practical syllogism is usually employed for the purposes of *explanation ex post actu* or *prediction* rather than used in the now-case. There is, accordingly, much more to be said about the conceptual conclusiveness of the practical syllogism.

2.2.   To make some further progress let us see whether we can make the schema of practical syllogism a logically valid "theoretical" inference schema. Technically we can do this as follows. Schema (13) can be made logically valid by adding the following generalization as a new premise:

(L)   Whenever A intends to bring about E at t and considers the
           doing of X no later than t' necessary for this, and if normal
           conditions obtain, then A will do X no later than when he
           thinks the time t' has arrived.

The generalization (L) should be understood in the sense that normal
conditions obtain at least right before and during t', which is the latest
time for acting. (L) can be regarded as implicitly quantifying over several
things, e.g., over agents, intention contents (or rather ends E), actions,
and times. But the logical form of (L) in fact contains still other quantifiers.
Thus, using our earlier symbolism of Chapters 2 and 6, I suggest that the
practical syllogism *(13)*, when turned into a valid theoretical inference
schema, has the following generic logical form (using the notational
simplification of Chapter 2 that the verbs have only three arguments and
that the circumstance-predicate T has only two arguments):

(14)   (P1) $(Ey)I(A, (Ex)$ (Mod-Verbs (A, o, x) & $T(o^*, x))$, $y)$
            (P2) $(Ez)B(A, (x)(Ey)$ (Mod-Verbs (A, o, x) & $T(o^*, x) \rightarrow$
                  Mod′-Verbs′ (A, o', y) & $T'(o'', y))$, $z)$
            (P3) $N(A)$
            (L) $(y)(z)(I(A, p, y)$ & $B(A, q, z)$ & $N(A) \rightarrow$
                  $\overline{(Eu)\ (\text{Mod′-Verbs′}\ (A, o', u)\ \&\ T'(o'', u)))}$
            (C) (Eu) (Mod′-Verbs′ (A, o', u) & $T'(o'', u))$.

Here $N(A)$ says that A is in normal conditions. In (L) I have used p and q
to represent the second arguments of I and B (of (P1) and (P2)). Our
formalization does not explicitly mention the times t and t'; they are
assumed to be imbedded in the predicates T and T'. (L) does not explicitly
quantify over the agents A nor over the propositional contents (p and q)
nor over the actions (Mod-Verbs, Mod′-Verbs′), although it might. That
is not, however, essential for our present purposes.

   Schema *(14)* leads to the agent's action (Mod′-Verbing′) in its conclusion
rather than merely to the agent's setting himself to perform this action.
This can be regarded as justified by assuming that one can pack into the
N-conditions everything that distinguishes setting oneself to do from doing.
Our broad notion of preventing will be assumed to do that job.

   With these simplifications and qualifications, *(14)* is a "theoretical"
counterpart or translation of the practical inference schema *(13)*. *(14)*
represents a logically valid inference in first-order predicate logic, and

thus it would seem that practical inference, in the descriptive sense that we have been discussing, is, after all, only a species of ordinary logical inference.

What is the status of the generalization (L)? It says that for A's every intending and believing (with the appropriate contents) there exists an appropriate action following them, given that A is in normal conditions. Thus, for example, whenever A intends to illuminate the room and thinks that flipping the switch is necessary for this, he performs a flipping of the switch, given normal conditions.

(L) can be regarded as representing a *conceptual truth* within or relative to the framework of agency. It just makes a conceptually true statement about certain (generic) intentions, beliefs and actions. The normal conditions it speaks about must be interpreted so as to respect this. Perhaps they can be thus understood so as to *also* keep them epistemically respectable at least in some cases. But I doubt that they could *always* be specified *ex ante* except in the general and rather vague terms we have been using (see Tuomela (1976c) for this). Expecially, I doubt that one can specify in more detailed terms the notion of preventing so as to make it perform the work we have given to it. We shall soon see what happens when some "logical pressure" is put on the claim that (L) is a conceptual truth.

It is important to notice that although (L) is a generalization (it generalizes explicitly over singular episodes) it is not a specific covering *law* (as it is a conceptual truth about agents). Thus we do not have here anything like a covering law explanation or prediction at all. The role of (L) is merely to exhibit part of what lies "at the back" of the practical syllogism (*13*). It should be noticed that it says nothing explicit about the *causal* connection we have claimed to be intimately (conceptually) involved in practical inference. Thus, there is more at the back of (*13*) than what (L) lays bare.[1]

To illuminate the conceptual character of (L) and, what amounts to the same in our construal, of the schema (*13*), we consider an instantiation of (*14*). Assume that we use the inference pattern predictively and have substituted a name a for a specific intending and b for a specific believing:

(*15*)   (P1*) $I(A, (Ex) (\text{Mod-Verbs } (A, o, x) \ \& \ T(o^*, x)), a)$
         (P2*) $B(A, (x)(Ey) (\text{Mod-Verbs } (A, o, x) \ \& \ T(o^*, x) \rightarrow$
                  $\text{Mod'-Verbs' } (A, o', y) \ \& \ T'(o'', y)), b)$
         (P3*) $N(A)$

(L) $(y)(z)(I(A, p, y)$ & $B(A, q, z)$ & $N(A) \rightarrow$
       (Eu) (Mod'-Verbs' (A, o', u) & $T'(o'', u)))$
_____
(C) (Eu) (Mod'-Verbs' (A, o', u) & $T'(o'', u))$.

Now a critic might argue against the conceptual validity of (L) as follows. Suppose the agent Mod'-Verbs'. Then the event (episode) picked out by the variable u may be taken to be different from the events named by a and b. But there cannot be logical (including conceptual) connections between logically distinct events. Therefore (L) must be contingent.

But this argument is wrong. There are several counterarguments to be made here. First, it can be argued against the critic that the variable u runs over entities (events) c such that a (possibly together with b) forms the initial part of c. We have earlier discussed in what sense this could be the case. Roughly, it holds if a (or a cum b) can be regarded as naming an effective intending (trying, in our terminology) and c a *"full"* action token such that the singular *overt* action temporally succeeds the causally effective trying-episode. But as we have emphasized, our notion of trying here is conceptually connected to Mod'-Verbing' as it is even functionally defined in terms of the latter. Even if it were possible to somehow redefine the concept of trying independently of action concepts in the future (e.g., in neurophysiological terms) that would mean a change in its meaning, and in any case that would have almost nothing to do with (L) as we now have it.

I think the above answer suffices to reject the Humean criticism in cases when a really names a trying-event. But that can be the case only in the "instantaneous" uses of the practical syllogism, i.e., in the now-cases. But perhaps the main point of the Humean criticism was just that cases with time-gap exist and are even typical. Indeed, here lies its force. Our Humean is completely right in that the premises and the conclusion of (L), and hence of (*13*), can be instantiated as ontologically separate; viz. c may be taken as distinct from a (and b). Thus, the events (episodes) a and c are logically independent in that sense. However a *as an* intending and c *as an* action are still conceptually connected, for the universals (the concept of intending and the concept of action) are here conceptually connected. Taken this way, the Humean critic's premise is simply wrong.

Let us now ask whether the antecedent of (L) could be instantiated without its consequent being instantiated or, alternatively, whether the premises of (*13*) could be instantiated without its consequence following. If this were a possibility then the premises of a practical syllogism could be said to be conceptually connected without logically *entailing* the conclusion. For instance, von Wright (1971) and (1976b) holds such a position, but I do not. I think that we can and should understand the normal conditions in (L) and in *(13)* so that (L) cannot thus become falsified and *(13)* be made an invalid schema of inference. (Notice, I naturally accept that in *(13)* (P1) and (P2), *simpliciter*, could be true and (C) false, but that is of course not at stake here.)

I will here give one more argument to support my position. It goes via a *normative* version of the practical syllogism. Consider the following schema corresponding to *(13)*:

(16)   (P1) A intends to bring about a certain end E at t.
           (P2) A considers that unless he does action X no later than at time t' ($t' < t$), he cannot bring about E at t.
           (P3) Normal conditions obtain.
           _____
           (C) A ought to do X no later than when he thinks time t' has arrived.

Schema *(16)* is (at least syntactically) identical with *(13)* as far as its premises are concerned. Its conclusion, however, is a normative statement (in a sense). It specifies a practical necessity or norm for A vis-à-vis the premises. I think we must accept *(16)* as representing a valid inference schema.

One way of arguing for the validity of *(16)* is the following "metalinguistic" one. We may back *(16)* by claiming that it represents a conceptual truth about the phrase 'intend', expressible as a norm (rule) of roughly the following kind:

(NI)   One ought to use 'intend' so that *(13)* is true in all circumstances.

Schema *(16)* could now be regarded as a kind of material counterpart of (NI). The "ought" in its conclusion would reflect in the agent's mind the "ought" of (NI), which is a rule supposedly taught to A and (hopefully) internalized by A.

What I am after here is that (NI) could be taken as a "linguistic" norm in *our* speech community (and culture). Every "full fledged" member of this speech community is assumed to have internalized (NI) in the strong sense that he *obeys* it or that it *guides* his behavior. This kind of internalization presupposes a causal mechanism, I claim. The norm in question has not been properly internalized unless it is causally effective in producing behavior. What is needed here is a notion of *normative* or, as we shall call it, *purposive* causation, which preserves conceptual connections. This notion will be analyzed in Chapter 9.

If what has been said above is on the right track we have a new dimension to take into account when discussing the conceptual validity of (*13*). This is the *internalization* dimension. It gives us a new way of defending the validity of (*13*). For suppose its premises are instantiated but its conclusion fails to be. Then we can say that *A* is not a full fledged member of our speech community or has not yet properly learnt (NI), or that he is momentarily disturbed or "irrational" or something like that.

I emphasized earlier that it does not seem possible to *ex ante* clarify the notion of prevention (in our broad sense) in more detailed terms. The factors which distinguish, say, Nixon's becoming the president from his setting himself to become the president are so many and so various that no informative *ex ante* characterization seems possible.

We are, however, dealing with a new aspect of the situation when we take into account the normative dimension. If we accept (*16*), then we can discuss what distinguishes the truth of '*A* does *X*' from the truth of '*A* ought to do *X*'. *A* would obviously do *X* in a *deontically perfect* "world" or under deontically perfect circumstances. In other words, an ideal member, and a "really" full fledged member, of our community would do *X* whenever he ought to do *X*. If now the premises of (*16*) are true in some situation but the agent does not actually do *X*, we may blame *A*'s lack of "rationality" (in *this* sense). If we could add into the "normal conditions" of the practical syllogism that *A* always does what he ought to do, then we would have bridged the present gap, which, so to speak, distinguishes our actual circumstances from deontically ideal ones. But such a new assumption is obviously too strong to be accepted in general.

It seems that the gap between "ought to do *X*" and "doing *X*" cannot be bridged merely by adding new "purely" descriptive premises or qualifications. There is, however, one more possibility that I find viable. It is to require that the agent's episodes, which instantiate the premises, normatively or purposively cause the agent's doing *X* (or at least his attempting to do *X*, in case the world does not cooperate). This strategy just adds a *dynamic* element into the picture, so to speak, while making no changes in the conceptual situation as characterized by a "descriptive" practical syllogism (such as (*13*)). This is what we shall defend in the next section and in Chapter 8, and that will be seen to bridge (in part) the present gap between "competence" and "performance" (in linguists' terminology).

3. PRACTICAL SYLLOGISM AS A SCHEMA FOR UNDERSTANDING BEHAVIOR

3.1. The view that practical syllogisms, such as those represented by, for instance, (9) or (13), are logically conclusive or binding has been discussed extensively by von Wright (1971, 1972, 1976b). He has also emphasized the pivotal role of the practical syllogism in "understanding" behavior as intentional action. As my above views are in some respects different from von Wright's it will be useful to note what von Wright has to say about the logical conclusiveness of the practical syllogism. Furthermore, our discussion of von Wright's views below will also bring up some interesting points related to the practical syllogism as a vehicle of *understanding* behavior and to the notion of *intentionality*.

In von Wright (1971) it is claimed that the "tie" between the premises and the conclusion of a practical syllogism is *logical*. (Sometimes von Wright also says 'conceptual', and it seems that 'logical' must be taken broadly so as to be synonymous with the former phrase in this kind of context.) Von Wright's aim is to give an argument which shows that the tie is indeed logical.

The basic idea of the argument is this. If one can show that it is not (logically) possible to verify or falsify the premises and the conclusion independently of each other, then the tie between them is *logical* (rather than *empirical* or *causal*, the notions which von Wright contrasts it with). Before going into any details let me say two things. First, von Wright's argument rests on the premise that if it is not (logically) possible to verify the premises and the conclusion independently, then there must be a conceptual connection between them. This sounds like confusing meaning with evidence, even if only the *logical* rather than something like a *factual* possibility of verification seems to be under consideration. I think that, for instance, two theoretical scientific propositions could be conceptually independent even if they were not independently testable in any circumstances. Meaning is not reducible to testability nor even to truth conditions.

Another problem with von Wright's idea of proving logical connection in terms of verification is that it seems circular as an argument. If one thinks, following von Wright, that the nonexistence of independent verification does entail conceptual connection, one cannot yet epistemically use this idea without circularity. This is, roughly, because "meaning comes before truth". The truth of some X cannot be ascertained (i.e., verification is impossible) before one has a conceptual grasp of X. The concept of evidence for X presupposes the concept of X, we might also say. (For a longer and somewhat different argument to show the circularity of von Wright's argument see Martin (1976).)

Despite the above criticisms one can in a useful way examine von Wright's discussion of verification and take it as an illustration of, rather than as an argument for, the conceptual connection. Von Wright summarizes his discussion concerning verification as follows (von Wright (1971), pp. 115–116):

(a) "The verification of the conclusion of a practical argument presupposes that we can verify a correlated set of premises which entail logically that the behavior observed to have occurred is intentional under the description given to it in the conclusion. Then we can no longer affirm these premises and deny the conclusion, i.e., deny the correctness of the description given of the observed behavior. But the set of verified premises need not, of course, be the same as the premises of the practical argument under discussion."

(b) "The verification of the premises of a practical argument again presupposes that we can single out some recorded item of behavior as being intentional under the description accorded to it either by those premises themselves ("immediate" verification) or by some other set of premises which entail those of the argument under discussion ("external" verification)."

From the cited passages and the accompanying context we can accordingly extract two claims (see von Wright (1971), p. 115 and also (1972), p. 49). Using some symbolism we can state these theses as follows (variables and constants are not syntactically distinguished in our notation):

(17)   (a) If C then $(EP)[PS(P, C)]$
        (b) If P then $(EC)(EP')[PS(P, C) \lor (PS(P', C) \& P'$ entails $P)]$

(18)   The singular action truly described by C is *intentional* if and only if $(EP)[PS(P, C)]$.

((18) is not meant to be an *analysis* of intentionality, but it is still a true claim in von Wright's theory.) Here P and P' represent conjunctions of (potential) premises of some practical syllogism and C represents an action description, which could be a consequence of some practical syllogism. $PS(-, -)$ represents the relationship in which the premises and the conclusion of a practical syllogism stand.

(18) claims that the schema of practical syllogism captures exactly the class of intentional action descriptions. Accordingly, a practical syllogism always assumes that its conclusion C describes an intentional action (or an action as intentional).

In von Wright's account the action, say X, which C attributes to A is intentional if and only if *(17)* is true. *(17)* can then be read as giving necessary and sufficient conditions for A's intentionally doing X, too.

We may now consider whether *(17)* is acceptable. My general answer is that it is not quite acceptable; I will give my arguments below. However, *(17)* is so close to what I take to be a correct analysis that after a discussion of the faults of *(17)* it will be rather easy to give a corresponding acceptable analysis.

Let us first consider condition (a) of *(17)*. It says that if C is true then suitable premises must exist to satisfy the relation *PS*. However, there are some difficulties with (a).

The first problem with (a) is that not all actions are intentional (as described); and some actions are not intentional under any descriptions (see Chapter 10). For instance, I may nonintentionally wave my other hand while writing this, and this action does not seem to be intentional under any description. As the practical syllogism concerns only intentional actions (cf. *(18)*), obviously C must be restricted to descriptions under which the action considered is intentional.

Suppose next, that A, for instance, intentionally hummed a melody, pushed a button, or passed the salt to his neighbour. In these cases there is an intention, a *telos*, *in* the act, we may say, although there is perhaps no further intention, i.e., no further goal, involved. The action was, so to speak, performed for its own sake, we might think. If this is accepted there is no *full* set of premises of a practical syllogism to match C, it would seem. A just intended to do X and did X, no further intention and no (non-trivial) belief was present, it may seem. This might be taken to show that (a) is false unless "mutilated" practical syllogisms with trivialized second premises are accepted. However, I don't regard this argument as forceful because one can think that i) the mutilated practical syllogism referred to only appears mutilated because the agent was not, or we were not, *aware* of his further intentions and of the relevant belief; ii) perhaps some other, completely different practical syllogism could be constructed to match C (we might consult Freud for advice).

Another difficulty with (a), which is a graver one, is that it may sometimes be satisfied only through a more "liberal" kind of conduct plan than the ordinary practical syllogism. Suppose that we see A running towards the train station. We assume that A's running (from the university to the station, say) was intentional. However, we are unable to find a practical syllogism of the ordinary kind (*13*) (or (*9*)) to match this action. We assume in normal fashion that A intends to catch a certain train. However, A does not consider his running to the station *necessary* for catching the train. He might have taken a tram or a taxi. Instead he considers that his running will very likely generate his catching the train. Perhaps he also considers that it is the cheapest and healthiest way. We may thus assume that the above intention and belief did activate him to run in this case. Nevertheless, there is no *ordinary* kind of practical syllogism to match the conclusion, it would seem. Thus it seems that (a) is too strong. Thus, we can only generally require the existence of a conduct plan which, through its being activated, purposively causes A's running, as we shall say.

Let us next consider condition (b) of (*17*). It requires that given some premises P there will be a related true practical syllogism $PS(P, C)$ or $PS(P', C)$. (One minor thing to be noticed here is that I don't see why it is required that, within external verification, P' entails P. It should suffice, I think, that P and P' only be related through the fact that they lead to the same action, described by C.) Suppose now that we are given a set of premises P. Does there always exist a related true practical syllogism as (b) requires?

Let us consider a problematic example that von Wright has discussed (cf. von Wright (1971), pp. 116–117). Suppose an agent has decided to shoot a tyrant. He thinks that it is necessary to pull the trigger of his revolver when standing in front of the tyrant. Let us assume that all of the so-called normal conditions of (*13*) are satisfied. However, when the time (or latest time) for acting comes, nothing happens. The agent does not even set himself to pull the trigger. No effects of paralysis or anything related are found – the normal conditions hold true. We assume that the normal conditions are specifiable in detail here and that they have been made epistemically respectable. In this example we are invited to make the somewhat difficult and very hypothetical thought experiment that all the premises of the practical syllogism in question are indeed *true* (and not, e.g., only *regarded* as true).

What should we say here? Von Wright himself leaves the case as a kind of mystery. He just says that we must accept that nothing happens in this example and that this shows that, despite the existence of a logical or conceptual connection between the premises and the conclusion of a practical syllogism, "the premises of a practical inference do *not* with logical necessity entail behavior. They do not entail the 'existence' of a conclusion to match them" (von Wright (1971), p. 117). "It is only when action is already there and a practical argument is constructed to explain or justify it that we have a logically conclusive argument. The necessity of the practical inference schema is, one could say, a necessity conceived *ex post actu*" (von Wright (1971), p. 117).

I agree that one might perhaps accept the tyrant shooting example as a kind of weird conceptual possibility and that the practical syllogism only represents some kind of *ex post actu* necessity. But I think that we must say something by way of explanation and clarification here and that we cannot leave the matter as a mystery. Both the puzzling example itself and the notion of *ex post actu* entailment are badly in need of clarification.

Let us first consider a point related to the *ex post actu* entailment of a conclusion C by some premises P in a practical syllogism. Any notion of (logical) entailment must presumably contain as part of its meaning that the entailment fails if P is true and yet C is false. Let us thus try to falsify an alleged *ex post actu* entailment of a C by a P when C and P are thought to be related by a practical syllogism. P is assumed true. Now, if C is true the entailment should be taken to hold, according to von Wright. If, however, C is false and no action occurs (as in the tyrant killing example) we are not allowed to infer that the entailment does not hold. So we seem to be left with a completely nonfalsifiable and empty notion of entailment. von Wright might want to reply here either that it is a mistake to speak of entailment in anything like the above kind of truthfunctional sense at all or that my argument should rather be taken to demonstrate the conceptual character of the practical syllogism, viz. that the practical syllogism is an interpretative schema rather than a pattern of inference. I think both of these hypothetical replies are somewhat unsatisfactory for, as I will argue, one can regard the schema both as a truthfunctional inference schema and also as a schema for interpreting and understanding behavior.

I would like to regard the notions of conceptual (or logical) *connection* and conceptual (logical) *entailment* as representing two aspects of one idea. They are not to be separated as von Wright does. (His attempt to use the notion of *ex post actu* entailment may be regarded as an unsatisfactory attempt to connect the, for him separate, notions of logical (conceptual) connection and logical (conceptual) entailment.) There are two considerations to be brought up here.

First, we consider condition (b) of (*17*). Given true premises *P*, can a conclusion *C* fail to be true? Even if we indeed accept the tyrant shooting example as showing this, there is still something to be said to defend condition (b). Though its consequence would then have to be regarded as false, the establishing of its falsity seems to be possible only if something like the principle (b) of (*17*) is regarded as true. I mean, roughly, the following by this paradoxical statement. The agent behaves in some way or other all the time (except perhaps when unconscious). How do we now know in our example that the potential killer does not pull the trigger? Well, we *see* it. We said earlier that nothing happened with the trigger. But of course the agent was behaving in one way or another all the time. Perhaps he just stood there; or perhaps he pulled a newspaper out of his pocket and started to read it calmly. In any case we conceptualize his behavior in terms of the framework of agency, and this conceptual framework, internalized by us, contains (*17*) (or, rather, a closely related principle, as we shall see) as a kind of basic principle for *understanding* behavior as action.

In a certain dialectical sense then, putative counterexamples to the logical validity of the practical syllogism can (*logically* can) be found to be counterexamples only if the practical syllogism is regarded as valid (viz. as not having any counterexamples). I think that von Wright has driven himself into this paradoxical, although perhaps not completely intolerable, situation by accepting *both* the "mutual verification" (or "conceptual connection") thesis and the tyrant shooting example (as a counterexample).

My solution to the puzzling situation is as follows. While I accept the descriptive practical syllogism as logically conclusive (in its form (*13*) or (*14*)), I do not regard the tyrant shooting example as a counterexample to the logical conclusiveness of the practical syllogism. Considering that example as it was described earlier, I think nothing *conceptual* is missing in it. What is missing are the underlying dynamics: there is no *causal* connection between the intending and the action. That is, the agent's intending did not turn into a causally effective trying at all in this example, and this suffices to explain away the example as a counterexample. To put this differently, the agent was not fully "one of us" or a "full-fledged" member of our community at the time when the action was supposed to occur. We might also want to say that our broadly factual conception (theory) that all men are agents at any time was at best only approximately true of this man.

As I said, we might argue that the tyrant killing example is conceptually all right in that we could write out all the specific premises of the appropriate practical syllogism (such that these premises are true). Then, when no trigger pulling occurs, we need not necessarily start doubting the premises; we have the alternative strategy of, so to speak, going to the metalevel to claim that some of the preconditions of the "whole game" were not satisfied. In my analysis this entails that purposive causation was lacking, and thus the whole conceptual schema does not function as it is assumed to (when really applicable).

When applied to condition (b) of (*17*) our results amount to saying that the tyrant shooting example does not falsify it, since the general preconditions for applying the framework of agency, and hence of (*17*), to that example were not satisfied. In fact, I regard (*17*) (b) to be true. What is more, I think even the following strengthening of it to be true:

(b')   If P then $(EC)(PS(P, C))$.

That is, "direct verification" of premises should be enough from a *logical* point of view. If (b') were false, it would be so presumably because of unrealized intentions. But intentions are dispositions and our normal conditions, if correctly analyzed, specify everything needed for an actualization of them, I argue. Thus unrealized intentions do not fall within the scope of the practical syllogism and hence of (b'). In saying this it must be emphasized that P indeed be a true premise conjunction of a potential practical inference.

Concerning (*17*), we may then conclude that its condition (a) has to be replaced by a broader condition as discussed earlier; and its (b) is true but may be strengthened to (b').

3.2.   How about intentionality and formula (*18*)? Roughly, an action (described by a description C) can be said to be *intentional* if and only if the agent *meant* by his behavior to bring about the result (which he in fact brought about). Intentionality (in this semantical sense) is a notion intimately related to *meaning* (and *expressing* and *reference*). More generally, we can then say that an item is intentional if and only if it makes reference to or expresses something (cf. Chapter 3).

We may compare the intentionality of actions to reading a text. Assume that we are given some spoken or written text produced by an agent. That we are dealing with a *text* involves that the sounds or marks (i.e., the physical bearers of meaning) are interpretable into such and such physical elements and relations between these elements. The resulting physical system can be interpreted (or articulated) as words and sentences. The resulting sentences *mean* something. What this sentence meaning consists of I cannot of course here try to clarify in any detail. Roughly speaking, given any (descriptive or non-descriptive) sentence we can ideally make the well-known Wittgensteinian distinction between the *sentence radical* and the *sentence mode* or *mood* (broadly understood), and say the following (cf. Stenius (1967)). Sentence radicals (as natural-linguistic objects) can be taken as the physical systems we spoke about. They are to be isomorphically mapped (by means of some "key") into something extra-linguistic (typically), i.e., into what they come to represent in the sense of picture theory. The sentence radicals are associated with "moods" or "modes" such as assertion, interrogation, persuasion, belief, etc. (Cf. Chapter 6 for our specific treatment of the "moods" of belief, intention, and want.) This relates to what the speaker *means* by his words: he may, e.g., intend that his words make a question or that they assert something.[2]

Thus we are dealing with two sorts of naturalistic elements here: 1) the physical system consisting of sounds, marks, etc., 2) the situation in which the speaker produces in some "mood" the elements of this physical system. There are accordingly two things to be interpreted here. Theoretically we have to get hold of the so called key (projection rule) related to the mapping of the sentence radicals, and we have to specify the mood. In reading and understanding some text and in understanding spoken words we thus perform complex "reading off" which logically (although not necessarily psychologically) involves the above elements.

The above gives a rough sketch of what understanding linguistic actions involves. In this sense of 'understanding' we understand only *intentional* actions, i.e., what the speaker or writer meant by his bodily behavior and the intended results of this bodily behavior. This is the idea of intentionality that the formula (*18*), accepted by, e.g., von Wright, tries to capture.

Is (*18*) adequate for this purpose? Not quite. As the example about the man intentionally running to the station shows, a wider type of conduct plan than the ordinary practical syllogism suffices here. Thus, if *C* describes an action as intentional, it need not be the case that *C* is the conclusion of a practical syllogism of kind (*13*).

How about the converse? Surely that holds true. But again, a wider type of conduct plan may suffice as well. In fact, in the example about the man running to the station, the mentioned causally effective conduct plan suffices to guarantee the intentionality of the man's running. It is essential here that the man's conduct plan is causally effective. For that involves that the man's trying (willing) to run to the station is dynamically involved here. This trying is the *causal* locus of intentionality, we may say. The trying has the content that the man by his bodily behavior *means* to perform the action of running to the station.

Notice that intentionality, analyzed as a semantical notion, needs some bearer. This bearer can be sounds, marks, or bodily behavior. But it can also be (what we now call) brain processes. If trying (willing) turns out to be a brain process it has got to be a *propositional* brain process, and a process by which the agent means (or may mean) something. If we can "read off" some propositional meanings from overt behavior, we can, at least as a logical possibility, read them off from the agent's brain processes, given our analogy account of mental episodes and given also that these mental episodes are found by future science to be brain processes. The causal locus of intentionality (i.e., trying) thus may serve as the bearer of intentionality (in the semantical sense), although only derivatively, as specified by the analogy theory.

We shall later in Chapter 10 give exact and detailed conditions for an action's being intentional. Here it suffices to say that 1) (*18*) is too narrow and that 2) a causal action theorist can comfortably combine causal and semantical criteria of intentionality. Thus we get, as a first approximation, the following condition:

(*19*)   A singular action *u* of *A* is *intentional* under a description *C* if and only if *A* performed *u* because of a conduct plan *K*, in which *u* is described by *C*.

The 'because of' in (*19*) is to be later explicated in terms of our notion of purposive causation.

3.3 Our investigation of von Wright's verification argument led us to the conclusion that the whole argument relies dialectically on what it was meant to prove. This kind of hermeneutical "*Vor-Verständnis*" is presupposed, in fact, by *any* scientific describing. It can be demonstrated especially clearly within inductive inference (this is what Niiniluoto and

Tuomela (1973) argue for in their detailed account). Our preceding discussion, connected to the verification argument, can be related closely to the framework Niiniluoto and Tuomela employ. In fact, if (14) is taken to be an adequate formalization of the practical syllogism, their monadic framework fits the practical syllogism exactly. I shall not here, however, bother to present my argument using the formalized version of the practical syllogism, since that is not needed for my present purposes.

Let us consider a somewhat artificial example where the practical syllogism is used for the purposes of verification and confirmation. Suppose John has a wife who always needs to know where John is going when leaving home. John almost always responds truthfully to his wife's request. This time, when asked, John utters: 'I will go to the bank'. Supposing that John speaks the truth, we assume there are two ways for John's wife to interpret what John says, given that she interprets the above utterance as a saying. She formulates the two following alternative practical syllogisms to interpret and understand John's linguistic action (I use the simple schema (9)):

(20)   (P1) John intends to go and get some cash to buy a bottle of whisky.
          (P2) John considers that this requires his saying *to his wife* that he will go to the bank.
          _____
          (C)   John says: 'I will go to the bank'.

(21)   (P1′) John intends to go and see his friend Tom who every day sits at the bank of the nearby river.
          (P2′) John considers that this requires his saying *to his wife* that he will go to the bank.
          _____
          (C′) John says: 'I will go to the bank'.

The two possibilities of interpreting John's saying are of course due to the ambiguity of 'bank'. (We may assume that in the *logical forms* of (P2) and (P2′) the ambiguity does not occur, as we may there have the words $bank_1$ and $bank_2$.) (20) and (21) thus correspond to two interpretations of what John means by his linguistic action, given that his utterance-behavior is indeed taken as a saying (and hence as an action) and that he does not lie.

Accordingly, John's wife, Jane, can be taken to consider the following mutually exclusive and jointly exhaustive hypotheses for interpreting

John's behavior:

(22) $h_1 = $ (P1) & (P2)
$\qquad$ $h_2 = $ (P1') & (P2')
$\qquad$ $h_3 = \ \sim h_1 \ \& \ \sim h_2.$

The third hypothesis h_3 is given in a roundabout way here, but it is assumed to contain such possibilities as: John is lying, John is using language weirdly, John is not (intentionally) saying anything, etc. Jane now assigns epistemic (or inductive) probabilities to her hypotheses in the following way. Let T represent the conceptual and factual "background knowledge" Jane has. For instance, T entails that John is to be regarded as an agent who is very fond of whisky. We may now assume that Jane's probabilities are initially as follows:

(23)   (a) $P_0(h_1/T) + P_0(h_2/T) + P_0(h_3/T) = 1$
$\qquad$ (b) $P_0(h_1/T) > P_0(h_2/T) \gg P_0(h_3/T).$

Suppose that the (positive) difference $P_0(h_1/T) - P_0(h_2/T)$ is so small that Jane thinks she cannot decide which one to regard as the correct interpretation. So she gathers more evidence. In principle she can ask John various questions, observe his further behavior, etc. Let us assume, for simplicity, that she only asks which bank John will go to. John answers: 'I will go to the river bank'. Now Jane uses this crucial piece of evidence, call it e; she takes $P_0(e \ \& \ T/h_1) \ll 1$ but considers that $P_0(e \ \& \ T /h_2)$ is practically equal to one. We notice that considered *separately* John's two utterances (viz. his first and second answers), which both involve essentially the same intepretative problems, need not get one's inductive inference far at all, whereas jointly they may give rise to great changes in probability values. Thus in our present situation we may compare the difference $P_0(h_1/T) - P_0(h_2/T)$ with $P_0(h_1/e \ \& \ T) - P_0(h_2/ e \ \& \ T)$. Given our assumptions, it follows by means of a simple application of Bayes' theorem, that the value of the latter difference is smaller than that of the former. We may assume here that it is so much so that now $P_0(h_2/e \ \& \ T) \gg P_0(h_1/e \ \& \ T)$. Jane would probably, then, inductively conclude that John will go to the bank of the river.

Our simplified example shows that evidence changes epistemic probabilities and that such a change depends on 1) the assignment of initial probabilities and, in general, the selection of the measure P_0, 2) the background assumptions and "*Vor-Wissen*" T, 3) the evidence e, and 4) the conceptual framework in which the hypotheses h_i, the evidence e, and the background assumptions T are stated, and 5) the principle of "probability kinematics" used (here Bayesian conditionalization). It should be emphasized that none of these groups of factors 1)–5) are either "purely" empirical or "purely" a priori (see Niiniluoto and Tuomela (1973)).

In Niiniluoto and Tuomela (1973) it was argued that the probability measure P_0 will in general depend on some *broadly factual* considerations (on two extra-logical parameters), which the whole inductive procedure logically presupposes. It was also claimed that conceptual change very often affects the inductive situation. For instance, new background knowledge, explanatory theories, or empirical evidence may involve the introduction of new, perhaps theoretical, concepts, and that would involve a change in the probability measure used. Thus, instead of using simple Bayesian kinematics, we would change our probability measure P_0 into P_1, P_2, and so on.

As a special illustration of their theory of conceptual enrichment within induction Niiniluoto and Tuomela (1973) consider Goodman's (1955) well known grue-paradox. A version of Goodman's paradox could easily be formulated in the context of interpreting and explaining behavior as well. Thus, instead of Mr. Grue and Mr. Green discussed there, we would now have Mr. Reflex and Mr. Action. Mr. Reflex would describe behavior in non-intentional physicalistic terms, whereas Mr. Action would use intentional terminology. Similar general considerations that bear on the dispute between Mr. Green and Mr. Grue apply to the dispute between Mr. Reflex and Mr. Action as well (see Niiniluoto and Tuomela (1973), esp. pp. 189–192).

4. Extended Uses of Practical Syllogism

4.1. We have been emphasizing the importance of the practical syllogism (of type (9) or (13)) as a vehicle for *understanding* behavior. I think the deepest significance of the practical syllogism lies just in its use as an interpretative schema. When applied to the *explanation* of behavior the practical syllogism is somewhat insufficient, as will be seen in Chapter 8. The practical syllogism is also, accordingly, too restrictive as a conduct plan and as a psychologist's explanatory device.

We shall below discuss some extended uses of the practical syllogism (or, in fact, a liberalized version of it), and in that context some general comments on intention generation (or transferral) will also be made. To begin, we consider a simple example.

Suppose the agent's intention is to ventilate the room before time t. How does he carry out this intention? He, perhaps not fully consciously, devises a conduct plan, and starts by decomposing the original principal goal-intention into the suitable subintentions he has the means for realizing. Thus, the agent A also relies on some of his beliefs in forming his conduct plan. Perhaps he thinks that his opening the window will *generate* his ventilating the room. Suppose further that he opens the window best by pushing a button in front of him. These steps can be expressed by means of a kind of practical syllogism. We use a version of the practical syllogism where the time for action is not yet present, i.e., we speak of intention transferral. The second premise of the practical syllogism is now formulated in terms of action generation. We assume that in our present context it entails that the generating action is *necessary* for the generated action. For simplicity, we omit an explicit mention of the normal conditions. We get:

(24)   (PS$_1$) A intends to ventilate the room no later than at t.
                 A considers that his opening the window no later than at $t - n$ will generate his ventilating the room at or before t.
                 ─────────
                 A intends to open the window no later than at $t - n$.

           (PS$_2$) A intends to open the window no later than at $t - n$.
                 A considers that his pushing the button no later than at $t - n - k$ will generate his opening the window no later than at $t - n$.
                 ─────────
                 A intends to push the button no later than at $t - n - k$.

The practical syllogisms (PS$_1$) and (PS$_2$) are logically valid on the basis of what we have said earlier. Notice that the logical validity of the inference is not affected by the fact that the notion of generation in the second premises involves more than merely a necessary condition.

What we have above is a kind of small conduct plan: A's pushing the button (factually) generates A's opening the window, and the latter generates A's ventilating the room. Before discussing its nature in more detail, let us see how A carries out his plan.

According to his action plan A has decided to act no later than at $t - k - n$. When time gets close to that, say at $t - k - n - 1$, A will make his intending effective: he intends *now* to push the button at $t - n - k - 1$. We may write this final step as a degenerated form of a practical syllogism:

> (PS$_3$) A intends now to push the button now.
>   Therefore, A pushes the button now.

The *now* in (PS$_3$) is assumed to be just $t - k - n - 1$. Normally, as button pushing is a kind of basic action for A, his body and the external world co-operates with his trying and he succeeds in pushing down the button. (Strictly speaking, we shall argue in Chapter 10 that it is actually the bodily movements involved in button pushing that are describable as A's basic action as we shall use this technical term.) What is central here is that button pushing is the kind of basic action for A which requires no further means to be carried out. Thus, we have decomposed the original goal (the action of ventilating the room), to be called the *total* (intended) goal, into subgoals and ended up with a minimal subgoal – the performance of a basic action.

We have here a sequence $\langle$PS$_1$, PS$_2$, PS$_3\rangle$ of three practical syllogisms, which seems to correspond rather closely to our previous notion of an action plan. More generally, we consider sequences $PS = \langle$PS$_1$, PS$_2$, ..., PS$_m\rangle$ of m practical syllogisms and define:

(25)   A sequence $\langle$PS$_1$, ..., PS$_{k-1}$, PS$_k$, ..., PS$_m\rangle$ of practical syllogisms is a *PS-sequence* if and only if

  (1) $\langle$PS$_1$, ..., PS$_{k-1}$, PS$_k$, ..., PS$_m\rangle$ contains a "descending" series of goals related to each other by appropriate generational beliefs such that (i) PS$_1$ uses in its first premise the total goal; (ii) PS$_2$, ..., PS$_{k-1}$, PS$_k$, ..., PS$_m$ use in their first premises subgoals such that, for any k, the goal used in PS$_k$ is the goal appearing in the conclusion of PS$_{k-1}$;

  (2) the conclusions of the practical syllogisms PS$_1$, ..., PS$_{k-1}$, PS$_k$, ..., PS$_{m-1}$ are statements about intentions (to do certain goal actions);

  (3) the conclusion of PS$_m$ is about an intention to do a basic action.

This notion of a PS-sequence is meant to define a hierarchy of practical syllogisms such as exemplified by our above PS$_1$, PS$_2$, and PS$_3$. The clauses of (25) should be understandable without much additional clarification since the notion of a *PS*-sequence is a most simple one. It assumes that there is a *total* end or goal (an action) which is refined into finer and finer subgoals the last of which is just an action the agent can perform without doing anything which generates it. If, in (3), the intention to do this basic action is an intending *now*, we could also say that the agent tries (or sets himself) to do that action.

The subgoals are related *linearly* by generation (R_1 or R_2). A *PS*-sequence thus has, as its content, a linear graph of goal actions. Thus, it may occur as a *part* of a conduct plan structure (cf. Section 1 of this Chapter). A conduct plan structure may contain several branches, we recall. It need not be a tree even, as several action sequences may merge together. In any case, a *PS*- sequence has a linear structure and is connected with only one basic action (and hence only one action in the unifier's sense, cf. Chapter 2).

There is one important thing to be emphasized in connection with the form of the elements of a *PS*-sequence. It is that I have used the notion of action generation in the second premise of the syllogism. In our example we stipulated that generation in those cases entails that the generating action is a *necessary* condition of the generated action. But if we are interested in the formation of a cognitive psychological theory about action planning we may consider dropping the stipulation of necessary condition (even in any relative sense). We shall below sometimes use the term 'practical syllogism' to include such *liberalized* versions in which generation need not imply such necessity.

Why can we so liberalize the practical syllogism? If the necessity requirement is omitted, the practical syllogism ceases to be logically (conceptually) valid. It follows, too, that the practical syllogism ceases to be merely meaning analysis of the psychological concepts it contains. But assuming that we are interested in how people actually think and act we must of course be free to make moves like this. Still we shall continue to use our psychological concepts according to the original practical syllogism, for this we certainly can do. An agent can intend an end without there being any action of his which he considers necessary for the end, *ceteris paribus*; and the agent may perform an action which he conceives to be merely sufficient or likely for that end (cf. Chapter 6). We may thus also note that it is a clearly *factual* hypothesis that people typically act according to a certain action plan or a *PS*-sequence in our liberalized sense. (I think it is a hypothesis which is rather well supported.)

4.2. As we have emphasized, our notion of a conduct plan allows for the possibility of several basic actions occurring in it. Even if we shall not create an explicit theory about that case, we shall discuss one interesting situation here. If an agent builds a house, buys a car, or runs a mile there are normally several basic actions involved in such behavior contexts. These basic actions may be generationally *parallel* as in somebody's doing something with her left hand and (simultaneously) something else with her right hand in order to jointly bring about something. But, instead of or in addition to this, actions may also occur in a *sequence* (cf. driving a car or building a house). This case is also allowed by our notion of conduct plan. But in that case we really have a connected *sequence of action plans*, to use our terminology of Section 2 of this chapter. Within the framework of the practical syllogism, we correspondingly may have to deal with a connected *sequence of PS-sequences*.

If one considers behavior as a sequence of actions, this involves that behavior can be looked upon as a *process* — as a connected chain of action-events. One central principle of connection in that kind of behavior sequence is that of (negative) information *feedback*. Feedback from the results of action is what controls directed behavior and serves to correct action plans to conform with reality.

We shall below analyze the role of such feedback from the point of view of action theory and, for simplicity, be mostly concerned with what is usually called *activity*. Among activities we have talking, running, playing chess, and perhaps even non-intentional activities like sleeping. We shall, however, below restrict ourselves to activities in which one can be engaged intentionally. Thus, for instance, sleeping and falling down are excluded.

One can approach activities from several points of view. They can be analyzed as sequences of movements, for instance. But they can also be regarded as composed of achievements (performances). Thus running a mile or running for half an hour clearly have an achievement or result aspect. Or consider singing. It involves emitting sounds of a certain type and in a certain way and manner. Singing is intimately connected with satisfying some standards, and thus is characterized by achievement. It seems to me that this way of analyzing activities as complex achievement-actions is more important from the point of view of the science of psychology than the others — such as treating them as mere movement sequences or as unanalyzed activity-processes (cf. Chapter 10 for an exact treatment of complex actions).

Let us now consider the almost classical example of hammering activity discussed by Miller, Galanter and Pribram (1960). They analyzed hammering as a hierarchical activity consisting of lifting and striking the hammer such that this process was controlled by a feedback mechanism related to "reality" (position of nail). Let us thus consider Figure 7.1 (cf. Miller, Galanter, and Pribram (1960), p. 36):

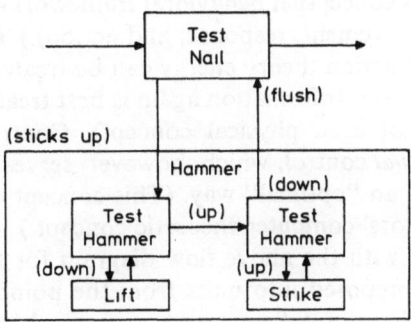

Figure 7.1

I think it is clear without a detailed further explanation what this simple diagram is supposed to represent. The plan analyzes hammering as a sequence of actions ("subroutines") of two kinds: striking the nail with the hammer and lifting the hammer. After each striking and lifting the position of the hammer is tested, and after each hammering, the position of the nail is tested, and so on. This testing procedure controls hammering in reaching its goal, that is, in getting the nail flushed. Hammering is thus treated as a goal-directed process controlled by the above kind of testing procedure which involves both internal bodily feedback and external information about results. The authors call the control mechanism a test-operate-test-exit unit or TOTE unit. (We have here in fact three TOTE units one of which is superordinate to the two "lower" ones.)

From the point of view of action theory there are many problems in the above analysis. One problem concerns what the arrows in the above diagram exactly represent. (Above we took them to represent information in a general unanalyzed sense.) Miller, Galanter, and Pribram (1960), pp. 27–29, consider three alternatives: 1) energy, 2) information, 3) control. Energy is to be taken as a physical concept. Information is in these authors' account somewhat unclearly regarded as "correlation over the arrows". Control again does not seem to be any more than "the order in which the instructions are executed" (p. 29). This sense of 'control' is what the authors finally opt for. I think that, in spite of some conceptual vagueness in the presentation, these authors are after something important.

It is central to notice that one's analysis and use of the notions of energy, information, and control have to be connected to one's conceptual framework. (Miller, Galanter, and Pribram somewhat confusedly use the elements of different conceptual behavioral frameworks in this connection (behavior as, e.g., movement, response, and action).) Thus we claim that in the connection of action theory energy can be treated as an underlying *physical* aspect of action. Information again is best treated as *propositional* information (and not as a physical concept). Control can be treated primarily as *conceptual* control, which, however, serves to connect action-plans with reality in an "optimal" way. (This concept of control is richer than the above authors' computer-theoretic concept.)

Another problem with the above flow diagram for hammering is that it is incorrectly decomposed into units from the point of view of action theory (though perhaps not from, e.g., a neurophysiological point of view). What we would like to have is the decomposition of the total action (activity) of hammering of the nail into suitable subactions. What could those subactions be? How are they connected? And how does feedback enter here? Can we perhaps apply the practical syllogism in an interesting way to these problems?

Let us consider the following practical syllogism (PS), which assumes that the total intended goal of the agent A is to flush the nail:

(PS)   (P1) A intends to flush the nail.
             (P2) A considers that his hammering the nail a sufficient number of times will generate the flushing of the nail.
             (C) A sets himself to hammer.

I have omitted time considerations and all the normal condition-assumptions in order not to unnecessarily complicate our present issues. In premise (P1) we have A's intention to bring about something—to flush the nail. The second premise speaks about the activity of hammering. But we can also conceive of hammering as an action, or can we? It seems that hammering the nail can plausibly be understood in the manner of Miller, Galanter, and Pribram (1960) to consist of repeatedly striking the nail with the hammer, each striking of course presupposing that the hammer has been lifted. Even if it is partly a contingent matter, we may perhaps assume that striking the nail with the hammer *and* lifting the hammer is a non-generated and in *this* sense a *basic* action for a typical agent. If so, the activity of hammering simply consists of a sequence of action tokens of the same type.

Our practical syllogism (PS) need not be conceptually valid as the agent need not consider hammering *necessary* for nail flushing. If the agent thinks so, the practical syllogism is logically valid. But irrespective of this (PS) may be equally effective in guiding the agent's behavior.

Now consider our diagram on hammering. If what I just suggested is acceptable we have to modify the diagram a little. For suppose that we are interested in an agent's making an action plan for hammering and that the striking-lifting pair is his minimal behavior-unit (basic action). Testing occurs after each basic action (or perhaps only, say, after each n basic actions).

What is this testing concerned with in terms of our framework? Testing of course means comparing achieved result with the ultimate intended result. Thus, testing presupposes *feedback*, i.e., information fed back to the agent from the external world, so to speak. In this case, the information fed back is primarily reflected in the second premise of our above practical syllogism. After each hammering-token the agent records how high up the nail sticks. Then he, so to speak, inserts this information into the second premise, and goes on to act. Thus I am suggesting the following scheme.

Prior to starting the hammering the agent will have to operate with an unspecified practical syllogism such as the above (PS). It is in fact a scheme for a sequence of connected hammering-tokens. But this sequence can be completely specified (e.g., as to the number of required basic actions) only *ex post actu*, i.e., after the total action (activity) has been performed. For its second premise can be determined only on the basis of direct commerce with reality right before the action token is performed. Thus, after, say, five hammering-tokens the agent's operative practical syllogism might be:

(PS$_6$)   (P1$_6$) A intends to flush the nail.

  (P2$_6$) A considers that in order to flush the nail according to (PS) it is necessary to perform a hammering-action now.

  (C$_6$) A sets himself to perform a hammering-action now.

(Again I have omitted time considerations except for the 'now' which refers to the time moment after A's fifth hammering-token.) We see that the plan (PS) is the operative practical syllogism during the whole activity of hammering. If the 'sufficient number of times' mentioned in the second premise of (PS) is *afterwards* found to be n times we are dealing with n

hammering-tokens $a_1, a_2, \ldots, a_n$. Each syllogism (PS$_i$) concerns one of these tokens, with our above example being concerned with a_6. Now, according to our previous analysis, this sequence PS$_1, \ldots,$ PS$_n$ is a *PS*-sequence, which, moreover can be said to be "controlled" by (PS) and "ordered" by perceptual feedback, so to speak. Feedback information tells the agent after each hammering token a_i that (P2$_{i+1}$) is acceptable. Notice that in some cases the agent might be able to quantitatively measure roughly how much the nail sticks up each time. Then we might use this information in the second premise. Thus (P2$_7$) might contain the information that the nail still sticks up $\frac{1}{8}$ of an inch. In such a case A might use, or act according to, the premise

> (P2$_7$) A considers that as the nail sticks up $\frac{1}{8}$ of an inch it is necessary for him to perform a hammering action now.

Then the premise (P2$_i$) would presumably be different in the case of different action tokens, and this would make clearer the nature of the *feedback loop* related to the second premise of (PS). This forms the basis of our TOTE-unit (test nail-hammer-test nail-exit).

We have seen that feedback gives information on the basis of which activity proceeds. Normally feedback concerns only the second premise in the above kind of cases. But it of course takes into account breakdowns (e.g., the hammer getting broken) as well. In such cases the implicit normal condition-assumptions of the practical syllogism do not any more hold, and hence the agent must go over to a different conduct plan. This indicates that the *function* of feedback is primarily that of *control* through information. Within our framework this means that the control in carrying out the conduct plan (generated by the global practical syllogism (PS)) functions (or can function) so that this carrying out is appropriate with respect to objective (physical and social) reality.

What I have said above is of course a simplified sketch but it still shows how activities may be fitted into our general framework (also see Chapter 10). Notice that my particular assumptions concerning how to decompose an activity into basic actions and TOTE-units are not very central, provided that the activity in question can be understood to have *achievement* character. Thus, basic hammering-actions might be construed differently, and testing need not (perhaps) occur after each basic action. But our purpose was only to show how the action theoretic framework and the cybernetic TOTE-framework can be fitted together; thus we can leave all the further details undiscussed here.

One could naturally imagine much more complex examples on the basis of Figure 7.1. Thus, we might think that in our analysis each practical syllogism PS_i is suitably connected to other actions (through its goal action). For instance each PS_i could thus be a member of some other PS-sequence. Hence, we would be dealing with a connected *sequence of PS-sequences* (or some other kind of conduct plans). Examples related to *problem solving* may be mentioned as examples of complicated but intelligently connected conduct plans (cf., e.g., Newell and Simon (1972); also cf. our discussion in Chapter 10).

NOTES

[1] Our generalization (L) could perhaps be modified so as to better fit the purposes of a causal theorist. (L) says that, in normal circumstances, for any psychological events (or short-term states) y and z there exists a singular action u of a certain kind. But one may argue that that is not enough. For given a certain intending and believing not *any* exemplification of the action will do. It has to be "the one" brought about by the activated psychological background. I think this argument has some force, especially if we view the situation from neurophysiology and accept that all psychological events have material carriers (cf. Chapter 9). If the agent has tried to do something (and thus activated his brain), the action inevitably follows, given that the body functions normally and the external world cooperates. So, as in typical causal laws, a kind of *coordinating predicate* may be needed (cf. Kim (1973) and Tuomela (1974a) for a discussion of it). Let this predicate be $P(y, z, u)$. Given y and z it coordinates them with the singular event u, whose existence is guaranteed by (L).

Instead of (L) we would now use

(L') $(y)(z)(I(A, p, y)$ & $B(A, q, z)$ & $N(A) \rightarrow$
            $(Eu)(\text{Mod}'-\text{Verbs}'\ (A, o', u)$ & $T'(o'', u)$ & $P(y, z, u))$.

Notice that I am still not claiming that (L') has to be a *specific* causal law, even if it now contains the coordinating predicate (cf. Chapter 9).

[2] Technically, the (literal) meanings of both (merely) descriptive and non-descriptive (especially "performative") sentences can be analyzed in terms of truth conditions, it seems. For such a theory, see especially the unified account by Lakoff (1975).

CHAPTER 8

EXPLANATION OF HUMAN ACTION

1. ACTION-EXPLANATIONS

1.1. The philosophical problems related to the explanation of actions are very intimately connected to those pertaining to the characterization of intentional actions and to the understanding of behavior as action, as will be seen soon. Thus, although the aim of this chapter is to clarify what it is to explain actions, our discussion below will be a rather direct continuation of the discussion begun in the previous chapter.

When one discusses the explanation of actions one can either emphasize the differences or the similarities of action explanations with explanations in the physical sciences. It is, however, rather obvious that of the action theories proposed so far causal theories come closest to "real" science and to "real" psychology, viz. what is currently being done in the natural sciences and in psychology (especially modern cognitive psychology). As will be seen below, our view on action explanations is indeed close to what can plausibly be said about "ordinary" scientific explanations. There are two differentiating factors I would like to mention here, however. The first is that, while the "hard" sciences are primarily concerned with the explanation of laws and regularities (as contrasted with particular facts and events), within action theory we are primarily interested in explaining particular actions rather than action-laws or the like. (Of course, even particular actions can normally be explained only as instantiations of their action types.)

Another feature that distinguishes action explanations, and for that matter psychology and the other social sciences, from what is the case in the physical sciences is that the subject matter of investigation of psychology is in an important way different from that of, say, physics. Briefly put, agents follow *rules* in a sense that, for instance, stones or electrons do not. This difference in the object of investigation is also important in the context of explanation, and it has to be properly accounted for. All this is of course no news; the main news will be arguments for the causal character of rule-following and our theory of *purposive causation* (as contrasted with mere causation). This theory takes the above difference into account in an important sense.

We shall start our discussion of action explanations by some general remarks on scientific explanation, to set the stage for what is to come. Many kinds of views have been expressed as to the nature of scientific explanation. Let us mention a few ideas. First, to explain an event (or a state of affairs) could be to cite its cause or determinant. Secondly, to explain an event described by a certain statement D could be taken to be giving a suitable informative redescription D' of the event. Thirdly, to explain an event described by D could be taken to be to cite a statement D', which contains a law of nature as its part, such that D' logically entails D. Fourthly, it may be argued that a statement D' explains an event-describing statement D if D' conveys understanding and information concerning D. All these ideas, and many others, have been explored either separately or explicitly connected to some of the others. What we are going to say about explanations below will take into account all the mentioned aspects of explanation, at least to some extent. We shall start by discussing a so far unmentioned idea of explanation, according to which explanations are some kind of answers to appropriate queries (especially why-queries).

Suppose an event occurs. We describe it by a certain statement D. We may then query why, how, when, etc. the event occurred. Our query is a speech act and it may be assumed to "contain" a *question statement* such as 'Why D?'. We shall below often call such question statements simply questions. They are to be distinguished from *queries*. A query typically has more content than that involved in the question sentence we relate to it. (We shall below pretend, for simplicity, that a unique question sentence is associated to each query, although this is of course not strictly true.)

Suppose John opened the window. I ask: Why did John open the window? Suppose further that in one case John just stumbled and the window opened due to that non-action. In another case John opened the window intentionally in order to ventilate the room. This highlights the fact that one and the same question sentence may be contained in quite different queries.

To illustrate essentially the same point differently, we quote Collingwood's famous example: "For example, a car skids while cornering at a certain point, strikes a kerb, and turns turtle" (Collingwood (1940), p. 304). Consider now the question "Why did the car turn turtle?" From the car driver's point of view the cause of the accident probably was cornering too fast. From the country surveyor's point of view the cause was a defect in the surface or camber of the road. From the motor-manufacturer's point of view the cause was defective design in the car. Finally, from a vulgar-marxist's point of view the cause might have been the capitalistic system (which has led to the manufacturing of such small private cars). These different answers to the above question demonstrate that, in fact, four different queries may thought to have been made. Each query contains the point of view, aspect, interest as well as presuppositions which the answer implicitly or explicitly reflects. (When we speak of queries below we shall assume that they completely account for the point of view, interest, etc. in question.)

Our preliminary thesis about the explanation of singular events now becomes the following. An explanation of a singular event is a (partial or complete) answer to an explanatory query concerning the event. The explanatory query in question is typically a why-query and it is then, in part, expressed by a question sentence of the form 'Why D?'. Explanatory questions may, however, take more complicated forms, and other question words than 'why' may of course occur in them. In general, the more of the content of a particular query is packed into the question sentence, the more complicated that sentence naturally becomes.

Even if the above remarks would help to clarify what an explanatory query is, it does not yet help much in clarifying the notion of explanation. For that we still need an account of what an *answer* to an explanatory query is. Let us here say in a preliminary way that an answer to a query expressed by 'Why D?' should at least *convey understanding* concerning D (or what D expresses). We shall later (in Chapter 9) return to the problem of answerhood in some detail. Another problem we will briefly discuss later is the degree of objectivity (i.e. independence of speaker, occasion, etc.) a notion of explanation characterized in terms of answerhood can and should have. Before going into these problems in detail, it is useful to briefly discuss the explanatory queries and questions typical of the context of action-explanations.

1.2. Suppose a man goes to a bank, signs a check, and pays his debt with it. This is a description of something that goes on in the bank at that time. It may also be considered to give some kind of explanation of the man's action of signing the check, for it refers to the man's effective purpose (to pay his debt).

There are at least two central features to be noticed in this example even without elaborating it further. One is that the description of the explanandum and the description of the explanans must fit together in the sense of belonging to the same conceptual framework — whatever that exactly means. Another feature is that the explanation of the action is teleological in that it somehow refers to the agent's anticipated future course of behavior (as viewed by him).

The agent's behavior in the bank can be described by means of different conceptual frameworks, and in many ways within each framework. Such different conceptual frameworks cannot fruitfully be mixed together, if at all. We may, for instance, describe the agent's behavior in the "molecular" terms of neurophysiology. Then, we would speak about muscle movements and the like and would explain such explananda by telling a suitable neurophysiological story about the agent's bodily goings on. Or we may describe the agent's behavior in the "molar" terms of stimulus-response psychology. In that case behavior is often functionally described as a response to some stimuli (and not as "pure movement" as philosophers often claim). Such responses are explained by reference to stimuli only (extreme behaviorism) or to stimuli and "intermediating" states (neo-behaviorism). It is very common, however, that stimulus-response psychologists mix different conceptual frameworks in that they try to investigate all of man's activities, be they mere movements, reflexes, habitual behavior or intentional actions, within one artificially created conceptual framework (cf. e.g. Hull's theory). I am of course not against creating new technical conceptual apparatuses for theorizing in psychology. It must, however, be taken into account, as we did in Chapters 3 and 4, that psychological concepts are strongly culture-laden. Concept formation in psychology has tended to neglect such conceptual social features when inventing new technical concepts (see Taylor (1964) for good illustrations of this).

In this book our main concern is intentional action. Thus we will say next to nothing about reflexes and reflex-like behavior. A few remarks will be made concerning non-intentional actions (i.e. actions which are non-intentional no matter how described) when discussing action tokens$_1$ in Chapter 10. Much of habitual and routine-like behavior is analyzable by means of intentional actions. Such behaviors thus belong to the scope of our investigation in principle, although very little will explicitly be said about them below. In any case, since the nature of intentional action and its explanation is a very central topic in psychology, we are fully justified in restricting our scope.

How do we explain intentional action? Recall our example of the agent's signing the check. As soon as we have decided to describe the agent's behavior as an intentional signing of the check, we do not, at least in the first place, ask explanatory questions appropriate only to muscular movements or reflexes and "mere" responses. Our queries instead involve questions related somehow to the *teleological* element in the action performed.

Suppose an agent A performed an action of kind X. There are now several appropriate queries and questions relevant to the explanation of this singular action. To approach them, let us try to pack much of the information contents of these queries into their contained questions. We may then consider the following list of appropriate explanatory questions:

(1)   Why did A do X?
(2)   What was the purpose (intention, aim, end, goal) of A's action?
(3)   What was A's motive for his doing X?
(4)   What was A's reason for his doing X?
(5)   Which of A's wants (or other conative attitudes) brought about his doing X?
(6)   Which of A's beliefs (or other doxastic attitudes) brought about his doing X?

Since we are going to interpret their contents broadly, these questions form a representative list of appropriate explanatory questions that can be asked in the case of an arbitrary intentional exemplification of action type X.

The first question is of course a very general one. It accepts as answers the citing of both *causes* and *reasons* for actions, whereas (4) seems to accept only the latter. Question (2) accepts a large class of answers. For our present purposes the small differences in nuance between 'purpose', 'intention' (representing the "goal" of an intending), 'aim', 'end', and 'goal' can be ignored. All genuine answers to (2) can be paraphrased in the form:

(7)   *A did X in order to do Y.*

Here Y is A's goal action for the purpose of which X was performed. (As in our discussion of the practical syllogism in the last chapter we could have alternatively spoken of achieving an end-event or -state E which is the *result* contained in the action Y.) The in-order-to relation is a kind of prototype of a teleological relation. Actions X which are not done for any further purpose Y, if there be such, can also, as a limiting case, be teleologically explained in the sense of (7).

How about our other explanatory questions? The concepts of motive and reason as used in the context of action-explanation are rather vague and ambiguous. Among possible motives I will include below at least emotions, feelings, sensations and personality traits. Thus, actions may be motivated by, e.g., love, fear, pain, and patriotism. Among reasons I will count proattitudes and doxastic attitudes which produce intentional actions. Thus, it will be seen that questions (5) and (6) are contained in (4) in that an answer to (5) or (6) in effect can be transformed into an answer to (4). (Note: (5) also takes into account duties, obligations, norms, etc. as they come in through *extrinsic* wants.)

It should be noticed here that (5) and (6) presuppose that wants and beliefs on the occasions in question did bring about the actions. Here 'bring about' is a (broadly) causal phrase. It is undeniable that this or related causal phrases are used in the context of action-explanation. Although this can be taken, prima facie, to give some aid and comfort to a causal theorist, a pre-Jonesian (or neo-Wittgensteinian) can still claim against a causalist that such locutions carry no ontological commitments.

If A did X with or out of a certain motive or with a certain reason, the corresponding *explanatory* claim seems to be statable as:

(8) A did X because of Y.

Here Y gives the motive or reason, as the case may be. Explanations in terms of productive wants and beliefs can also be given by means of claims of the type (8). For, if this kind of want- and belief-explanations are indeed reason-explanations, these cases reduce to the above case without further ado; cf. (14) below.

As why-questions are typically answerable in terms of claims of the type (8) we seem to be able to handle all proper action-explanations in terms of answers of the kind (7) and (8). But I claim more. I argue that (7) can further be reduced to (8). What is still more, I claim that the force of 'because of' is *causal*. More specifically, it is concerned with so-called *purposive causation*.

If the explanatory questions (1)–(6) (and the corresponding queries) would *exhaust* the types of proper explanatory questions that can be asked in the case of intentional actions, then we could at least make the following claim about action-explanations: All proper explanations of intentional actions are *intentional-teleological* (IT-explanations, for short). In other words, any proper explanation of an intentional action which answers one or more of the questions *(1)–(6)* is an IT-explanation. The word 'teleological' indicates that such an explanation refers to a *goal* of the agent. The word 'intentional' again specifies that the agent's *intendings* are somehow involved, although not necessarily referred to in the explanation.

Given our broad notion of reason (to be discussed below) explanations given in terms of the agent's reason will exactly amount to intentional-teleological explanations, given that *(8)* is satisfied. Thus, all intentional-teleological because of-explanations of intentional actions are in this sense reason-explanations; and, as claimed above, all reason-explanations are purposive causal explanations, given that they satisfy *(8)*.

Before arguing for our strong theses, let us mention a couple of weaker explanatory queries which need not satisfy *(8)* as we here interpret it. Let us first consider a query reflected by the following question:

(9)   What does *A*'s doing *X* express?

Put in different words, we may ask for a (further) meaning, sense, or significance of *A*'s doing *X*: *what* indeed was it *A* did in doing *X*? Here we are asking for (additional) understanding of *A*'s action *X* and for understanding in a different sense than above in *(8)*. The meaning or sense asked for is, furthermore, not only meaning in some strict semantic sense but in a pragmatic sense, broadly understood.

Question *(9)* may get a partial answer from adequate answers to our earlier explanation-seeking questions, especially those answered by explicitly citing reasons for acting. But something more seems to be involved here. What this more is, is hard to say in exact terms. Speaking very generally, it is required that the agent's action be approached from how the agent "sees" or conceives of the situation in which he acts, from what the situation pragmatically means to him. The agent then expresses, by his behavior, his aims and beliefs. How the agent does all this is strongly relative to cultural and social factors. But he does it so as to make intelligible communication possible within his (speech) community. A good illustration of the "hermeneutic" and culture-laden aspects of action is provided by *linguistic* actions ("speech acts") like promising or thanking.

In Chapter 7 we discussed the use of the practical syllogism as an *interpretative* schema. Clearly, it takes into account part of the above aspect. Presumably, an advocate for the practical syllogism such as von Wright or Stoutland would take it not only to provide a fully adequate answer to (9) but to regard (9) as representing the central and perhaps the only explanatory query to be asked in the context of action-explanations (cf. Stoutland (1976b)). I agree that the intentional-teleological explanation of intentional action presupposes that behavior be understood as action and that the practical syllogism (though in a liberalized form) is central for that (cf. Chapter 7). Redescribing action by means of further practical syllogisms may also convey important explanatory understanding. Granting all this, there is something more to be queried, which is what answers of type (8) are meant to account for.

There is a plausible reading of the explanatory answer (8) in which the explaining factor Y gives (in some sense) *sufficient* grounds for the action (X). That is, (8) seems to require that the explanatory factor in some strong sense *produces* the action. Given the explanans, the explanandum has got to be true. There are, however, situations in which it is appropriate to explain (in a *weak* sense) an action simply by answering a question of the following general type:

(10)   How is it possible that A did X?

Question (10) can be answered by citing some necessary conditions for action. The problem is of course how to find more interesting necessary conditions for A's doing X than, for instance, his having been born or his brain containing a certain sufficient amount of oxygen. It would seem that, for instance, ability may serve as an interesting explanans here. Thus, e.g., A's having an IQ of 155 may explain A's solving a certain puzzle. It does it the better, the worse people with a lower IQ fare with this problem. It seems that how possible-explanations can be accounted for and clarified in probabilistic terms. The more probable an explanans makes the explanandum, relative to the prior probability of the explanandum, the better is the explanans (see the discussion in Chapter 11 below and in Chapter 7 of Niiniluoto and Tuomela (1973)). As we shall see in Chapter 12, cases of probabilistic causation can be taken to give answers of kind (8) while they at the same time answer a how possibly-question of kind (10).

One type of how possible-explanation worth taking up here, even if we shall not discuss it further, is the explanation of action by means of sociological factors such as social position or class of the agent. To take a simple example, one may claim that agent A was elected member of parliament because he was sufficiently rich to afford a good campaign. Here, the explanatory factor is assumed to be a necessary, but not a sufficient, condition of the explanandum. Let me remark, however, that often these kinds of necessary social conditions are used to explain *failure* of action. For instance, A may have wanted to become a teacher but was too poor to afford his education.

Sometimes a distinction is made between normative and non-normative or "descriptive" explanation of action. Normative explanation refers, in its explanans, to some moral or legal *norms*, or the like, whereas descriptive explanation does not. Thus, one might try to explain somebody's going to church or somebody's stopping his car at a red light by citing an appropriate norm. This seems to be a clear case of a reason-explanation. The central explanatory factor here is, nevertheless, the agent's "*obeying*" the norm or rule in question. In discussing what is involved in the agent's "obeying" or "internalizing" a norm, in "following" a rule, or acting "under normative pressure", etc., in a *particular* explanatory situation, the non-descriptive aspects of the norm or rule in question do not play a central role. (Instead we have to speak about e.g. the agent's wants and intentions to follow the rule.) Thus, I take the distinction between normative and descriptive explanations in the above sense to be a relatively uninteresting one.

2. CAUSALITY AND INTENTIONAL-TELEOLOGICAL EXPLANATION OF ACTION

2.1. According to our causal theory of action, the basic ontological pattern of explanation is the following. A change (e.g., an action non-functionally described) in a substance (the agent) is explained by another change (effective intending cum believing, i.e. an inner event activating a conduct plan) in the substance such that the first change is caused by the latter change (non-functionally described). Under their normal functional descriptions, the latter type of changes are conceptually connected to the first type of changes. This ensures the tightness we need to have between mental causes and the actions they have as their effects.

The basic pattern of explanation in a non-causal pre-Jonesian theory of action (such as von Wright's) is completely different. According to its explanations are mainly answers to questions of the form (9) such that no causal production of a change in the substance by another change in the substance is involved. In fact, there is no requirement as to what the ontological side of an explanation is to be like. The sole concern is with interpreting and reading off behavior (or, rather, what the agent thinks is his behavior) in terms of the practical syllogism.

Our causal theory of action is broadly factual in its nature in that science, in principle at least, is assumed to tell us what kind of mental processes and episodes there are. Still, our commitment to a certain pattern of explanation is rather to be called metaphysical (although I do not want to draw a sharp line of demarcation between the metaphysical and the factual). In order for our pattern of explanation to work, the appropriate mental episodes must of course be there and so must the relevant causal connections. Although we have earlier given both conceptual and empirical arguments for our causal theory of action (see especially Chapter 4), some specific further arguments will still be given in the context of action-explanations. In fact such additional arguments will be most welcome when defending our causalist theses about action-explanations, to which task we now turn.

Let us summarize our general theses about intentional-teleological action-explanation as follows (for arbitrary explanations E):

(T1)   If E is a proper explanation of an intentional action which is an answer to at least one of the questions (1)–(6), then E is an IT-explanation.

(T2)   If E is an in order to-explanation of an intentional action, then E is a because of-explanation.

(T3) E is an IT-explanation satisfying (8) if and only if E is a reason-explanation.

(T4) E is a purposive causal explanation of an intentional action if and only if E is an intentional-teleological because of-explanation of that action.

A few words about the contents of these theses are now in order. First consider (T1). I will not here really try to give an explicit definition of a *proper* explanation of an intentional action. In fact, our theses (T1)–(T4) just serve to give a partial explication of it. Intuitively speaking, I can, however, say that *proper* explanations are meant to cover action-explanations in terms of *proximate mental factors* – roughly as typical common sense explanations in terms of such factors are given. Thus, our term 'proper' serves to exclude e.g. Skinnerian explanations in terms of external stimuli. But on the other hand we have to require in (T1) that E answers at least one of (1)–(6), otherwise we might have to accept, for example, an agent's merely having a certain kind of power of imagination (a how possible-explanans) as a proper explanans of his intentional action.

It should be noticed that as such (T1) does not explicitly speak about causation at all. As we characterized IT-explanation earlier it does not yet require that the explanatory intendings be causally productive. (That additional requirement is made explicit especially in (T4).) Unless this extra requirement is made, the converse of (T1) does not hold (e.g. answers of the kind (9) disqualify it).

(T1) has been introduced mainly to emphasize the importance of IT-explanation. (But we shall not, nor need we, spend much energy in trying to make it more explicit.) There is a related thesis which also emphasizes the importance of IT-explanation: (T1') Every intentional action admits an IT-explanation. What we are going to say about IT-explanation below serves to support about equally well both (T1) and (T1'). For every intentional action has a proper explanation, I claim, and such an explanation is an IT-explanation.

The meanings of (T2) and (T3) should be relatively clear as they stand. (T2) may be regarded a consequence of our causal theory of action and hence it is a broadly factual hypothesis. (T3) will be seen to be true in part due to our explication of the notion of reason.

(T4) explicates the force of because-of according to our causal theory of action. As will be seen, probabilistic purposive causation is to be included here, although we shall not explicitly discuss it in the present chapter. It should be noticed here that not all because of-explanations of behavior (e.g. of the behavior that an action ontically "consists of") are because of-explanations of an intentional action. A because of-explanation of an action will be understood here to always be a proper answer at least to question (1).

Our above theses are obviously closely interconnected. Equally obviously it is hard to *conclusively* show any of them to be true or false. Furthermore, at least in the case of (T2) and (T4), our central theses, no conclusive a priori arguments exist. Their truth or falsity is ultimately to be decided a posteriori on scientific grounds. Nevertheless, one can try to make these theses sound plausible and likely also by means of a priori arguments, and this we shall try to do. On the whole, our discussion of theses (T1)–(T4) will concentrate on showing that purposive causation *is* involved in action-explanations.

We have earlier characterized the intentionality of actions by reference to suitable liberalized and extended practical syllogisms, viz. conduct plans. Our above claims about action-explanations amount, by and large, to claiming that all and only *because of*-explanations of intentional actions are purposive causal explanations (in terms of some appropriate conduct plans). Even if we have characterized conduct plans functionally by saying that they are explanatory entities and even if the intentionality of actions has been characterized by reference to them, our above claim is not completely circular. This is because of two considerations. First, the exact details of action-explanations still can and will contain additional non-tautologous information. Secondly, the possibility of retreating from functional descriptions of causes and effects to non-functional ones in the case of each singular explanation shows that explanations in terms of activated conduct plans are informative in a sound sense.

Our basic intuition about action-explanations is very simple. We claim that all *direct* action-explanations satisfying (8) are to be given by explicit or implicit reference to an *intending* which was made effective and which purposive-causally produced the overt action to be explained. The explanatory intendings, embedded in a conduct plan of the agent, may be generated or activated by a number of different factors, depending on the case. Thus *intrinsic* and *extrinsic wants* may generate intendings (see Chapter 6). *Beliefs* (e.g., perceptual beliefs) may generate or activate intendings. Similarly, *emotions* (e.g., envy), *feelings* (e.g., fear), and *sensations* (e.g., pain) activate intendings in those cases where they lead to or produce *intentional* action. Explanations in terms of *purposes, goals, aims, ends*, etc. refer either to the contents of proattitudes or doxastic attitudes, and hence they are included in the above cases. As will later be seen, *reasons* are also to be explicated in terms of psychological attitudes. I believe the above cases exhaust the so-called *proximate* psychological determinants of intentional actions and, hence, their direct productive explanatory factors. There are of course *distant* determinants also, such as various environmental events and general societal factors such as social class. In this book we shall not, however, be concerned with such distant factors.

2.2. Let us now proceed to a discussion of the inadequacy of non-causal action explanations. We shall discuss simple kinds of intentional-teleological explanations and try to argue that they essentially involve a causal component.

Let us consider our example about the man who went to the bank to pay his debt. We explain his action of signing the check teleologically by saying that he did it in order to pay his debt. This intentional-teleological explanation refers to a further purpose of the agent. It is now easy to suggest that the practical syllogism serves to explicate intentional-teleological explanations. In fact, von Wright (1971) suggests that to give a teleological explanation for an action is to construct an ordinary kind of practical syllogism (cf. schema *(13)* of Chapter 7) for it (von Wright (1971), pp. 98–100). Accordingly, the methodological role of the practical syllogism is considered pivotal in the social sciences: "It is a tenet of the present work that the practical syllogism provides the sciences of man with something long missing from their methodology: an explanation model in its own right which is a definite alternative to the subsumption-theoretic covering-law model. Broadly speaking, what the subsumption-theoretic model is to causal explanation and explanation in the natural sciences, the practical syllogism is to teleological explanation and explanation in history and the social sciences" (von Wright (1971), p. 27). Let us consider von Wright's suggestion in some more detail.

Here, we may accept that an explanation given in terms of the practical syllogism at least normally is an acceptable teleological explanation of an intentional action. The first premise of the practical syllogism gives the further purpose for which the means-action in the conclusion is performed. Are all teleological explanations of intentional actions then explanations by means of some practical syllogism? Von Wright discusses this question and, after some discussion, gives an affirmative answer to it.

He considers the possibility that there are several alternative actions which the agent considers necessary for a given end. Suppose there are two such actions X and Z. Then the agent has a choice. The practical syllogism only explains why the agent does X or Z, but it does not explain his doing X as opposed to Z. It is possible, however, that there is another practical syllogism explaining the agent's choice, for instance, in terms of X's being more effective, cheap, etc., than Z for the end. But there need not exist a reason, and hence such a practical syllogism, for *every* choice, von Wright argues. It is therefore a matter of *contingent* fact whether such an additional practical syllogism can be found, and that is not enough according to him.

Von Wright also considers the possibility that the agent does not consider any means-action to be *necessary* for the end. We may now be able to explain the agent's action (at least if he considers it to be *sufficient* for the end). Still such an explanation is *incomplete* as it is not *logically conclusive* in the sense the ordinary practical syllogism is (von Wright (1971), p. 100). Thus, von Wright thinks he can conclude that all proper (i.e. conclusive) teleological explanations of intentional actions are explanations in terms of practical syllogisms.

Von Wright's claim that the teleological explanation of action coincides with explanation by means of a practical syllogism is wrong, I argue. The class of teleological explanations of actions neither contains nor is contained in the class of explanations in terms of the practical syllogism. Furthermore, von Wright's account of both teleological explanation and of explanation by means of the practical syllogism lacks an important component, viz. causality. What is more, von Wright has strongly exaggerated the role of the practical syllogism, even if it were understood causalistically in the way argued in Chapter 7. We will now proceed to a more detailed discussion of these remarks.

Let us consider a simple example of explanation in terms of an emotion. An agent A goes to the cellar of his house. We explain this by saying that he is afraid of thunder-storms and lightning strikes and that right before he goes to the cellar it was thundering heavily. A's action was intentional, we suppose. We may now try to exhibit this explanation in a more complete form as follows:

(11) A is afraid of thunderstorms.

A intends to save his life.

A considers that unless he goes to the cellar
he will not save his life.

A goes to the cellar.

(11) resembles the unqualified form of the ordinary practical syllogism (schema (9) of Chapter 7). But in addition it contains a new premise which describes A's emotion out of which he acted. In the second premise the hidden intention is assumed to be the intention to save one's life — it might have been something less solemn as well (e.g. the intention to escape the sound of thunder).

We can proceed in the same way as in the last chapter by reading schema (9) as really being of kind (7.13). Then (11) becomes logically conclusive in the sense schema (7.13) was argued to be. But does it not follow then that the first premise of (11) is superfluous? If so, our original explanation sketch seems not to work as it explains A's action by reference to his emotion. However, it is only because the emotion activated A's intention that we can refer to the latter in our explanation.

Should we now say that the practical syllogism is too *narrow* in that it excludes the intentional-teleological explanation (11) in the sense of requiring us to omit the first premise? I think one can defend an affirmative answer to this question. On the other hand, one may also claim that A's intention gives the (most) proximate explanatory factor while the activating emotion represents a more distant explanatory factor.

I shall here accept an affirmative answer to the above question. The explanatory conduct plan in the case of explanations by emotions (and, analogously, feelings and sensations) should (or, at least, may) be taken to include the "immediate activator" of the agent's intending. Taken in this wider sense it has more *explanatory power* than (*11*) considered without the premise describing the agent's emotion, for the emotion may be compared with a standing factor which explains *several* analogous situations in the agent's life.[1] One can construct examples to show that *feelings* and *sensations* play an explanatory role similar to emotions (cf. *A* was in pain, had a throbbing sensation, etc.).

A further argument for the explicit introduction of emotions and related factors into the immediate intentional-teleological explanans of an intentional action is that they may account for the *style* or *manner* with which the action was performed. For instance, an angry person often closes the door, or says good-bye, etc. in a different manner than a calm person.

Emotions, feelings and sensations sometimes lead to behavior which is not intentional (cf. jumping out of pain). We shall not consider their explanatory role in such cases here.

Going back to the narrowness of the ordinary practical syllogism as an explanatory schema, there is yet another argument. Suppose, alternatively, that in our example the agent did not think that going to the cellar was *necessary* to save his life. Maybe he only thought that going to the cellar *or* entering his car was required (cf. our discussion in Chapter 7 of the man trying to catch a train). Or maybe he thought there was nothing which was necessary for him to do in that situation. Perhaps he thought that going to the cellar would only slightly enhance the attainment of his goal. Does (*11*), modified in one of these ways then, only give us an *incomplete* explanation (cf. von Wright (1971), p. 100)? No, it definitely does not, I argue, as long as a factor moving the agent to act is mentioned in the explanans. Mere logical conclusiveness is not per se interesting from the point of view of explanation.

When an agent forms an intention he also, typically, in some general sense chooses his means for reaching his goal. Here, he is not a priori bound by any conceptual barriers concerning what is analytic and what is synthetic about the means-goal relationship or concerning the exact form of his practical inference. Thus, he does not have to worry about the logical conclusiveness of the practical syllogism. We as scientists and methodologists must be satisfied with recording how the agent decides to act and how he, as a consequential matter of fact, acts. Questions about the form and adequacy of teleological explanation are separate from the question about the validity of any related practical syllogism. The only a priori restriction for the content of the belief in the second premise of the practical syllogism is a requirement of believed generation (in a wide sense including probabilistic generation), which we used earlier in the liberalized version of Chapter 7. The rest is a factual problem to be solved by scientists.

Why is the practical syllogism too *wide* even when considered in its original logically conclusive form? As it is normally stated, it does not require that the agent's intention in any factual sense brings about or *causes* the action in question. This is something we have been repeatedly emphasizing. It is now proper to summarize our arguments for the causal character of practical inference and to bring up some new arguments.

A. The most fundamental argument for the presence of a dynamic (i.e. causal) element in practical inference comes from how we have construed the concept of a mental episode. That is, the version of the functionalist analogy theory of mental episodes that we adopted and discussed in the earlier chapters of this book (especially in Chapters 3 and 4) *conceptually* introduces mental states and events as something causing behavior in suitable circumstances. Thus, for instance, to intend that *p* means, roughly, being in a dispositional state with the structure *p* (or with the structure that *p* has) such that this state, given suitable internal and external circumstances, will cause the bodily behavior needed for the satisfaction of the intention. To take an example, consider a man who intends to stab someone and who in fact stabs him. This stabbing was intentional, in the typical case, only if the man's intending brought about or caused his stabbing.

To summarize, we have in our account dispositional states as well as events that actualize them, the latter acting as efficient *causes* of behavior. At the same time there is a *conceptual* (semantical) connection between concepts representing the (possibly complex) mental disposition, on the one hand, and, on the other hand, both the generic action which ultimately satisfies the intention and the generic means-actions which are related to the intention via an ordinary practical syllogism.

Let us now consider an agent who has formed an intention to illuminate the room. His deliberation makes him — through intention transferral — form an *intention to flip the switch*. When the time for flipping the switch arrives, this intention becomes a causally effective intending or *trying* (as we have technically called it) to flip the switch (see Chapter 6).

The mental event of trying is characterized functionally in terms of what it causes. Thus, a trying to raise one's arm is a singular occurrent event which, the world suitably cooperating, causes the bodily events necessary for the satisfaction of the intention. In our example it causes the arm's going up which constitutes the "observable" or "overt" embodiment of the arm-raising action. We recall that a trying is an intentional mental act, but it is not an action caused by another trying; trying to try is, if anything, only trying. Also notice that I have not required a trying to be a fully conscious event.

It is important to see that each *singular* claim concerning the causation of arm rising by a trying is a *synthetic* claim. For in the case of any such singular causal claim, the trying and the movement (or action, if you like) are ontically *distinct* events that can be redescribed so that 1) every trace of intentionality as well as 2) the reference to generic logical connection between the intention and the corresponding action disappears from the description.

We do not at the present time know how to make such a redescription for the corresponding *generic* events nor whether it is more than a conceptual possibility. If the generic notion of trying were, after all, physically redescribable (e.g., in the sense of being contingently identical with some physical property) our generalization (L) of Chapter 7 which, in effect, says that the concept of trying (effective intending) is semantically connected with the corresponding action concept might lose its status as a semantically valid principle (although that does not have to happen). What *will* happen here is to a great extent up to future neurophysiology.

The point I have been trying to make in this argument for causation is of central importance, for it means not only that whenever an agent performs an action there is an underlying causal bodily process involved but that, in addition, this causal nexus is *conceptually* built into the very notion of intentional action (cf. Chapters 3 and 10). It is a conceptual truth that there is a causal connection between the premises of an ordinary practical syllogism and its conclusion. More exactly, *each* singular episode (satisfying the premises), non-functionally described, causes the occurrence of a non-functionally described singular episode satisfying the conclusion. Consider still my greeting somebody by my raising my arm. My intending to greet him becomes, through intention transferral, my effective intending to raise my arm *now*. In case I intend-out-loud I may say: "I will raise my arm now". We know how this utterance can be non-functionally described in physicalistic terms. The inner episodes of intending or willing are to be regarded as analogously construed and as analogously non-functionally describable (cf. Chapters 3 and 6). As it should be rather clear how to give the overt action of arm raising a non-functional description, we have here the elements entering the causal relation.

If one analyzes singular causation in terms of a backing law account, as we will do, one thing has to be noticed. It is that while there is a causal connection in the case of each exemplification of the premises and the conclusion of a practical syllogism, different backing laws may be involved in different exemplifications. The different exemplifications are, however, conceptually connected (cf. Chapter 7). The causal connection between the premises and the conclusion represents, and is also a precondition of, full-fledged conceptual activity (here: practical thinking leading to action).

B. My other arguments for the causal character of practical inference are ultimately related to the main argument above. Still, they are sufficiently independent to warrant a separate discussion. The following argument indirectly relates practical inference to the causalist theory of inference.

We may argue on the basis of "our" conception of inference that something causal must be involved in inferring. To take a simple example from ordinary logical inference, the fact that the premises p and $p \to q$ logically imply q has *normative* character. The basic norm here is: One *should not* infer $\sim q$ from the premises p and $p \to q$. A person cannot be inferring if he concludes $\sim q$ from the above premises. Furthermore, under epistemically normal conditions an agent *should* infer q given the premises p and $p \to q$.

Thus, in order to be able to make logical inferences, an agent must *internalize* at least some logical principles. This is to be understood in the sense that the agent must be able to *use* some principles without deliberating on their validity. As Lewis Carroll and others have pointed out, *modus ponens* is one of those most basic logical principles which an agent must have internalized in order to be able to make inferences. For if a person does not already infer according to modus ponens this principle cannot be taught to him by formulating it to him in the metalanguage, since that formulation cannot be understood without using modus ponens, etc. A regress is involved here. The same point can be made about *universal instantiation* as well.

We may now argue that the internalized use of such normative logical principles as modus ponens is based on the fact that there is a *causal* connection between a person's thinking of the premises and his thinking of the conclusion. (This holds for the overt actions of thinking-out-loud as well.) The causal connection in question is "ontologically" prior to deliberation and inferring.

The above remark applies also to practical inference. Analogously with the cases of "theoretical" inference, there are cases of *non-deliberative practical inference* in which the agent has more or less completely internalized the inference pattern so that only a minimal amount of voluntary control is exercised in the inference process. To give a simple example of such a practical inference without deliberation, something which is an action token$_1$ but not an action token$_2$ in our later terminology, consider the following. An agent A infers-out-loud or otherwise reports to us that he has inferred the statement p_2 from the premises p_1, $p_1 \rightarrow p_2$, and $p_2 \rightarrow q$. We construe the following simple explanatory syllogism (exemplifying scheme (*13*) of Chapter 7):

(*12*) A intended to deduce the statement q from the premises p_1, $p_1 \rightarrow p_2$, and $p_2 \rightarrow q$.
    A considered that in order to deduce q from the premises p_1, $p_1 \rightarrow p_2$, and $p_2 \rightarrow q$ it is necessary to first deduce p_2 by means of modus ponens.
    Normal conditions obtained.
    _____
    A inferred-out-loud p_2 from the premises by means of modus ponens.

In this example A did not deliberate on whether or not to use modus ponens — he "just inferred". There does not seem to be any other good way to account for this than to simply say that a *causal* inference process took place here.

C. The next argument for causality involves making a distinction between the *justification* of an action and the *explanation* of an action. One may justify a person's doing an action by citing a reason (want, desire, intention; see our later account) that one has even if one did not act on or because of that reason. However, one cannot explain an action by citing a reason unless one acted because of it (cf. Davidson (1963)).

Thus, in a certain situation (e.g., when biking) I may have two reasons for raising my arm: I have a reason to signal for a right turn and I also have a reason to greet somebody, both actions performed by my raising my right arm. Suppose I only acted on one of these reasons. I greeted my friend but decided to continue along the same street one more block before turning right (or perhaps I forgot to signal). I thus only formed an intention to greet my friend and not to signal, although I might have had a good reason to signal (perhaps the street was blocked ahead of me).

Our causal theory handles this situation by saying that the causally effective intending was the greeting-intention. In fact, no intending to turn right even existed here. I acted *because of* my reason to greet my friend, and the force of this because of-relation is causal (cf. our (8)). It is the causal pattern that is explanatory here, in the first place. It does not suffice to say that my arm rising *is* greeting or that it *is not* signalling for that is only "mere" redescription, failing to account for the difference in acting on one reason and not on another. A causal theory of action gives a good non-verbal solution to the problem at hand, and this can be taken to give clear support to the causalist view.

D. A central criterion of adequacy for scientific explanation is that the explanans should give reasons for expecting the explanandum to be true. An explanatory pattern which fits this requirement is the cause-effect relationship. It accounts for how the explanandum-episode comes about, and it certainly gives good reason to think that an explanandum-episode is to be expected whenever the explanans is true.

Of course, I am not saying that all explanations should refer to deterministic cause-effect relationships. There are non-causal nomic connections and there are, conceivably, cases of indeterministic causation. What I want to emphasize specifically here is that there are important cases of action-explanation which do not become intelligible unless a reference to an underlying dynamism — a pattern of producing — is made. A good case in point is the tyrant-shooting example we discussed in Chapter 7. There we argued that what was missing was causality. The pattern of the practical syllogism is explanatory in the sense of affording understanding (cf. our question (9)), but as such it is merely a conceptual pattern. If it is to be used to explain events in the real world, it must also be ontologically connected to the real world. Given that actions are considered as events when using the practical syllogism for explanation, if we would only restrict ourselves to redescribing one and the same event (perhaps by referring to its context of occurrence) we would gain less understanding than if we, in addition, refer to causal connections.

E. An argument for the causal character of practical inference, which continues our present line of thought, relates to the following problem posed by Wittgenstein: What is left when we subtract from the agent's arm raising his arm rising? The aspect of this problem that I want to emphasize here is: What features qualify events and processes as *results* of (intentional) actions? (Here I use 'result' in the sense of von Wright (1971).) Thus: what distinguishes my arm rising when it occurs as a result of an action from a case of its "merely" occurring (e.g., as a reflex)?

A non-causal theory might try to give such "verbal" answers as: "One uses different ways of conceptualizing the cases regarded as different" or "Only in the case of intentional action, as distinguished from 'mere bodily movement', does the agent's behavior express his intention to raise his arm". Our causal theory, on the contrary, looks for an ontically and dynamically real solution to the problem. We thus refer to the mental act of trying, which accounts for the voluntary control and purposive causation present in the case of intentional arm raising but not in mere arm rising. Thus, if a trying-event purposively causes the arm raising (here non-functionally described as the bodily component of that action), the arm rising counts as a result of the full action of arm raising. In general, reference to trying or effective intending gives only a partial characterization of the notion of a result of intentional action. It seems to give a sufficient condition but not a necessary one, for not every intentional action is preceded by a trying to do specifically that action (type); cf. Chapter 10.

F. The next argument for the causal character of practical inference and for the need of a causal theory of action can best be given by means of an example. We consider a situation in which the agent A intends to see a certain person B. He learns that that is possible only if he goes to his friend E's party at the latter's summer cottage. He also considers that his only chance to have a sauna bath that night is to go to E's sauna. So A forms an intention to take a sauna bath while visiting E. (A takes his towel with him, etc.) So we have the following two practical syllogisms both with true premises and a true conclusion (I state only the central premises):

> (13)   (PS1) A intends to see B.
>   A considers that unless he visits E he will not see B.
>   ———————————————————————
>   A visits E.
>   (PS2) A intends to have a sauna bath.
>   A considers that unless he visits E he will not have a
>   sauna bath.
>   ———————————————————————
>   A visits E.

Here, the intention expressed by the first premise of the second practical syllogism need not be regarded as conditional on the intention in the first premise of (PS1), although it might at first sight seem to be so.[2] The intendings in these two practical syllogisms can be regarded at least as conceptually independent. Furthermore, we invite the reader to conceive of the situation so that the intention in the second practical syllogism, contrary to the situation in the first practical syllogism, does not generate the means-belief. Rather the belief (in conjunction with the intention in (PS1)) generates the "sauna-intention". Thus, while it is *true* (or assertable) that

> (a)   if A had not intended to see B he would not have visited E,

it is *false* that

> (b)   if A had not intended to have a sauna bath he would not have
>   visited E.

If we accept a conditional analysis of causation (e.g. Lewis' (1973b) or a related account), we are at least technically, but, I argue, here also intuitively, entitled to claim that in the case of the first practical syllogism there is a *causal* connection between the intention (cum belief) and the action, whereas in the case of the second practical syllogism there is not. This, if accepted, explains the difference in the explanatory power in these two cases; I do not see how a non-causal account (such as von Wright's) could handle this case satisfactorily. (Also see the relevant treatment of competing intentions in Chapter 9.)

G. As has been emphasized already, causal theories of action are continuous both with common sense and current scientific practice. A causal theory has *unifying* and *explanatory* power that a non-causal theory such as von Wright's lacks (cf. our earlier citation on the role of the practical syllogism). One aspect that has not yet been sufficiently emphasized in this context is that our causal theory gives an account of the bodily behavior involved in the action in the special sense of giving a causal account of the action *non-functionally* described (cf. point A. above). In this sense, we may say that our causal theory also accounts for the *possibility* of action.

Above we have argued for the presence of a causal component in the context of action-explanation. What has not yet been adequately emphasized is that this causation cannot be, so to speak, "mere" causation. It must be *"purposive"* (or "normative" or "final") causation. We need such purposive causation to account for the idea of intendings (cum beliefs) causing the action in practical inference. The basic idea of purposive causation is that it is something that (in a sense) *preserves intentionality* or *purpose*. Thus, if an agent acts on a certain conduct plan (e.g., the premises of a practical syllogism) the actions caused by his intendings have to satisfy the conduct plan, given that "normal conditions" obtain. If an agent forms an intention to go to the theater at 8 p.m. tonight, this intention should cause him to satisfy the intention (purpose) by causing whatever bodily behavior is needed to satisfy that purpose. Hence, it should not e.g. cause him to go to a restaurant at that time. The conceptual connection between an intention and the means-action, through which its satisfaction is attempted, must be preserved. The dynamic element of control needed here is provided by the mental act of trying (active intending) by means of which the agent exercises his causal power. (A stronger motivation for the need of purposive rather than mere causation will be given in the next chapter.)

We have above, in effect, discussed and argued for our thesis (T2) which turns in order to-explanations into causal because of-explanations by claiming that a causal element must be present in all IT-explanations. The force of the because of-relation is not only causal but purposive-causal, as just pointed out. This is what our thesis (T4) claims, too.

We have not yet discussed reason-explanations (thesis (T3)) in detail. Let us turn to them next.

2.3.   What kind of entities are reasons for acting? We saw above that purposes or aims can be such reasons. Thus, conative attitudes must be closely related to reasons. How can we make this more precise?

In a recent paper Myles Brand distinguishes between two different types of account of the notion of reason: 1) *want* theory and 2) *value* theory (see Brand (1970)). Generally speaking, a person's (intrinsic) wants are reasons according to the first account, whereas a person's (overall) interests and obligations are reasons according to the second account. Brand himself argues that value theory gives a correct account of reasons, and he gives the following formulation for it (Brand (1970), p. 939): "For every person A, every action x, and every r, r is a reason for A's performing x if and only if there is a state of affairs y such that:

(1) $r =$ its being in A's overall interest that y obtain, *or*
             $r =$ its being A's obligation that y obtain; and

(2) A believes that his performing x makes y obtain, *or*
             A believes that his performing x is a causally necessary condition for y's obtaining."

In Brand's definition a reason consists of a *propositional content* operated on by a certain kind of conative attitude, viz. *extrinsic* want in our terminology. But such a restriction to extrinsic wants seems to me incorrect. Surely intrinsic wants can be reasons as well. My reason for buying a detective novel may simply be my intrinsic want to read it, even if it would better serve my overall interests to spend my time writing a philosophical paper instead. Thus, Brand's condition (1) seems too restrictive. In addition to extrinsic wants intrinsic ones must be taken into account as well.

How about condition (2), then? Here, we have the same problem as with the second premise of the practical syllogism (cf. above and Chapter 7) and with defining intending (see Chapter 6). The correct solution to all of these cases, I believe, is to require only action-*generation* without specifying for all cases that the generating action x is either necessary or sufficient for y. What it will be in each particular case depends on the agent and the context. We cannot legislate a priori about the notion of reason more than to require generation in a rather vague and abstract sense.

Could our liberal version of the practical syllogism discussed above and in the preceeding chapter be used to define the notion of reason? It seems that it cannot since its first premise only concerns intentions. Any kind of conative attitudes may serve as reasons. An agent may perform an action because he wants, wishes, hopes, or feels obligated to do it, and each one of these conative attitudes may in principle represent his reason.

It seems to me that beliefs and related doxastic attitudes also can serve as an agent's reasons for doing something. For instance, an agent may be driving his car towards a certain place in a foreign area when he notices a sign informing drivers to this place to make a right turn. Here the agent's noticing, which is a perceptual believing, can be said to be the agent's reason for his turning right.

The notion of reason thus seems to include both conative and doxastic attitudes (together with their contents, of course). Next we have to notice and emphasize the distinction between *a reason* for acting and *the agent's reason* for his acting. The first notion just refers to a possible rationale for a man's acting and can be any kind of *propositional content* an agent may strive for or believe in (cf. the *justification* of action discussed above). But we are here after the explanatory reason, i.e. the reason *for* or *on* or *because of* which the agent acts.

Let us now employ our earlier machinery and use the predicate P for a *conative* attitude in general and B for any *doxastic* attitude. Then I propose the following characterization for an agent's acting for a reason for a case were an agent A performed a singular action x:

(14) ρ was A's *reason* for doing x
        if and only if there are statements q and q' such that:
        (1) a) $\rho = \cdot(Ez)P(A, q, z)\cdot$, where q is an action-description, or
            b) $\rho = \cdot(Eu)B(A, q', u)\cdot$, where q' is either a statement about action generation or an action-description; P is a predicate representing a conative attitude and B a predicate representing a doxastic attitude;
        (2) A believed that there is a singular action y which x indirectly or directly purposively generates such that y satisfies q (in the case of (1)a)) or q' (in the case of (1)b));
        (3) A acted on ρ.

In this characterization clauses (1) and (2) can be taken to define what it is for an agent to have *a* reason for acting (the rationalization or justification aspect). Clause (3) in effect adds that, indeed, ρ was the reason on which or because of which A in fact acted.

The reason is in our definition taken to be a dot-quoted third person proposition to the effect that A wanted (etc.) that q or that he believed (etc.) that something q' (cf. our earlier remarks relating the explanatory questions (5) and (6) to (4)). This is exactly how it should be; e.g. A's reason for raising his arm perhaps was that A *intended to greet me* (q describes A's greeting here). Or his reason may have been that he believed that his raising his arm generates his greeting me here. (This generation is what q' describes.) A's reason for raising his arm might be his (having the) belief that his raising his arm satisfies his want to raise his arm. Therefore q' need not always be concerned with action generation in the strict sense. Statements a) and b) in (1) are just the kind of statements which we have taken to be elements of conduct plan descriptions.

Clause (2) in effect serves to show the relevance of action x to action y. (Note: in the typical case (1)b) trivially guarantees the truth of (2).) This relevance can in simple cases be reduced to direct generation, i.e. $R_i(x, y)$; $i = 1, 2$, according to A's conduct plan. For instance, this suffices for basic actions. But in the case of arbitrary *complex* actions we need *indirect purposive generation*. That is, we require that A believed that x (indirectly) generates y (or the corresponding types, rather) according to his conduct plan into which the reason-statement (i.e. (1) a) or b)) belongs. Anticipating our treatment in Chapter 9, we accordingly explicate (2) for the general deterministic case by

$$(2') \qquad IG^*(x, y).$$

The relation IG^* includes both direct and indirect generation. Furthermore, it implicitly refers to the existence of a conduct plan, which must here be one containing the mentioned reason statement.

Clause (3) finally requires that A acted *on* his reason. This is a central requirement in view of what has been said earlier in this chapter. We accordingly construe (3) in terms of purposive causation: A's reason ρ purposively caused his doing by his bodily behavior whatever is required by x (as an exemplification of some X, according to q). If now, for instance $\rho = \cdot(Eu)W(A, q, u)\cdot$, then the purposive cause, or at least part of the purposive cause, would have been the particular wanting state which was here playing the role of an intending.

More generally, we might want to distinguish between the agent's *full* reason and his *partial* (or, perhaps, *immediate*) reason. (*14*) is explicitly concerned with partial reasons only. The agent's full conduct plan would then give the agent's full reason. (Notice, that in clause (1) of (*14*) we may take the 'or' in the inclusive sense, which allows both the conative and the doxastic attitude to be jointly taken as A's reason.)

On the basis of what we have said about reasons for acting it is clear that reason-explanations, which explicitly refer to the agent's effective conative and doxastic attitudes, are IT-explanations. For reason-explanations are purposive-causal explanations, which involve the agent's activated conduct plans and hence his effective intendings – explicitly or implicitly.

But it also seems plausible to think that all IT-explanations of intentional actions satisfying (*8*) are reason-explanations and that our thesis (T3) holds in both directions. For if explanations of intentional actions in terms of goals, aims, purposes, intentions, wants, obligations, beliefs, emotions, feelings, and sensations indeed do implicitly or explicitly refer to effective intendings, then all these cases represent reason-explanations in the sense of (*14*). We have earlier argued that all these cases do rely on effective intendings. It thus seems that IT-explanations satisfying (*8*) cannot fail to be reason-explanations (in our above sense). In saying this we must assume that IT-explanations do escape the problems due to so-called wayward causal chains. That this is indeed the case will be argued in Chapter 9.

Our theses (T1), (T3), and (T4) make use of the somewhat vague notion of an IT-explanation. We have tried to capture the class of IT-explanations in part through the notion of intentionality and in part through an examination different types of teleological explanation. According to our characterization of intentionality all intentional actions contain an implicit or explicit reference (although perhaps an opaque one) to some intending. We have now been claiming in effect that this intending or some related intending (e.g. one connected by means of the practical syllogism) can be used for *explanation* of the action in question. This explanation is in part related to, and at least it presupposes, "hermeneutic" understanding of behavior in terms of the conceptual framework of agency. In part it is related to the production or coming about of behavior. Here our causal theory refers to *purposive* causation in terms of intendings (plus the relevant beliefs).

In the light of above considerations our theses (T1)–(T4) should seem plausible. We have above concentrated on (T2) and (T4), which are central to our causal theory of action. (T1) and (T3) were taken into account only to give some kind of completeness to our account. They are not so central for our purposes, and we have therefore not spent so much space and energy on them.

As an additional clarification we may finally try to sharpen in exactly which way our account makes action-explanations rely on effective intendings. Consider the following claims:

(i) In the case of every intentional action there is an effective intending (i.e. trying or willing) which purposively causes it. (This effective intending need not be an intending to perform an action of *that* type.)

(ii) In the case of every intentional action we in our IT-explanation of that action explicitly or implicitly refer to a purposive-causal effective intending when explaining it.

(iii) In the case of every intentional action it is the case that every reflective person (having the relevant concepts) upon reflection believes or at least is disposed to believe that there is an effective intending which purposively causes it.

Thesis (i) is a broadly factual thesis which our causal theory can be taken to make *prima facie*, at least if we give our theory a strict interpretation. Whether or not (i) is in fact true is supposed to be decided by future science. Notice, however, that presently we do not need for our action-explanations as much as the *ontological* thesis (i), for (iii) will suffice for our purposes. In fact (iii), which is a, seemingly weaker, thesis about our *conceptual scheme*, is what most of our arguments for the causal character of action-explanation concern in the first place. There is, however, *as such* (and without further assumptions) no clear entailment relation between (i) and (iii) in either direction, although ultimately we would like to see these theses coincide in truth value. In any case, I accept (iii) as a kind of "ideal type" for our present rational reconstruction of explanation.

As to thesis (ii), which is a thesis about *explanation*, our theory of explanation does not (quite) commit us to its truth (cf. here our claims in Chapter 9, too). This is basically because we do not require a *direct* reference (neither explicit nor implicit) to effective intendings when IT-explaining intentional actions. Whether or not an *opaque* or indirect reference is invariably made will depend somewhat on the notion of opaque reference (and redescription of events) in the contexts where we in our explanations opaquely refer to trying-events under, e.g., neurophysiological descriptions. But in any case, as will be seen in next chapter (especially from clause (1) of (9.5)), no explicit or implicit direct reference to intendings need to be invariably required in action-explanations. For we shall construe (complete) explanations as suitable nomological arguments and they need not always straightforwardly speak about intendings.

NOTES

[1] In fact, it is not only because the fear explains a greater variety of actions but also because it here gives a *deeper* explanation of the agent's action of going to the cellar (and the other actions it was just claimed to explain) and because it explains his *activated* intending to save his life (and the relevant activated intendings in other analogous cases).

[2] Kim (1976) has discussed the so called Sturgeon-counterexamples against the explanatory power of the practical syllogism. In these counterexamples one of the two

intentions is subordinate to the other one, and the subordinate intention does not have the same explanatory power as the other one, as with our case. However, in the case of such Sturgeon-counterexamples it is plausible to argue that the subordinate intention is in fact a *conditional* one. In our example it would mean that the first premise of (PS2) would say that A intends to have a sauna bath, given that he is visiting E. Then the second practical syllogism is no longer valid. This is von Wright's (1976a) reply to Sturgeon-examples. However, our above example is not to be read as a Sturgeon-example with an implicit conditional intention.

Kim (1976) tries to answer Sturgeon's counterexamples by restricting the cases of allowed intention-generation. As his conditions are not causal and as they are easily shown not to be exhaustive, we shall not discuss them here.

CHAPTER 9

DEDUCTIVE EXPLANATION AND
PURPOSIVE CAUSATION

1. DEDUCTIVE EXPLANATION

1.1. In the last chapter we started our discussion by claiming that explanations can fruitfully be viewed from the point of view of querying and answering. Explanations can thus be understood as partial or complete answers to appropriate queries (especially why-queries). It is now appropriate to return to this view and to clarify it in some more detail.

If explanations are answers we must clarify the concept of an answer. Answers are best discussed by starting from queries. An explanatory situation will consist of the questioner (explainee), the answer-giver (explainer), the question statement (containing the explanandum as its descriptive component) and the answer-statement (explanation). If we think of questioning and answering in general it becomes obvious that an answer depends on the background knowledge and situational beliefs of the questioner (explainee). In the scientific context this entails that we must think of the explainer and the explainee as members of some scientific community sharing a certain paradigm ('paradigm' taken, e.g., in the sense of Kuhn's (1969) 'constellation of group commitments').

It is now clear that the relation of explanation can be taken to have at least five arguments:

(*1*) $E(E, L, E', T, B)$, i.e., E explains L to E' by T, given B.

Here E represents the explanatory relation itself and it reads 'explains'. E and E' stand for the explainer and the explainee, respectively. L is the explanandum statement and T the explanans statement. B stands both for the general kind of paradigmatic background assumptions involved and also for the situational presuppositions possibly present.

The explanans T is an answer-statement supposed to answer the question

(*2*) $?L$,

where ? typically stands for 'why' (i.e. (*2*) then reads 'Why L?'). Notice that, although '$?L$' is a question statement representing the explanatory query, it need not exhaust the information content of the query. Similarly, the explicit answer-statement T need not exhaust the information content of the corresponding answer (i.e. the speech-act of answering).

We shall below analyze the explanatory relation E by regarding E, E', and B as fixed (as a kind of "boundary conditions"). We thus largely abstract from the "speakers" E and E' and "occasion" represented by B, so to speak; we try to keep the notion of explanation objective in this sense of fixed "pragmatic imbedding". We shall not analyze these fixed elements further here, nor shall we discuss in any detail the price we possibly have to pay for this restriction (see Tuomela (1976b) for a brief discussion on this). Our main concern below will then be the two-place explanatory relation $E(L, T)$ and the notion of answerhood, i.e. the conditions E and T must satisfy in order for T to qualify as an answer to '$?L$'.

We shall below concentrate on why-queries and why-questions. This is justifiable if the central explanatory questions (8.1)–(8.6) discussed in Chapter 8 are answerable by causal because of-statements. What this involves is that questions (8.2)–(8.6) are in a sense subsumable under (8.1). Let us consider this briefly by means of our earlier example of the man A who paid his debt in the bank.

What was the purpose (intention, aim, end, goal) of A's signing the check? His purpose was to pay his debt; he performed the action because of his purpose to pay his debt, we may answer. Even if question (8.2) explicitly asks only for A's purpose, the explanatory query behind this question concerns the effective purpose because of which A acted. The answer to question (8.2) will have to contain, explicitly or implicitly, some kind of causal statement, and hence such an answer will also be an answer to the broader question (8.1). Similarly, behind questions (8.3)–(8.6), we found a causal element which makes all of these questions answerable by means of statements which also qualify as answers to (8.1). Thus, in technical terms, the question

(3)   Why (Ex) (Signed $(A$, the check, $x)$ & In (the bank, $x))$?,

which consists of the explanandum statement operated on by the why-operator, becomes (at least partially) answered by each explanatory answer to (8.2)–(8.6).

Most explanations of actions are in a sense only partial or incomplete. Mentioning merely an effective purpose or want or reason, for instance, may very well suffice for explanation. That is, singular actions are explainable by citing their singular causes or parts of such singular causes.

However, such explanations can be called partial or incomplete in the sense that they can be reconstructed to give complete nomological explanatory arguments, as will be seen.

Speaking in general terms, answers to questions should convey understanding concerning what the question is about. Thus we can put down the following obvious requirement:

(4)   If T is an explanans statement answering 'Why L?', then T must convey understanding concerning L.

(4) can be regarded as a conceptual truth in a double sense: both explanation *per se* and answerhood are conceptually connected to understanding. The obvious trouble with (4) is that the notion of understanding is not one of the clearest.

Whatever we think of understanding we can say at least this. If an explanatory answer gives good *reasons* to think that the explanandum is the case, then the answer conveys understanding concerning the explanandum. In other words, if an explanatory answer serves to give a good *argument* for the explanandum being the case, then the answer conveys understanding and information concerning the explanandum. What is more, we can say, for obvious and well-known reasons, that *nomological* arguments (i.e. arguments making essential reference to laws of nature) are especially good candidates for giving good reasons to think or to expect that the explanandum is the case.

We shall here accept without further argument the view that explanatory answers must convey information and understanding by making implicit or explicit reference to laws (cf. Tuomela (1973a)). An especially typical explanatory pattern is explanation by reference to *causal* laws. An explanatory argument should also satisfy various other desiderata. For instance, an explanation should give a proper amount of non-redundant information, and it should be non-circular. As I have elsewhere discussed these requirements in detail, no further elaboration and motivation will be given here (see Tuomela (1972), (1973a), (1976b)). The end result, which will be presented below in a summary way, is called the *DE-model* of explanation.

We have above discussed what might be called *conclusive* arguments, viz. arguments which give conclusive reasons to think that the explanandum is the case. We can accordingly speak of conclusive explanatory answers. A (complete) conclusive explanatory answer is a nomological argument which gives good and conclusive reasons to think that the explanandum is the case. That the argument is a good one is supposed to entail, among other things, that it conveys understanding concerning the explanandum. Though we shall discuss explanatory arguments which are not conclusive in Chapter 11, below we will restrict ourselves to conclusive ones.

As a summary of the above sketchy discussion, we propose the following analysis of conclusive answerhood in an explanatory context:

(5)   A statement T is a potential *conclusive explanatory answer* to the question 'Why L?' (relative to fixed E, E', and B) if and only if

    (1) T constitutes, or is a part of, the premises of some potential ϵ-argument for L;
    (2) T conveys understanding concerning why L is the case;
    (3) (2) is (at least in part) the case because of (1).

This definition employs the notion of an ϵ-argument. It is our technical concept which is equivalent to the formal notion of DE-explainability, to be discussed below. Condition (1) allows that T is just a singular statement rather than the whole nomological explanans. However, (1) contains the idea that if T is to count as an explanatory answer there must *exist* an ϵ-argument (which is a *nomological* argument) for L.

A few remarks will be made below on clause (2). Clause (3) is needed for the following reasons. Citing T must convey understanding concerning L because T gives a good argument for L (which is in part what (1) says). We wish to exclude cases where T conveys understanding (to the explainee) concerning L through some other mechanism. For instance, the explainee E' could gain *some*, perhaps slight amount of, understanding concerning L due to the fact that the explainer E is an authority in the field and that E shows by means of T that L fits a certain familiar pattern; yet E' might fail to gain this understanding through the fact that $\epsilon(L, T)$. We could put this in different terms by considering the statement:

(U)   If $\epsilon(L, T)$, then T conveys understanding concerning L.

First, (U) need not be invariably true. There might be occasions in which the formal conditions for the relation ϵ, to be discussed below, are true but (2) fails, nevertheless, to be true. The possibility of cases like this obviously depends very much on what (2) is taken to contain. Secondly, and this is practically what was previously said, (U) need not be taken to express a nomological and productive if-then relation. We wish to exclude paradoxes such as gaining understanding concerning L, given the citing of T, in the first place is due to some miracle or some other irrelevant factor. Part of what is involved in T's being an *explanatory* answer (meant in the sense of *scientific* explanation) is just that T conveys understanding via a suitable nomological argument, which is what (1) explicates.

Definition (5) concerns *potential* answers only. Here we refer to the usual potential-actual dichotomy in the context of models of explanation. If the premises and the conclusion of an ϵ-argument are *true (or approximately true)* we call it actual, otherwise it is called potential. Analogously we speak of potential and actual answers.

A few general comments on understanding in the context of action explanations can now be made. What is it to understand that or why something is so and so? To understand, for instance, that light comes on when the flip is switched, or to understand why the light comes on, is related to one's *knowledge* of the causal connections involved. The deeper the physical knowledge one has mastery of the better and deeper is one's understanding. Understanding is then, in general, connected to one's knowing something. Thus, it is also connected to one's *power to reason* in various ways and to perform various linguistic and non-linguistic *actions*. For instance, if I understand why the light came on in a standard case, I can reason about what happens when I flip the switch and about what happens when there is a blackout, and so on. I can also *produce* and *prevent* the light's coming on by appropriately operating with the flip. The above are the basic elements involved in the *causal pattern of understanding*.

While we have emphasized most the causal pattern of action explanations, there is also another pattern involved. This could be called the *hermeneutical pattern of understanding* (cf. question (8.9)). This is what is involved in understanding a text and in understanding intentional aboutness. More specifically, this pattern is involved in understanding behavior as intentional action, when "reading off" intentions from behavior, and when using the practical syllogism as a schema for interpreting behavior (cf. Chapter 7). As we have analyzed action explanations, both patterns of understanding are essentially involved. They are, furthermore, conceptually intertwined because in our analysis it is a conceptual truth that actions are purposively caused by suitable effective intendings (cf. Chapters 7 and 10 on this).

1.2.   Our next task is to clarify the notion of ϵ-argument in (5). In other words, we shall briefly present the so-called *DE-model of deductive* *explanation* developed in Tuomela (1972), (1973a) and (1976b). This model, which applies both to singular and general explananda, was originally developed on the basis of some principles of "informativeness", which fit very well together with what has been said about explanation earlier in this chapter. Very roughly, these principles amount to saying that an explanation should give a minimal amount of relevant information about the explanandum within each "quantificational level" (determined by the quantificational depth of the explanans-law), given, of course, that the explanans logically implies the explanandum.

Before going into specific problems of explanation, I still want to emphasize the nature of the enterprise we are engaged in. We think of singular deductive explanations as rationally reconstructed arguments of a certain kind. As argued above, they are arguments given by (idealized) explainers to (idealized) explainees in answer to the latter's questions concerning the reasons (*"Begründung"*) for the *nomic expectability* of what the (singular) explanandum statement describes.

Some general aspects of this enterprise must be taken up here. First, our analysis will rely on the notion of lawlikeness (or lawhood). Thus, at least some idea of how to pick lawlike statements (or statements with a high degree of lawlikeness) from among universal generalizations will be assumed (cf. Tuomela (1976a, 1976b)).

Secondly, our present enterprise adopts the working assumption that those deductive arguments which are *explanatory* have some common *logical* features over and above the fact that the explanans logically entails the explanandum. What ever other, non-logical, aspects explanations have we abstract from here (see e.g. Tuomela (1973a) concerning them). The DE-model of explanation is supposed to deal with *direct* and *minimal* explanations (although it accepts an idea of quantificational levels of explanation; see Tuomela (1972), pp. 386–387). Roughly speaking 'direct' means explanation by a direct covering under the explanans law. 'Minimal', again, refers to the fact that the covering law is to be minimal in the sense of not containing more general information about the world than the *type* of cases being explained need for their becoming covered (see Tuomela (1976b) for a broader and more detailed account).

Let us now proceed to the logical aspects of deductive explanation. One central requirement for deductive explanations is that they should not be more circular than "necessary". (Of course there must be some content common to the explanans and the explanandum; otherwise there could not be a deductive relationship between them.) As recent discussion on explanation has shown, the following kind of general condition has to be accepted: In an explanation the components of the explanans and the explanandum should be noncomparable. We say that two sentential components or statements P and Q are *noncomparable* exactly when not $\vdash P \rightarrow Q$ and not $\vdash Q \rightarrow P$. (See e.g. Ackermann (1965) and Tuomela (1972) for a discussion leading to the acceptance of this general condition.) Actually, our analysis of noncomparability needs some refinement, mainly because of the vagueness of the notion of a component. To accomplish this we use the notions of a sequence of truth functional components of an explanans and of a set of ultimate sentential conjuncts of an explanans (cf. Ackermann and Stenner (1966)).

A sequence of statemental well-formed formulas $\langle W_1, W_2, \ldots, W_n \rangle$ of a scientific language $\mathscr{L}$ is a *sequence of truth functional components* of an explanans T (theory plus "initial condition" statements) if and only if T may be built up from the sequence by the formation rules of $\mathscr{L}$, such that each member of the sequence is used exactly once in the application of the rules in question. The W_i's are thus to be construed as tokens. The formation rules of $\mathscr{L}$ naturally have to be specified in order to see the exact meaning of the notion of a sequence of truth functional components of a theory finitely axiomatized by a sentence T. A *set of ultimate sentential conjuncts Tc* of a sentence T is any set whose members are the well formed formulas of the longest sequence $\langle W_1, W_2, \ldots, W_n \rangle$ of truth functional components of T such that T and $W_1 \& W_2 \& \ldots \& W_n$ are logically equivalent. (If T is a set of sentences then the set Tc of ultimate conjuncts of T is the union of the sets of ultimate sentential conjuncts of each member of T.) We may notice here that although by definition the Tc-sets of two logically equivalent theories are logically equivalent they need not be the same. Also notice that there are no restrictions which would exclude, e.g., the use of a causal or nomic implication in $\mathscr{L}$. Still we do not below need such an implication in our object language.

Now we are ready to state a better version of the noncomparability requirement for a Tc of a theory T constituting an explanans (cf. Tuomela (1976b)): For any component Tc_i in the *largest* set of truth functional components of T, Tc_i is noncomparable with the explanandum.

In addition to this condition, we require that the explanans and the explanandum of an explanation are consistent, that the explanans logically implies the explanandum, and that the explanans contains some universal laws.

Finally, there is a nontrivial logical condition for our explanation relation, called ϵ, which guarantees that an explanans provides a proper amount of relevant information. This is condition (5) below. (The reader is referred to Tuomela (1972), (1973a) and (1976b) for a discussion of its acceptability.)

Now we can state our model of explanation (termed the DE-model in Tuomela (1973a)). Let T be a statement, Tc a set of ultimate sentential components of T (or actually a conjunction of components in the context where $\epsilon(L, Tc)$), and L a statement to be explained. As we have associated a certain Tc with T, we can use $\epsilon(L, Tc)$ and $\epsilon(L, T)$ interchangeably. We say:

(6) $\epsilon(L, Tc)$ satisfies the logical conditions of adequacy for the potential *deductive explanation* of (singular or general) scientific statements if and only if

(1) $\{L, Tc\}$ is consistent;

(2) $Tc \vdash L$;

(3) Tc contains at least one universal lawlike statement (and no non-lawlike universal statements);

(4) for any Tc_i in the largest set of truth functional components of T, Tc_i is noncomparable with L;

(5) it is not possible, without contradicting any of the previous conditions for explanation, to find sentences $S_i, \ldots, S_r (r \geq 1)$ at least some of which are essentially universal such that for some $Tc_j, \ldots, Tc_n (n \geq 1)$:

$Tc_j \& \ldots \& Tc_n \vdash_p S_i \& \ldots \& S_r$

not $S_i \& \ldots \& S_r \vdash Tc_j \& \ldots \& Tc_n$

$Tc_s \vdash L$,

where Tc_s is the result of the replacement of $Tc_j, \ldots, Tc_n$ by $S_i, \ldots, S_r$ in Tc, and '$\vdash_p$' means 'deducible by means of predicate logic, but not by means of instantiation only, and without increase of quantificational depth'.

Condition (5) is not quite unambiguously formulated as it stands. The reader is referred to Tuomela (1976b) for its clarification (note especially the conditions (V) and (Q) discussed in that paper) and also for an alternative interpretation of '$\vdash_p$'.[1] See Tuomela (1972) for a detailed discussion of the formal properties of the notion of explanation that this model generates. Here it must suffice to make the following general remarks only.

In the above model of explanation, an explanandum may have several explanantia differing in their quantificational strength (depth). On each quantificational level, however, only the weakest explanans-candidate qualifies. Our model thus generates an explanation-tree for each explanandum such that the explanantia in different branches may be incompatible whereas the explanantia within the same branches are compatible and increasingly stronger.

More exactly, the DE-model has the following central logical properties (see Tuomela (1972)):

(7) (a) $\epsilon(L, Tc)$ is not reflexive.

(b) $\epsilon(L, Tc)$ is not symmetric.

(c) $\epsilon(L, Tc)$ is not transitive.

(d) If $\epsilon(L, Tc)$ and if, for some Tc', $\vdash Tc' \rightarrow Tc$ (assuming not $\vdash_p Tc' \rightarrow Tc$), then $\epsilon(L, Tc')$, provided every Tc'_i in Tc' is noncomparable with L.

(e) $\epsilon(L, Tc)$ is not invariant with respect to the substitution of either materially or logically equivalent explanantia or explananda.

(f) If $\epsilon(L, Tc)$ and if, for some T' such that $\vdash T \equiv T'$, T and T' have identical sets of ultimate sentential components (i.e. $Tc = Tc'$), then $\epsilon(L, Tc')$.

(g) If $\epsilon(L, Tc)$ and for some L', $\vdash L \equiv L'$, then $\epsilon(L', Tc)$ provided that, for all Tc_i in Tc, Tc_i and L' are noncomparable.

(h) If $\epsilon(L, Tc)$ and $\epsilon(L, Tc')$, then it is possible that Tc and Tc' (and hence the corresponding theories T and T') are logically incompatible.

What especially interests us is property (e). The lack of linguistic invariance exhibited by it shows or expresses the fact that explanation is a pragmatic notion: How you *state* your deductive argument may make a great explanatory difference.

There are various *epistemic* conditions which can be imposed on the explanans and the explanandum in the case of *actual* explanation (viz. when the explanans and explanandum are (approximately) true). Here the explanandum must be known, or at least believed, to be true. We may also want to ideally require that the explanans is believed to be true or at least approximately true. In any case, we must require that the explanans is not believed to be "completely" false. We shall not here discuss these epistemic requirements nor any further methodological or metaphysical components of explanation (see, e.g., Tuomela (1973a) for a brief discussion). When we speak of DE-explanation or of an ϵ-argument we below, for simplicity, mean an (actual or potential) argument satisfying the above conditions (1)–(5). The conjunction of these conditions can thus here be taken as necessary and sufficient for the truth of '$\epsilon(L, Tc)$'. Nothing else needs to be required for our present purposes.

We have now given our clarification of what a conclusive explanatory query involves, relative to a fixed "pragmatic embedding". Our basic thesis concerning action-explanations, here, is this: *Conclusive IT-explanations are complete or incomplete ϵ-arguments which convey suitable understanding concerning why the explanandum is the case.* A more informative discussion of incomplete (especially singular) explanations requires a treatment of singular causation, to which we now turn.

2. Purposive causation

2.1.   One central factor to be clarified in discussing causal accounts of action is naturally causation itself. It is rather unfortunate that most causal theorists have neglected a detailed treatment of causation altogether (cf. Goldman (1970)). I have elsewhere developed an account of singular causation by reference to explanatory backing laws (see Tuomela (1976a)). In Tuomela (1975) it was applied to problems of action-causation to yield a concept of directed or *purposive causation*.

Below I will briefly summarize my theory of ordinary event-causation. This sketch is complemented by an account of the important notion of purposive causation which we will need later.

The type of backing law account I have developed starts with the idea expressed in the following partial analysis or "definition":

(8)   Event a is a deterministic *cause* of event b only if there are singular statements D_1 and D_2 such that D_1 and D_2 truly describe a and b respectively, and there is a true causal law (or theory) S which jointly with D_1 deductively explains D_2 (but D_1 alone does not explain D_2).

Given our above account of deductive explanation, we immediately get much informative content in (8). We introduce a two-place predicate '$C(a, b)$' which reads 'singular event a is a (direct) deterministic cause of singular event b'. (Note that we are discussing only *contributory* event causes here.) We then give the following truth conditions (in a wide "assertability"-sense of truth) for $C(a, b)$, assuming that a and b have occurred:

(9)   The statement '$C(a, b)$' is true if and only if there are true singular statements D_1 and D_2 and a true causal law (or theory) S (possibly also including singular context-describing statements) such that D_1 and D_2 truly describe a and b, respectively, and the following conditions are satisfied:

(1) S jointly with D_1 gives an actual explanation of D_2 in the sense of the DE-model, viz. there is a Tc for $T = S \& D_1$ such that $\epsilon(D_2, Tc)$ in the sense of the DE-model;

(2) a is causally prior to b;

(3) a is spatiotemporally contiguous with b.

Our earlier discussion has clarified the content of clause (1). Clauses (2) and (3) cannot be discussed in any detail here (cf. Tuomela (1976a)).

What kind of formal properties does $C(-, -)$ have? Solely on the basis of the logical properties of ϵ as well as (2) it is easy to list a number of them. For instance, it is asymmetric and irreflexive, and it fails to be transitive. One feature to be especially noticed, is that the truth of '$C(a, b)$' is invariant with respect to the substitution of identicals in its argument places. In this sense '$C(a, b)$' is extensional.

On the basis of the definition of $C(a, b)$ we immediately get a conditional:

(10)   '$D_1 \vartriangleright\!\!\rightarrow D_2$' is true if and only if there is a true theory S such that $\epsilon(D_2, S \& D_1)$.

If S is a causal theory and if the conditions (2) and (3) of (9) are satisfied, then $\vartriangleright\!\!\rightarrow$ expresses a *causal* conditional: 'If an event a-under-D_1 would occur, then so would an event b-under-D_2'; or simply 'a-under-D_1 caused b-under-D_2', for actually existing a and b (as in fact assumed).

Truth in this definition may be considered relative to a "model" or "world" constituted by the explanatory situation in question, viz. typically (a part of) the *actual* world.

Are our causes in some sense *sufficient* and/or *necessary* for their effects? We immediately notice that if $C(a, b)$ then $D_1 \vartriangleright\!\!\rightarrow D_2$, in our above terminology. As D_1 describes a then it follows that a cause is sufficient in the strong sense of the truth of the above conditional. But, as we allowed that the backing statement S may contain singular statements, D_1 may be only a *part* of a sufficient singular condition for D_2.

However, it is a *necessary part* of such a condition because of the properties of ϵ. Thus, if $S = L \,\&\, D_3$, where L is a law and D_3 a singular description, then '$D_1 \,\&\, D_3 \rhd\!\!\rightarrow D_2$' is true with respect to the backing law L. Furthermore, it follows from the properties of the explanatory relation ϵ that $D_1 \,\&\, D_3$ is a *minimal* sufficient, and even $\rhd\!\!\rightarrow$ -sufficient, condition for D_2 with respect to L.

More generally, L might have an antecedent which consists of not only one such minimal sufficient condition but instead a *disjunction* of disjoint minimal sufficient conditions. Then we could say that a singular cause a under the description D_1 is a necessary but insufficient part of an unnecessary minimal sufficient condition of b under D_2. This is essentially what is known as an INUS-condition (cf. Mackie (1965) and the analysis of it in Tuomela (1976a)). Continuing the above example, we assume that '$D_1 \,\&\, D_3$' is true and thus the only exemplified sufficient condition of D_2. In this sense a-under-D_1 can here be said to be an *ex post facto necessary* condition, but this only holds if *overdetermination* is barred (cf. Tuomela (1976a)). (To require $\sim D_1 \rhd\!\!\rightarrow \sim D_2$ would be a stronger possible alternative here, too.)

There is no good reason to think that the form of a causal law (theory) has to be fixed *a priori* (see Tuomela (1976a)). Therefore, not much more can be said about deterministic causes than that they are (at best) necessary parts of minimal sufficient conditions, perhaps relative to a given "causal field" of possible causal factors (cf. Scriven (1972) for a related discussion and similar conclusions).

Another thing to be said about our notion of singular cause is that it gives only *a* cause and not *the* cause. What is called *the* cause depends on the various pragmatic aspects of the situation (cf. Collingwood's example cited earlier in Chapter 8). We shall below return to this problem.

2.2. Our above notion of "mere" causation is not yet what our causal theory of action needs. As we have repeatedly emphasized, we need *purposive* (or final) causation, which is something that preserves intentionality or purpose. This invariance property will be achieved through a relativization to an appropriate conduct plan description.

We shall not give more arguments here for the need of purposive contra mere causation (cf. Chapter 8). Instead we will proceed "deductively" and present our technical account of purposive causation. The discussion following it will, however, give further motivation and support for our analysis.

Let us thus go to the heart of the matter and formulate a tentative characterization of purposive causation. We consider an action token u performed by an agent A. (In our later terminology of Chapter 10 we are considering an action token$_1$.) We assume that u has the internal structure of a sequence of events $\langle t, \ldots, b, \ldots, r \rangle$. Each event in this sequence is assumed to be spatiotemporally contiguous at least with its predecessor and its successor event. Presystematically speaking, t is going to represent the event or part of the event activating A's conduct plan. In the case of intentional action it will be a trying ($=$ effective intending, willing) cum (possibly) the relevant believing. The event b will represent the maximal overt bodily component in the action u, and r will be the result of the action. We recall that the result of an action is, roughly, the overt event intrinsically connected to the action in the way a window's becoming open is connected to the action of opening the window. What (if anything) is needed between t, b, and r is up to future scientists to find out. From the philosopher's conceptual point of view, it then suffices to represent u by $\langle t, b, r \rangle$. Furthermore, we shall below often simplify matters and speak of the composite event u_o which is understood to have been obtained by suitably composing b and r (e.g., in the sense of the calculus of individuals).

Let $u = \langle t, u_o \rangle$. Then we introduce '$C^*(t, u_o)$' as a two-place predicate to be read 't is a direct purposive cause of u_o'. We "define", relative to an agent A:

(11)   '$C^*(t, u_o)$' is true if and only if

    (a) '$C(t, u_o)$' is true; and

    (b) there is a conduct plan description K such that in K t is truly described as A's trying (willing) by his bodily behavior to exemplify the action which $u(= \langle t, u_o \rangle)$ is truly describable as tokening according to K, and u_o is represented in K as a maximal overt component-event of u; t and u_o are assumed to belong to the domain of a conduct plan structure which satisfies$_2$ K.

Regarding the notions that occur in (11), we shall later try to better clarify what a maximal overt part of an action is. Thus, we shall discuss both the nature of b and r. Notice that at least u_o need not belong to the domain of the structure as an event separate from $u = \langle t, u_o \rangle$, though it may.

The notion of satisfaction$_2$ means, roughly, ordinary model-theoretic satisfaction *and* a requirement that A has a correct "wh-belief" of the overt "external" entities in the domain of the conduct plan structure satisfying K (we now use K as a constant) in the model-theoretic sense. Here wh-belief means belief about what (or which or who, as the case may be) the entity in question is. Speaking in other terms, the conduct plan K must in a stronger sense than the ordinary model-theoretic one be *about* the objects in the domain of its model. We shall below return to this important problem.

What our above definition comes to saying is that the trying t purposively causes u_o only if t under the description D_1 that it has in K causes (in the ordinary sense of event-causation) the event u_o as described by a statement (D_2) in K. Notice that the backing law justifying the truth of '$C(t, u_o)$' of course can employ quite different predicates from those appearing in D_1 and D_2. The essential content of the above definition — over and above the causation requirement — is that t can indeed be truly described as a trying (activated intending) to bring about an action of a certain type such that u_o is describable as an overt part of a token of that action type.

Is the above definition of $C^*(-, -)$ intuitively justifiable? Let us first consider whether the definiens or analysans gives a necessary condition for purposive causation. I can see no serious objections to that, in view of what has been said earlier. Let us briefly illustrate this by a simple example. If an agent intends to ventilate the room and considers it necessary to open the window to do that, then he proceeds to open the window, given that normal conditions obtain. In our analysis purposive causation gets involved as follows. The agent's intending to ventilate the room becomes effective and "develops", through intention transferral, to his trying (= effective intending) to (now) do, by his bodily behavior, whatever is required to open the window. This takes place through a purposive causal process. The agent's trying purposively causes the required bodily behavior. Now clearly we can analyze the situation by saying that there is a practical syllogism (and hence a conduct plan) involved, as the very wording of the example indicates. The causal process takes place in accordance with this practical syllogism, and this can be taken to show that the definiens of (11) is satisfied.

How about sufficiency, then? Could it be possible that an event t causes another event u_o such that these events somehow accidentally or spuriously satisfy an otherwise appropriate conduct plan K so that the agent cannot be said to *intentionally* act *on K*? Or to put it slightly differently, assuming that any intentional action involves acting on *some* suitable conduct plan, could it be that t and u_0 spuriously satisfy a conduct plan K' while they "genuinely" satisfy another conduct plan K? If so, we might choose K' in our definition and incorrectly infer that '$C^*(t, u_0)$' is true.

I am not quite sure what to say about the above kind of situation, partly because the intuitive notion of 'acting on a conduct plan' is not very clear. But perhaps one could think of Freudian cases where a person might in a sense be said to act on two *incompatible* conduct plans — on one which is his own rationalization and on another true one which is the psychoanalyst's construction. Or perhaps some schizophrenic person with a "split" mind, or a "split-brain" patient, could be said to act on two incompatible conduct plans in a still more perspicuous and convincing sense.

It seems that the above kind of problems can be handled. Let us thus consider what we call the Principle of Expression. It says, roughly, the following two things: (a) all singular mental events (or states) have some descriptive physical properties, and (b) if two singular events or states agree in all of their physical properties they cannot disagree in any of their mental properties. This Principle of Expression formulates a *weak version of materialism*, and it can be regarded as a broad kind of factual hypothesis. We shall below, without further discussion, tentatively accept this principle as a working hypothesis.

Now consider the case exemplified by our examples. If a trying t satisfies and activates a conduct plan K (let us write it as $K(t)$, viz. as if K were a predicate), then, in this case $\sim K'(t)$ is true for the other conduct plan K'. For the trying event concerned with K' the inverse statement holds. We may assume here that we initially have the following definite descriptions for t and t' (activated intendings *cum* believings such as in our (PS1) and (PS2) of *(8.13)*):

$$t = (\gamma x)(I(A, p, x) \ \& \ B(A, q, x))$$
$$t' = (\gamma x)(I(A, p', x) \ \& \ B(A, q', x)).$$

Our assumptions made above for this case now entail that t and t' are not identical; indeed, they disagree with respect to at least the complex mental predicates K and K'. Our Principle of Expression now entails that t and t' must also differ with respect to some of their physical properties (presumably brain-properties).

Concerning the Freudian case (and assuming that we are dealing with intentional action) it seems that — barring causal overdetermination — neurophysiological research would now, given the truth of Freud's theory, have to confirm our suspicion that the patient acts only on the plan attributed to him by the psychoanalyst, and the results would contradict condition (a) of our definition of $C^*(-, -)$. Thus, a closer investigation would probably show that after all $C(t, u_o)$ does not hold. This is our main solution to the problem of spurious purposive causation. To repeat, we think that cases of spurious purposive causation simply fail to be cases of mere event-causation.

In the schizophrenic case the analogous thing might happen. But investigation might also show that both $C(t, u_o)$ and $C(t', u_o)$ hold. Especially in the case of split-brain patients this seems to be an experimentally evidenced possibility. (The same action might require the joint working of both hemispheres. In the case of split-brain patients, this working would be co-operative only accidentally, but in those accidental cases we would have incompatible trying events t and t'.) Then, we simply conclude that both $C^*(t, u_o)$ and $C^*(t', u_o)$ hold, i.e. the agent indeed simultaneously acted both on K and on K' (assumed to be incompatible). This situation is of course not at all excluded by our earlier characterization of purposive causation. So far, then, our (11) seems adequate.

There is still another problem left. It concerns the existence of two *compatible* conduct plans K and K' such that both satisfy our definition of $C^*(t, u_o)$ although now only one of them should intuitively satisfy it. To see what is involved here, we consider our example (8.13) with the two practical syllogisms (PS1) and (PS2), where a causal connection was claimed to be present only in the case of (PS1).

According to our treatment of the Freudian case, we might simply try to say that (PS2) does not satisfy the requirement demanding the existence of a causal connection (i.e., $C(t, u_o)$ simply does not hold). This answer, of course, works equally well here as it does in the Freudian case. But now I am raising the problem about how to identify and classify the trying-events. In the case of incompatible conduct plans K and K', we were naturally led to infer that t and t' are not identical and that consequently there was something in the brain to be found to distinguish them. But in the present case we might come to think that there is only one mental event of trying (to visit E), call it t, which activates both K and K'. That is, we cannot find two incompatible descriptions of the tryings in (PS1) and (PS2), as was the case earlier. Any trying t satisfying (PS1) satisfies (PS2), and vice versa. (This is at least a conceptual possibility; t might, e.g., be a "total" brain state or event regarding its neurophysiological nature.) Then we would have $C(t, u_o)$ and the satisfaction of the requirements of conformity to conduct plan both for K (the premises of (PS1)) and K' (the premises of (PS2)). What should we do in this case?

It seems that we have to use something like *maximal* conduct plans in our definition of $C^*(t, u_o)$ for this case, in order to get hold of the right conduct plan. As we know, there is (formally) a common extension K'' of any two compatible conduct plans K and K'. Should we then not look for such a maximal consistent extension or for a theory which gives a complete description of the situation? This is practically what we shall do. We recall that in the present case there is one trying event t and one overt action u_o. Furthermore, lots of other events and states may be involved in a particular situation where an agent sets himself to do something. Let us now imagine the following.

Assume that we are in the possession of a finite language of thought (cf. Chapters 3 and 6). This language will employ all the concepts in the "framework of agency", and, in order to be learnable, it must be finitary in character (as, e.g., Davidson has forcefully argued). Now, we interpret all the extralogical constants of this (first-order) language into a domain of the singular actual events, states, etc., and objects present in *this* behavior situation. We call the resulting Tarskian set-theoretical structure $\mathscr{M}$. Let us assume that *in principle* we can, for all cases, determine $\mathscr{M}$. (Extremely difficult epistemological and methodological problems will normally be involved in that.) Given $\mathscr{M}$, we can discuss the *theory of $\mathscr{M}$* (i.e. $Th(\mathscr{M})$) with respect to the full language of the framework of agency (plus whatever else is required). $Th(\mathscr{M})$ is, as usual, defined as the set of all statements true of $\mathscr{M}$.[2] Now, as is well known, any theory T which is satisfied by $\mathscr{M}$ and which is *complete* (in the logical sense) is equivalent to $Th(\mathscr{M})$. What we shall do then is to modify our earlier definition by the requirement that the conduct plan K be complete:

(*12*)   '$C^*(t, u_o)$' is true (with respect to the behavior situation conceptualized by $\mathscr{M}$) if and only if

    (a) '$C(t, u_o)$' is true (with respect to $\mathscr{M}$); and

    (b) there is a complete conduct plan description K satisfied[2] by $\mathscr{M}$ such that in K t is truly described as A's trying by means of his bodily behavior to exemplify the action which $u(= \langle t, u_o \rangle)$ is truly describable as tokening according to K, and u_o is represented in K as a maximal overt component-event of u; t and u_o belong to the domain of $\mathscr{M}$.

It should be emphasized, for further reference, that this definition does not rely on any (exact) systematic notion of an action *type*.

Several derivative causal notions can now be defined on the basis of $C^*(-, -)$. We shall below define the notion of indirect "extra-linguistic" purposive causation and two "linguistic" notions corresponding to the extra-linguistic notions.

We thus start by defining indirect ordinary event-causation. Let $e_1, e_2, \ldots, e_n$ be singular events. Now we say that e_1 is an *indirect* deterministic cause of e_n (we write $IC(e_1, e_n)$) just in case e_1 is causally linked to e_n through a chain of direct causal relationships:

(13) $IC(e_1, e_n)$ if and only if $C(e_1, e_2) \& C(e_2, e_3) \& \ldots \&$
         $C(e_{n-1}, e_n)$ for some singular events $e_2, e_3, \ldots, e_{n-1}$.

We now go on to characterize indirect purposive causation. In principle we can define such a notion for any two mental events, and not only for a mental event and an action. (An analogous remark of course can be made for direct purposive causation.) But here we are mainly interested in how far an agent's causal powers extend. Thus, we want to investigate, for instance, whether effective intendings can be said to indirectly purposively cause or generate all of the agent's intentional actions (or rather, to use our technical terminology, their maximal overt parts).

We now let a singular action u consist of a sequence $\langle t, b, r \rangle$ (short for $\langle t, \ldots, b, \ldots, r \rangle$), where t represents a trying event and b represents a bodily event which in the present case will be describable as an action (cf. Chapter 10). The event r is the event of the coming about of the result state of the full action u. The event b will below be assumed to indirectly generate the event r. Thus, the composition (sum) of b and r (which we call u_o) is describable as an overt part of a generated action.

Now we get in an obvious way a notion of indirect purposive causation:

(14)   '$IC^*(t, r)$' is true (with respect to $\mathcal{M}$) if and only if
         (a) '$IC(t, r)$' is true (with respect to $\mathcal{M}$) such that t corresponds to e_1, b to e_2, and r to e_n;
         (b) there is a complete conduct plan description K satisfied$_2$ by $\mathcal{M}$ such that in K t is truly described as A's trying to do by his bodily behavior what u tokens according to K, and u_o (composed of b and r) is truly described as a maximal overt part of an action which is (directly or indirectly)

factually, and thus not merely conceptually, generated by some of A's actions in this behavior situation; t and u_0 are assumed to belong to the domain of $\mathcal{M}$.

Our definitions (12) and (14) do not yet cover all actions, for there are other more complex actions which will be discussed later. The notions of factual and conceptual generation R_1 and R_2 will be defined in exact terms below.

It should be noticed that in (14) we, in effect, require that u_0 occur according to the agent's beliefs (or other relevant cognitions), that is, we now compare u_0 with the contents of A's beliefs. Thus, technically speaking, we require that u_0 satisfies the proposition named by the singular term in the second argument of one of A's beliefs.

We next define two notions of linguistic purposive causation, both for the cases of direct and indirect causation. They are defined as follows:

(15)   '$D_1 \rhd\!\!\rightarrow D_2$' is true (with respect to $\mathcal{M}$) if and only if '$C^*(t, u_0)$' is true (with respect to $\mathcal{M}$) such that D_1 truly describes t and D_2 truly describes u_0 as required by clause (b) of definition (12).

(16)   '$D_1 \rhd\!\!\!\rhd\!\!\rightarrow D_n$' is true (with respect to $\mathcal{M}$) if and only if '$IC^*(t, r)$' is true (with respect to $\mathcal{M}$) such that D_1 truly describes t and D_n truly describes r as required by clause (b) of definition (14).

Our conditional $\rhd\!\!\!\rhd\!\!\rightarrow$ could also have been defined by means of the conditional $\rhd\!\!\rightarrow$. For $\rhd\!\!\rightarrow$ is essentially $\rhd\!\!\rightarrow$ together with the requirements for employing a *causal* backing theory and relativization to a conduct plan as required by clause (b) of our definition (12). In addition, the truth is explicitly relativized to a given model $\mathcal{M}$, assumed to represent an actual behavior situation. (Notice that in Tuomela (1976a) and in (10) above the relativization lies implicitly in the notion of an explanatory situation.) All the same, $\rhd\!\!\rightarrow$ entails $\rhd\!\!\rightarrow$, which again entails, and is indeed much stronger than, Lewis' variably strict conditional $\Box\!\!\rightarrow$, as shown in Tuomela (1976a).

It should be emphasized here again that the backing theory underlying our technical notions $C(-, -)$, $C^*(-, -)$, $\rhd\!\!\rightarrow$, $\rhd\!\!\rightarrow$, and $\rhd\!\!\!\rhd\!\!\rightarrow$ need not be in the vocabulary of the conduct plan K, nor need it be a psychological theory at all. Our account thus does not require that there even be any specific factual psychological laws at all. As will later be argued, ▷→ especially seems to be a good candidate for explicating singular causal explanations in, e.g., psychology and history. (It is so especially, if we do not require that the chain start with a mental action of trying, but allow for any kind of causal sequences with mental actions occurring some place in the sequence.)

Given our technical apparatus above, we can now formally define our notions of action generation which were discussed at length in Chapter 7. Our formal explications add nothing to the substantive content given to the concepts of action generation in that chapter. Rather, they serve to make the formal structure of these notions clearer and thus to save them from, e.g., the paradox that any action generates every other action; a paradox which mars Goldman's treatment (see Tuomela (1974b)).

Let us start our explication with the notion of *factual* generation '$R_1(e_1, e_2)$' which reads 'singular event e_1 factually generates singular event e_2':

(17)   '$R_1(e_1, e_2)$' is true (in a given situation $\mathcal{M}$) if and only if (with respect to $\mathcal{M}$) there are true descriptions D_1 of e_1 and D_2 of e_2 and a suitable true factual theory S such that $\epsilon(D_2, D_1 \& S)$.

What a *suitable* theory (law) is we shall not discuss any more here (a *causal* theory or conjunction of causal theories is a paradigm example (see Chapter 7)). We recall that it is required that S (semantically) entails that the relation R_1 is irreflexive and asymmetric. In the case where the backing theory S is true in our actual "world" and $(Ex)(x = e_1)$ and $(Ey)(y = e_2)$ we speak of *actual* R_1-generation, otherwise only of possible R_1-generation. (Obviously, we may also speak about direct purposive generation R^*, which is defined on analogy with direct purposive causation C^*.)

Completely analogously with our definition (13) of indirect causation (IC), we can define a notion of *indirect generation* (IG) by using R_1 instead of C in the definiens. A notion of *indirect purposive generation* (IG*) is obtained from our definition of indirect purposive causation (14) by using relation IG instead of IC in its clause (a). Since we will later often return to the relation IG* let us give its definition explicitly. (The formal definition of IG is so obvious on the basis of what we have just said that we will not state it explicitly.) In our definition of IG* below factual generation again refers to R_1 (as defined by (17)) and conceptual generation to R_2

(to be defined formally by (*19*) below):

(*18*) '$IG^*(t, r)$' is true with respect to $\mathcal{M}$ if and only if

    (a) '$IG(t, r)$' is true (with respect to $\mathcal{M}$) such that t corresponds to e_1, b to e_2, and r to e_n;

    (b) there is a complete conduct plan description K satisfied$_2$ by $\mathcal{M}$ such that in K t is truly described as A's trying to do by his bodily behavior what u tokens according to K, and u_o (composed of b and r) is truly described as a maximal overt part of an action which is (directly or indirectly) factually, but not merely conceptually, generated by some of A's actions in this behavior situation $\mathcal{M}$; t and u_o are assumed to belong to the domain of $\mathcal{M}$.

Let us now go on to conceptual generation. Our concept of conceptual generation R_2 is defined to hold between an event u and its two different but conceptually (semantically) related descriptions D_1 and D_2: '$R_2(u, D_1, D_2)$' may be read "u-under-D_1 conceptually generates u-under-D_2". (If D_1 describes u as an X and D_2 as a Y we might also speak, in terms of intensional structured events: u *as an* X'ing conceptually generating u *as a* Y'ing.) The bare bones of the relation R_2 then are as follows:

(*19*) '$R_2(u, D_1, D_2)$' is true (in a given situation $\mathcal{M}$) if and only if (with respect to $\mathcal{M}$) there is a suitable true (assertable) meaning postulate (or conjunction of meaning postulates) S such that $\epsilon(D_2, D_1 \, \& \, S)$, where D_1 and D_2 truly describe u.

Here the "suitable meaning postulates" are supposed to cover a wide variety of semantic postulates and rules, norms, etc. (see Chapter 7 for our clarification). $R_2(u, D_1, D_2)$ is assumed to be irreflexive and asymmetric with respect to the second and third arguments (due to S).

We can define a conditional $\bigodot_{\overrightarrow{2}}$, representing conceptual action generation, as follows: '$D_1 \bigodot_{\overrightarrow{2}} D_2$' is true if and only if '$R_2(u, D_1, D_2)$' is true. ($\bigodot_{\overrightarrow{1}}$ is of course analogously definable.) The conditional $\bigodot_{\overrightarrow{2}}$ corresponding to purposive conceptual generation is defined on the basis of $\bigodot_{\overrightarrow{2}}$ completely analogously with the suggested definition of $\rhd\!\!\rightarrow$ on the basis of $\rhd\!\!\rightarrow$ (cf. our above treatment).

2.3. In recent philosophical literature a need for something like a notion of purposive causation has often been felt (see, e.g., Chisholm (1966),

Goldman (1970), and Davidson (1973)). We shall now illustrate how our
approach can handle some problematic examples. First we consider an
example by Chisholm:

"Suppose, for example: (i) a certain man desires to inherit a fortune;
(ii) he believes that, if he kills his uncle, then he will inherit a fortune;
(iii) this belief and this desire agitate him so severely that he drives
excessively fast, with the result that he accidentally runs over and kills a
pedestrian who, unknown to the nephew, was none other than the uncle."
(Chisholm (1966), p. 30.)

Can our account explain why the man's killing his uncle was *not*
intentional? Well, it seems obvious that the man did not kill the uncle
according to and because of his conduct plan: there is no operative
complete conduct plan K such that we would have direct purposive
causation or indirect purposive generation of the action token of killing.
Still we may analyze the situation so that the killing is "merely" caused
by the agent's desire and belief.

Next we consider Goldman's example, modified by using 'intends'
instead of 'wants' (Goldman (1970), pp. 60–61). Suppose an agent A is at a
dinner party. A intends to offend the host and he believes that if he grimaces
as he eats the soup the host will be offended. But Oscar, A's practical·
joker friend, knows about A's intention and A's belief, but he is determined
to prevent A from intentionally offending his host. So Oscar puts the foul-
tasting stuff in the soup that makes him grimace. Here we may seem to
have a causal connection between A's conduct plan (intention and belief)
and A's grimacing. For his having that conduct plan caused Oscar to put
the stuff in the soup and this caused A's grimacing. Still, A's grimacing is
not intentional. A's grimacing may seem to correspond to his conduct
plan (contra the uncle-killing example) because the basic act in his plan
was grimacing itself.

But we must ask: What is A's operative conduct plan? Perhaps it was
only insufficiently and misleadingly characterized here. We seem to be
able to say that the causality in question does not preserve intentionality
(it is only "mere" causation $C(-, -)$), for A's grimacing was not caused
by his trying to grimace. We may even suppose that an inner trying-to-
grimace occurred. But still we say that it did not properly cause the
grimacing, for something abnormal happened and prevented the normal
causal chain from occurring. Thus, *ex post facto* we seem to be able to say
that the causation did not take place as intended.

Perhaps we then have to admit afterwards that the conduct plan indeed was incompletely characterized, though it seemed adequate *ex ante*. Thus, there was found to be an *unintended* model for the conduct plan, which was not, and perhaps could not be, anticipated. But why cannot we just admit that a conduct plan and a prediction concerning our actual world may be relative to some *ex ante* partly unspecified or even unspecifiable implicit beliefs and normal condition assumptions. Occasionally these beliefs prove to be incorrect and the normal conditions cease to hold; this is what happened in Goldman's grimacing example. Thus we can claim that, after all, A's grimacing did *not* take place according to and because of his conduct plan. Therefore it was not intentional nor purposively caused (or generated) either.

Of the above examples Chisholm's deals with so-called *internal* and Goldman's with *external wayward* causal chains. Let us once again illustrate, in a slightly different way, the problem of internal causal chains, which Davidson thinks is "insurmountable". We consider Davidson's (1973) example: "A climber might want to rid himself of the weight and danger of holding another man on a rope, and he might know that by loosening his hold on the rope he could rid himself of the weight and danger. This belief and want might so unnerve him as to cause him to loosen his hold, and yet it might be the case that he never *chose* to loosen his hold, nor did he do it intentionally." (Davidson (1973), pp. 153–154.)

Davidson also remarks that it will not help to add that the belief and the want must combine to cause him to want to loosen his hold, for there will remain the two problems of how the belief and the want caused the second want, and how wanting to loosen his hold caused him to loosen his hold.

In analyzing this example we must first ask whether the agent's loosening his hold really was an action of his at all. I think we can understand and explicate the situation so that it is an action token$_1$ in our later sense of (*10.5*). But it is clearly not an intentional action as it was not purposively caused by a trying of the agent, although it was non-purposively caused by his (mere) want (cum belief). The agent did not even form an intention to perform the action, and still less, of course, could he have had an effective intending. Now, since trying is missing, so is purposive causation and so is intentionality (cf. (*10.14*)). Naturally, the required purposive causal intention-transferral from the more general intention to get rid of the weight and danger to the intention to loosen his hold is also missing, as the appropriate intendings did not even exist.

Perhaps Davidson would regard my solution as verbal and somehow trivial. But I would like to rebut that for the following reasons: (1) Trying and intending are clearly conceptually different from mere wanting. (2) Since they are probably ontologically genuine entities, neurophysiologists and psychologists can, in principle, find out more about their difference in the future. (3) One has the kind of self-knowledge of one's (controlled) trying which is missing in the case of the (more or less uncontrolled) wanting in the example. (4) My type of solution gives almost all a philosopher can be expected to do; the rest is up to scientists.

Thus it seems that within our analysis it is possible to exclude "weird" and "wild" causal connections by indicating that such connections are not cases of direct purposive causation or of indirect purposive generation.[3] Of course a *strict* proof cannot be obtained due to the present state of the sciences of psychology and neurophysiology, for we lack much scientific information about the nature of conduct plans and of acting on them. Such information would be needed particularly for the methodological task of ascertaining which conduct plan an agent in fact acts on.

One problem related to our discussion above is how to handle *mistakes* in acting. By mistakes we here primarily mean erroneous actions, which do not occur according to the agent's conduct plan. Thus, consider an agent who intends to illuminate the room by flipping the light switch. Suppose, however, that A has confused the switch and the doorknob. He thinks he tries to flip the switch but ends up turning the doorknob. What kind of purposive causation, if any, is involved here?

One thing to be said first is that even if his action is not intentional when described as a turning of the doorknob it is still intentional under the description 'an action believed by A to be a flipping of the switch'. In giving this description we in fact bring out one of the tacit presuppositions included in the agent's conduct plan, namely that those of his beliefs which he acts on are correct (true). Perhaps we should also say, in our example, that his trying to flip the switch was incorrectly described as such. Rather, it should be described as a trying to flip what A believed to be the switch. We need the analogous requirement that A correctly believed that the doorknob is identical with the switch, i.e. that A correctly believed that the doorknob was the switch. But of course such a *correct* belief is impossible to have as the doorknob is nonidentical with the switch; hence he did not intentionally flip the switch. Thus, A's trying to flip (what he believed to be) the switch purposively caused his flipping what he believed to be the switch, but it did not purposively cause his turning the doorknob nor, of course, his flipping the switch.

Our account of purposive causation and generation guarantees a certain coherence—coherence in terms of the agent's way of describing the situation—between the cognitive subjective elements and acting. If we want to have "objectively correct" action descriptions as the consequents of our (linguistic) causal statements, the above kind of (active) correct beliefs must be required (see Subsection **2.4** for a further development).

We may here also say something about purposive causal *over-determination*. Let us thus change our example (*8.13*) where the agent A's visiting his friend E was discussed. Assume now that not only was A's intention to see B operative but that A's intention to go to E's sauna was effective as well. We assume that both of these intentions (cum the relevant beliefs) were sufficient for A's visiting E. Let I_1 describe the event of A's intending to see B cum the accompanying belief, and let I_2 similarly stand for a description of the premises of (PS2). Now, if V describes the common conclusion of (PS1) and (PS2) we can assume that the following causal statements hold true of our modified example:

(a) $I_1 \rhd\!\!\rightarrow V$
(b) $I_2 \rhd\!\!\rightarrow V$
(c)   not $(\sim I_1 \rhd\!\!\rightarrow \sim V)$
(d)   not $(\sim I_2 \rhd\!\!\rightarrow \sim V)$
(e) $\sim (I_1 \vee I_2) \rhd\!\!\rightarrow \sim V$.

It should be clear that the causal statements (a)–(d) are acceptable as true or assertable in the situation of our example. But also (e) is, provided we now speak of truth in a broad Sellarsian sense including explanatory assertability (see the discussion in Tuomela (1976a)).

The conclusion for cases of overdetermination then is that at least for *standard* cases (a)–(e) could all be accepted. Above we explicitly considered only compatible intentions. As cases with incompatible intentions (i.e. intentions which cannot simultaneously be satisfied) are not conceptually possible (cf. Chapter 6), we do not have to worry about them.

2.4. To clarify and illustrate some of the points made earlier in this chapter (especially on the semantics of conduct plan descriptions) as well as to make a couple of additional comments, we will discuss a simple example in detail. Suppose an agent A is at the university at 4.50 p.m., and wants to go home. He forms an intention to catch the 5 p.m. train and then comes to think that only by taking the taxi standing at the university building will he be able to achieve that goal. We may assume that the want here plays the role of an intending with a somewhat unspecific content. We may, but do not have to, think that A looks at the train schedule and then forms the more specific intention. We, as psychologists, may now partly characterize this situation by the following statements:

(i) A wants to go home
(ii) A intends to catch the 5 p.m. train
(iii) A is at the university at 4.50 p.m.
(iv) A considers that unless he takes the taxi he will not catch the 5 p.m. train.

Let us assume, for simplicity, that (i) is made true in this particular case by a singular state (or episode) a, (ii) by b, (iii) by c, and (iv) by d. (No specific assumptions of the factual character of these episodes is made here.) We can now state our description of the agent's *conduct plan* in a preliminary way as follows:

(i') $W(A, (Ex) \; Goes \; (A, home, x), a)$
(ii') $I(A, (Ex) \; Catches \; (A, the \; 5 \; p.m. \; train, x), b)$
(iii') $At \; (A, the \; university, c) \; \& \; T(c) = 4.50$
(iv') $B(A, (x)(Ey)(Catches \; (A, the \; 5 \; p.m. \; train, x) \rightarrow$
         $Takes \; (A, the \; taxi, y) \; \& \; T(y) \leq 4.52), d)$
(v') $G_1(a, b)$
(vi') $G_2(b + c, d)$.

The symbolism in (i')–(iv') is rather obvious. (In a finer treatment we would identify our expressions with Sellarsian dot-quoted predicates and singular terms, e.g., '.Goes.', '.home.', to get to the "real" deep structures.) The predicate $G_1(a, b)$ reads 'a generates$_1$ b' and stands for want-intention generation. $G_2(b + c, d)$ analogously reads '$b + c$ generates$_2$ d' and stands for intention-belief generation. Here $b + c$ is the complex event composed of b and c. Similarly, one could discuss, e.g., belief–belief generation and intention–intention generation. Whether these kinds of generation are or can be *causal* in character cannot be decided on a priori grounds. The factual character of the generational relations must be left as equally open as that of the episodes a, b, c, d and their ontological interrelations. Our generation predicates are to be analyzed in terms of backing laws closely in the manner of our singular causal predicate $C^*(-, -)$. Thus, in a sense, they are only shorthand for a longer story involving essential reference to some psychological (or neurophysiological) backing laws or theories (possibly stated in quite different terminology).

In any case the conjunction (i')–(v') represents a conduct plan or, better, our deep structure description of the agent's conduct plan. (This conduct plan is of course not yet a *complete* theory in our earlier technical sense, but only an essential part of such a theory.) A conduct plan description, we have said, is a psychologist's explanatory instrument, at least in principle. (i')–(vi') can be taken to explain (and predict, too) at least A's taking a taxi. We notice that the explanation or prediction of this (type of) action by means of the ordinary practical syllogism (7.9) would refer to (ii') and (iv') and, implicitly in the normal condition assumptions, to (iii'). Our conduct plan description, in addition, contains a want-description as well as two statements concerning "mental generation". As we shall see, an immediate or direct explanation of A's action, such as his taking a taxi, will refer only to the purposive cause of this action. Still, as this purposive cause will be an episode, or part of an episode, activating the full conduct plan, we prefer (with respect to direct explanation of action) to refer to a description of the purposive cause which is as informative as possible. Thus, it seems wise to include at least (i')–(vi') in the explanatory conduct plan. It is in part up to future psychology to delineate the concept of a conduct plan. This concept cannot be fully specified a priori.

If we use this conduct plan for explaining the agent's taking a taxi (suppose he indeed carried out the plan), then we may perhaps add the following statemens to it:

(vii') $I(A, (Ex) \text{ Takes } (A, \text{ the taxi}, x)$ &
          $T(x) \leq 4.52$ & $At(A, \text{ the university}, x), e)$

(viii') $G_3(b + c + d, e)$.

Here G_3 stands for intention transferral generation; it is possibly causal in character, and it represents the, so to speak, *factual* aspect or side of the *conceptual* truth that (ii'), (iii'), and (iv') together guarantee the truth of (vii'). It is now the intending e which becomes the agent's trying (to take the taxi) at the right moment. Let f represent (the *overt* part of) the agent's action of taking a taxi such that A is aware of f in this sense. We then probably have $C^*(e, f)$ here, and it gives the prima facie explanation of f.

If we, furthermore, assume that the taxi arrived at the station in time (A caught the train (g)), then the action f factually action-generated g.

What does the actualized model $\mathscr{M}$ for the "full" conduct plan description $K = $ (i′) & (ii′) & (iii′) & (iv′) & (v′) & (vi′) & (vii′) & (viii′) consist of? The problematic thing in answering this is, of course, what the singular terms in the second argument places of the predicates for propositional attitudes denote. Their semantics may be given in two phases, so to speak.

The first phase simply consists of specifying which singular entities the singular terms (names of deep-structure propositions) in those second argument places denote. The answer is that they "replicate" and in this way refer to other singular entities with the same propositional structure but which (in the sense of the analogy theory) "are" in the agent's mind or brain (and are, so to speak, extra-linguistic with respect to the language *we* use in describing A's mental life and behavior). We recall from our discussion in Chapters 3 and 7 that such propositional brain episodes or propositional mental episodes are in central respects similar to propositionally interpreted utterings and other overt behaviors. For simplicity, we take these entities to be just inscriptions of some suitable sort, i.e. inscriptions with the same propositional structure that the singular term has when regarded as a sentence. (They might be, e.g., inscriptions in terms of neurons and neural structures.)

As we must represent these inscriptions somehow in our semantics, we cannot but write an inscription such as **(Ex) Goes (A, home, x)** for the singular term '(Ex) *Goes* $(A, home, x)$', since the current state of science permits no more. To put this more exactly, we may "intensionally" interpret the singular term '(Ex) *Goes* $(A, home, x)$' (or, perhaps rather '·(Ex) *Goes* $(A, home, x)$·'; cf. footnote 6.6) in our model as the individual $o = \langle \mathbf{i}, p \rangle$ where $\mathbf{i}$ is the inscription **(Ex) Goes (A, home, x)** and p is the proposition that A goes home. But below we just "extensionally" use $\mathbf{i}$ in the domain: in the first phase of our semantics we only need an inscription $\mathbf{i}$ of which the sentence in question can be seen to be (under some "key") a picture or a "structural" replica.

In the second phase of developing semantics, we take these inscriptions to be *statements* with a certain structure, such as p above, we have already given them (by, so to speak, *fixing* the interpretational key for the inscription **(Ex) Goes (A, home, x)**). Thus, they are not any more treated as denotata of names at all but as (the agent's) propositional representations of the states of the external world. In this latter analysis one can in fact use one's favorite semantic analysis to do the job.[4] Notice that when we speak of an action being performed *in accordance with*, or *satisfying*, a conduct plan it semantically means that the action satisfies (*for the agent*) the content of a propositional attitude of the conduct plan and that the agent's conduct plan can be said to be *about* the performed action.

We do not here intend to give a fully detailed discussion of semantics. We shall therefore only discuss our example. Let us write down the domain **D** of our present structure $\mathscr{S}$, which we use in the first phase of building our semantics:

D =   {A, a, b, c, d, e, b + c, b + c + d, (Ex) Goes (A, home, x),
             (Ex) Catches (A, the 5 p.m. train, x),
             (x)(Ey)(Catches (A, the 5 p.m. train, x) → Takes (A, a
             taxi, y) & T(y) $\leq$ 4.52)), (Ex) Takes (A, a taxi, x) &
             T(x) $\leq$ 4.52 & At (A, the university, x), the university,
             4.50}.

Now the model $\mathscr{S}$ for the conduct plan description (1') & - - - & (8') simply becomes:

$$\mathscr{S} = \langle D, W, B, I, G_1, G_2, G_3, At, T \rangle.$$

Here and below we (exceptionally) use bold face letters for the set-theoretic relations corresponding to the italic linguistic ones, for we want to emphasize these differences in the present connection. Thus e.g. **W** represents a three-place relation interpreting the predicate *W*, and so on. Now we say that $\mathscr{S}$ satisfies$_1$ the conduct plan description (i') & - - - & (viii') if and only if $\mathscr{S}$ is a model (in the usual model-theoretic sense) of (i') & - - - & (viii').

But we still have to clarify in which sense the overt actions **f** and **g** satisfy (we say satisfy$_2$) or accord with the conduct plan, for it is in this sense that we spoke of satisfaction when defining purposive causation. It is obvious that it is the sense of satisfaction in which **g** at least satisfies the statement (*Ex*) *Catches* (*A, the train, x*). In other words, now we treat the attitudinal contents (second place arguments or the corresponding inscriptions, if you like, of propositional attitude predicates) as statements satisfiable (in a sense to be clarified) by whatever entities external to the agent are appropriate here.

This second phase of doing semantics can be incorporated technically into our treatment as follows. Given the occurrence of some events, such as **f** and **g**, which are thought to make the conduct plan satisfied$_2$ we form an *extended* conduct plan. In this particular case we add the contents of the relevant propositional attitudes as additional statements in the conduct plan:

(ix') $(x)(Ey)(Catches\ (A,\ the\ 5\ p.m.\ train,\ x)$
            $\rightarrow Takes\ (A,\ the\ taxi,\ y)\ \&\ T(y) \leq 4.52))$

(x') $(Ex)\ Catches\ (A,\ the\ 5\ p.m.\ train,\ x).$

Now **f** obviously will interpret y and **g** x so that (ix') and (x') come out true in our example. We could also have added the corresponding satisfaction-statement for (i) to obtain a *maximal extended* conduct plan description. But to make sense of purposive causation, we have to require the satisfaction$_1$ of only some extended conduct plan description.

In our example case we may denote by **D'** the set $\{$**f, g, the taxi, the 5 p.m. train, 4.52**$\}$ of real singular entities. Then the model $\mathcal{M}$ used in our earlier definitions of purposive causation becomes

$$\mathcal{M} = \langle \textbf{D} \cup \textbf{D', W, B, I, G}_1, \textbf{G}_2, \textbf{G}_3, \textbf{A, T, Catches, Takes,} \leq \rangle.$$

$\mathcal{M}$ obviously satisfies$_1$ (i') & - - - & (x') and it satisfies$_2$ the description (i') & - - - & (viii'). But satisfaction$_2$ here does not mean *only* that $\mathcal{M}$ satisfies$_1$ (the model-theoretic sense) the description (i') & - - - & (x'). The additional requirement is that (i') & - - - & (x') has to be, in a stronger sense, *about* those elements in the domain of $\mathcal{M}$ which are external to the agent (i.e. those in **D'** $\cup$ {**the university, 4.50**}). We shall return to this problem after briefly discussing another problem complex which leads to this problem of aboutness.

We have so far been quiet about difficulties which arise when one performs quantification into an opaque context, such as a belief-context. Do we really get into those difficulties here? It seems that an adequate answer to this question is twofold. Namely, it seems that in our conduct plan descriptions themselves we can avoid quantifying into an opaque context, for we can try to keep our descriptions strictly to how the agent "views the world" and not to make any additional existential assumptions. But, at least in justifying statements about psychological event generation, quantification into an opaque context may be needed.

For instance, our earlier example concerning the mistake about the switch might be such a case, and in our above example the justification of $C^*(e, f)$ (now dropping our above typographic convention) may conceivably involve a law in which quantification into a belief context occurs. For now, it may seem, we explicitly have to speak about the agent's belief *of* or *about* a specified and, normally, actually existing object. Thus, in our earlier example A believes something about the switch, which is assumed to exist in the actual world. That belief is then to be understood so that it, in an obvious sense, brings together the agent's subjective "belief-world" and the actual world. But, I would not call such descriptions concerning a belief of or about some object merely descriptions of the content of that belief, for they are also partly *our* commentaries on the agent's belief.

In our switch flipping example the agent tries ('tries' in our technical sense) to flip something *s*, which he believes or takes to be the switch. This is what is "minimally" meant by a conduct plan description such as '(*Ex*) *Tries* (*A*, (*Ey*) *Flips* (*A the switch*, *y*), *x*)' which reads '*A* tries to flip the switch'.

Suppose now that *A*'s trying to flip the switch purposively caused his flipping the switch. Then this singular causal claim might conceivably (though this is not perhaps very realistic!) be backed by the following law (cf. the (L) of Chapter 7 and footnote 1 of the same chapter on the co-ordinating predicate *P*):

(L′) $(r)(s)(y)(z)[(W(A, (Ex)$ *Illuminates* $(A, r, x), y)$ &
$B(A, (x)(Ev)(Illuminates$ $(A, r, x) \rightarrow$
Flips $(A, s, v))$ & $r =$ *the room* & $s =$ *the switch*$), z))$
$\rightarrow (Eu)(Flips$ (A, s, u) & $P(y, z, u))].$

Obviously, in applying (L′) we have to assume that the want can be redescribed as a trying to flip the switch so that the following holds true of *A*'s trying (*t*):

(V) $(Es)(Tries$ $(A, (Ex)$ *Flips* $(A, s, x), t)$ & $s =$ *the switch*$)$.

Thus, I claim that before we have a genuine case of purposive causation it must be the case that the agent-external objects which satisfy the agent's conduct plan must be such that the agent's wants and beliefs are *about* them, and this (semantically) requires the truth of statements of kind (V). But such statements do not belong to our conduct plan descriptions. They normally belong to our commentary of the situation and, at best, may be entailed by the justifying backing laws. This in effect is the case in our present example. But, on the other hand, it is essential to discuss them in the second phase of building our semantics, for there we just have to say what it means for a conduct plan to be *about* some entities external to the agent in such a way that this aboutness is *for* the agent (or according to the agent's intentions and beliefs as expressed in his conduct plan). In other words, our notion of satisfaction$_2$ must include this idea of intentional aboutness. We must be able to say about some singular external event (or state or object) whether the agent's conduct plan is about it, i.e. whether it is a (complete or partial) satisfier (for the agent) of his conduct plan.

It can plausibly be argued that an agent's conduct plan is in that sense *about* an existing entity, say s, if and only if the agent knows or has a (correct) belief about *what* (or which or who) this entity is (cf. Hintikka (1974)). In other words, an agent's belief or want is about an existing object (or event or state)s if and only if an appropriate statement of kind (V) is true of s and furthermore the agent has a correct "wh-opinion" about s (i.e. correct opinion about who or what or which s is). The agent has, so to speak, to recognize s for what it is. This seems all right as far as the semantics of the situation is concerned. Thus, we have one answer to the question in what sense a conduct plan has to be about the events which satisfy it in the cases of purposive causation and generation (e.g. in terms of $\triangleright\rightarrow$ and $\odot\rightarrow$). Notice here that *results* of intentional actions must be events satisfying our aboutness condition.

But we must also say something about the conditions under which an agent can be said to have the above kind of correct wh-opinion. To put the matter still differently, given a model of type $\mathscr{S}$, we are looking for an extension $\mathscr{M}$ of $\mathscr{S}$ such that 1) $\mathscr{M}$ satisfies$_1$ a suitable extended conduct plan description and 2) $\mathscr{M}$ contains only such agent-external individual entities concerning whom (or which or what) the agent has a correct belief or opinion. We thus want to exclude from the set of possible extensions of $\mathscr{S}$, as *unintended* (almost literally!) those models of kind $\mathscr{M}$ whose agent-external individuals do not satisfy the aboutness condition.

A clarification may be needed at this point. We have above required that the agent be aware of the "agent-external" entities (and their relevant properties as specified in the conduct plan) in the sense that the conduct plan in a strong intensional sense is about them (for the agent). What we have called agent-external entities above may in other examples involve, e.g., overt (and in this sense non-internal) parts of his body. Thus in the case of raising one's arm, the arm belongs to these "agent-external" entities.

Furthermore, we notice that even inner events and states (e.g. those in D) must be epistemically accessible and "known" to the agent in the sense that at least under a suitable *functional* description relating them to overt action he has knowledge of what goes on in his body when he acts, e.g., raises his arm.

We cannot here go into a more detailed investigation concerning the aboutness problem. This problem has been discussed in the literature to some extent (see especially the acute and informative paper by Morton (1975)). To illustrate the issue we still state some conditions for aboutness which seem to us correct and philosophically illuminating, though they perhaps could be more specifically formulated. Let us thus consider the statement 'If A believes that something is P then he will do Q to it', which is a generalization resembling our above (L'), though it is somewhat simpler.

If, in this generalization, A's belief and action are to be *about* an existing individual x (cf. our (L')), then the following conditions, which elucidate correct wh-beliefs, must be satisfied:

(1) A is in possession of information that is true of x such that this information will allow A to go about doing Q to it.

(2) A correctly believes that there is something of which this information holds and of which P is true, and

(3) this information is obtained by means of a process that normally can guide doing Q, and similar actions, to their end. (Cf. Morton (1975), pp. 5–10 for arguments leading to similar conditions.)

Even if it is not possible to discuss the problems involved in any detail here, we have to make one remark. Namely, if it were the case that condition (3) would always require a causal account of beliefs (e.g., perceptual beliefs) we would seem to go in a circle. But I do not think that is the case, for information can be obtained non-causally as well, although I think that perceptual beliefs require a causal account. Secondly, even if our account would be circular in this sense, that is not fatal as long as the circle is big enough. (We do not think that a reductionist analysis of causality is possible at all.) Furthermore, such circularity would not affect our main purpose, which is to clarify what is involved in purposive causation.

This brief discussion of the problems of quantifying into opaque contexts and the problem of aboutness will have to suffice here, even if many details would have to be added in a full treatment of semantics. As a conclusion for our example, we notice that the statements '$C^*(e, f)$' and '$IG^*(e, g)$' are now seen to hold true with respect to $\mathcal{M}$ on the basis of our above treatment and what we otherwise assumed about causality and generation in this example.

Our final remark concerns conditional intentions and the role of feedback. We might change the above example by changing (iv') and (vii') to include the following: A is assumed to believe that if a tram goes by on or before 4.51, taking it will suffice for his catching the 5 p.m. train, otherwise he will have to take the taxi standing outside the university building. Now we are dealing with two *conditional* intentions: 1) A intends to take the tram, given that it comes no later than 4.51; and 2) A intends to take the taxi, given that the tram comes later than 4.51.

This kind of a case is very important in principle, for it leaves a *gap* in the conduct plan to be filled out by the agent. The information needed may require new actions or activities to be performed by the agent. In this case the result of his perceptual activities (e.g., no tram in sight at 4.51) will be fed back as information to fill the gap in the conduct plan (cf. the discussion of related cases in Chapters 7 and 12).

In our conduct plan description, the obtained information would be inserted and we might be willing to assert that A's conditional intention 2) in conjunction with that perceptual information purposively caused his taking the taxi. This causal assertion would have to be justified in terms of a factual (psychological) backing theory, not in terms of e.g. any conceptual truths about conditional intentions.

3. ACTION-EXPLANATIONS RECONSIDERED

3.1. Earlier in this chapter we arrived at the general thesis that all conclusive intentional-teleological explanations of intentional actions can be regarded as suitable partially or completely explicit ϵ-arguments which convey understanding concerning why the explanandum is the case. There are still some theoretical problems connected to this thesis that need to be clarified. We shall also further illustrate the content of this thesis, although we have already devoted much space to various considerations relevant to our main thesis above in the previous chapter and in Section 1 of this chapter (recall especially our theses (T1)–(T4) of Chapter 8).

One problem that we have hardly touched on so far is the specific nature of the explananda of action-explanations. To be sure, we have indicated that they will in some sense be intentional actions. But what does that really involve? Suppose we are asked to explain why Tom opened the window at midnight. Here an action has occurred and we are asked to explain it under one of its descriptions, viz.

(20)   Tom opened the window at midnight.

What is the logical form of (20)? Given that (20) was in this case made true by a certain singular action, call it b, we may consider the following (cf. Chapter 2):

(21)   Opened (Tom, the window, b) & At (midnight, b).

However, (21) is not strictly speaking our proposal for the logical form of (20). Rather (21) is an instantiation of that logical form, which is:

(22) (Eu)(Opened(Tom, the window, u) & At(midnight, u)).

Formula (22), we have said, in a sense represents a generic action. It is also something which may occur in the consequent of an action-law (if there is any) and which thus serves as a prediction statement. (21) again is a description of a concrete singular action.

We may appropriately say that statement (21) represents our explanandum in this particular case. In a clear sense (21) explicates what could be meant by saying that the explanandum of an action-explanation is "a singular action described in a certain way". We could also say here, in ontic terms, that our explanandum is action b as described by (21), and we could even say that it is b as an opening of the window by Tom, which, moreover, occurred at midnight.

More generally, a singular explanandum of an action-explanation will be of the form

(23)   Mod-Verbs (A, o, b) & $T(o', b)$,

to use the formalization of Chapter 2. We take b to be a name or definite description of a singular action. The corresponding generic statement is of course

(24) (Eu)(Mod-Verbs(A, o, u) & $T(o', u)$).

There are several remarks to be made about our proposal that schema (23) be taken as a kind of prototype of a singular explanandum of action-explanations. First, we are concerned here with an *intentional* action, viz. an action intentional under the description given to it in (23). This intentionality is a *presupposition* of explanation and thus is not "part" of the explanandum (even if we can formally indicate this intentionality by an operator as in Chapter 2).

Secondly, our formalism is noncommitted as to the specific ontic nature of the singular events such as b (the substitution instances of u). We, however, assume that these singular actions, at least, have an overt bodily component. Whether they, being intentional, contain an ontically "real" trying component t, as a Jonesian theorist wants to have it, is a separate matter. As a Jonesian causal theorist I, however, do make that assumption as well.

Thirdly, although singular action-explanations are concerned with (23) rather than with (24) it is only "through" (24) that we explain (23). That is, we primarily explain concrete singular actions by showing that they are exemplifications of suitable generic actions, which are nomically connected to some other universals. As we said, (24) might conceivably occur as the consequent of a psychological law, and in this way (23) would become explained through (24). However, two important qualifications are needed here. One is that the predicate $T(-,-)$ be typically concerned with specific, e.g. spatiotemporal and circumstantial, factors (cf. "at midnight"). It does not seem likely that any psychological law would cover them. What this means is that we may typically have to be satisfied with explaining only the first conjunct of (23), which of course is the central part of the explanans. There are even further and more serious retreats that one may, perhaps, have to make, but let us leave them till later, so that the prospects for action-explanations do not, as yet, appear too gloomy.

We recall that our basic explanatory idea is that all intentional actions are to be explained by reference to their purposive causes. Now, explaining a singular action b by saying that b (under a certain description D_2) was purposively caused by an episode a (under description D_1) involves "idiosyncratic" information about the concrete singular events a and b. That is, we at least must know that a was causally prior to b and that a was (sufficiently) spatiotemporally contiguous with b. This kind of information need not be embedded in the causal law backing the singular causal claim in question nor need it be found in the conduct plan description relative to which the causal connection is purposive. In this sense, an "ontic" explanatory claim of the form

(25) $C^*(a, b)$

may go beyond a linguistic explanation given by reference to universals only (as is the case when the explanation of (23) is reduced to an explanation of (24)).

We have considered statements of kind (23), or rather the first conjuncts of them, as the main explananda of action-explanations. We just added that, in the sense explicated, we may equally well speak about a singular action b (under the description given to it by (23)) as our main explanandum. However, there are also other types of explananda, for instance *conditional* explanandum statements.

We have been assuming all along that explanatory queries determine what is to count as a plausible explanation. For queries involve the interests and presuppositions of the questioner (explainee). Part of that is reflected in the explanatory question statement (e.g. 'why L?') contained in the query. One especially interesting way by which the query becomes reflected in its contained question statement is through *emphasis* (or stress). We may change an explanandum statement by changing its emphasized parts; thus, it is seen that in a sense one question statement can correspond to several different queries.

To see what is involved in the contexts of emphasized explanandum statements, we consider our earlier example (*20*). As we stated it, no parts of it were meant to be particularly emphasized. Let us now examine this same sentence by emphasizing different parts of it. Marking emphasis by italics, the following explanandum statements may be considered:

(*26*)   Tom opened the window *at midnight*.
(*27*)   Tom opened *the window* at midnight.
(*28*)   Tom *opened* the window at midnight.
(*29*)   *Tom* opened the window at midnight.

Linguistic emphasis is a strongly context-relative phenomenon and it has many different communicative uses. We shall below only be concerned with emphasis as *contrast*, for that seems to be the most interesting sense of emphasis in the context of explanation. By contrast-emphasis I mean roughly the following (clarified in terms of (*26*)). The speaker's interest is concerned with the time of Tom's opening the window. He may ask: When did Tom open the window? It is presupposed here that Tom did in fact open the window, but the questioner wants to know the time of the action. What is more, the situation described by (*26*) is *compared* with another one of Tom's (possible or actual) actions of opening the window, one occurring at some other time (e.g. in the morning, when Tom normally opens the window). How come Tom opened the window at midnight *rather than*, for instance, in the morning?

Let us write the indicated presuppositions of (*26*)–(*29*) by means of a given-clause. Applying our standard formalism we get (cf. (*21*)):

(*26′*)   Opened (Tom, the window, b) & At (midnight, b), given (Ex) (Opened (Tom, the window, b) & At (x, b))

$(27')$ Opened (Tom, the window, b) & At (midnight, b), given (Ex) ((Opened (Tom, x, b) & At (midnight, b))

$(28')$ Opened (Tom, the window, b) & At (midnight, b), given (EX) (X'ed (Tom, the window, b) & At (midnight, b))

$(29')$ Opened (Tom, the window, b) & At (midnight, b), given (Ex) (Opened (x, the window, b) & At (midnight, b)).

In the formalizations the given-clause always contains an existential quantifier. The questioner wants to gain understanding concerning the mentioned value of that quantifier when the statement occurs in a question. Thus, when asked *why* (26) is the case, the *time* of the action is under consideration, and information concerning it especially is asked for. Analogous comments go for (27)–(29).

In contrast with (26)–(29), an unemphasized statement such as (21) does not have a given-clause implicitly attached to it. An answer to a question of why (21) is the case must give truth grounds for the whole statement and not only to a special singled-out part of it.

Is the 'given' in (26')–(29') a truth-functional connective? Can we, for instance, paraphrase it as a material implication? We would then get, for (26'):

$(26'')$ (Ex)((Opened, the window, b) & At (x, b)) → Opened (Tom, the window, b) & At (midnight, b).

I do not think this is a plausible alternative at least in our present context. We can hardly expect (26'') to be a conclusion of any scientific explanation at least in the sense of its representing an instantiation of the consequent of some law.

However, consider a related case (cf. Niiniluoto (1976)). Assume b is a piece of copper which has expanded (E), when heated (H). Then, we might consider either

(30) $E(b)$, or

(31) $H(b) \rightarrow E(b)$ (i.e. $E(b)$, given $H(b)$)

as our explanandum. (30) could be explained by saying that b was heated, and that all copper expands when heated. (30) and (31) are of course very different as explananda, for (31) asks for an explanation of a connection which in fact serves to explain (30). In the typical case (31) can be taken to "represent" a *law* – b is an *arbitrary* sample of copper. Thus (31) really represents the law

$$(31')   (x)(C(x) \ \& \ H(x) \rightarrow E(x)),$$

where C stands for copper.

We conclude that how 'given' is to be interpreted depends on the context. In the case of $(26')$–$(29')$, it cannot be represented by a material implication ($\rightarrow$), and hardly by any connective at all. We may still consider the possibility that the given-clause is put in the explanans. If this were plausible then 'given' would obviously be explicable by our explanatory conditional $\rhd\!\!\rightarrow$. But this is surely an implausible suggestion. No nomic connection relating the presupposition (given-clause) of, say, $(26')$ to its antecedent can be expected to obtain.

The remaining suggestion now is that the given-clause should not be included in the explanation at all. That is, it is neither strictly a part of the explanandum nor is it a part of the explanans. It is a presupposition in a genuine sense, and its function is to "direct" the selection of the explanans, given an explanandum (such as (26)). The presupposition operates by eliminating possible explanantia. How does it do this? It does this by means of the following informativeness principle: The explanans must be informative not only with respect to the explanandum but also *relative to the presupposition the given-clause expresses.* Furthermore, what is regarded as a presupposition in a given context of explanation is not itself in need of explanation in *that* context. Hence, an explanation should not add to our understanding concerning the presupposition but rather concerning the explanandum, *given* the presupposition (i.e. concerning the explanandum "minus" the presupposition, so to speak).

As analyzed here, problems of emphasis do not then seem to present any particular difficulty for our approach. They only impose another informativeness criterion for our explanations. Thus, if explanations are conclusive (partial or complete) answers to why-questions and if such answers are (parts of) ϵ-arguments conveying understanding concerning why the explanandum is the case, we require something more of this "conveying of understanding" in the cases of explananda with emphasized parts.

Our above analysis of emphasis can be connected to the problem of how to select *the* cause from among several contributory causes. We recall from our earlier discussion that both mere and purposive singular causes are typically only parts of total (productive) causes. They may be partial in a purely ontic sense (e.g. by being parts in the sense of the calculus of individuals) but in any case they are partial causes in the following sense. If $C(a, b)(C^*(a, b))$, then $D_1 \rhd\!\!\to D_2(D_1 \rhd\!\!\to D_2)$, where D_1 describes a as exemplifying a universal P and D_2 describes b as exemplifying another one Q. We recall from Section 2, that P is typically a necessary part of a minimal sufficient condition for Q. Each such sufficient condition may contain several such necessary parts qualifying as contributory causes. Can we call any one of them *the* cause? If the sufficient conditions are mutually exclusive (as we have assumed) then we only have to consider, in our present situation, the sufficient condition which was realized. Thus, the possible candidates for *the* cause of b are the instantiations of all the necessary parts of that sufficient condition. The singular event a is one of them.

We may now reason as follows. A singular event, which in the above sense is a candidate for being the cause of b, is in fact to be regarded as *the* cause of b just in case it differentiates between the actual situation which included the occurrence of the effect b (call this the *effect* situation) and a certain actual or hypothetical situation with which the effect situation is being compared (call this the *comparison* situation). The problem of determining *the* cause divides naturally into two parts: 1) the problem of how to select the comparison situation, and 2) the problem of how to determine the cause, given a certain comparison situation. We shall not discuss these problems here in more detail. Problem 2) has been discussed in a sophisticated manner and in great detail by Martin (1972) and I believe his treatment is basically workable.[5] Concerning problem 1) we may recall Collingwood's example discussed in Chapter 8. Our present discussion of emphasis throws some light on that problem. Obviously the presuppositions in our (26')–(29') are intimately related to the selection of a comparison situation. They serve to (partly) determine what in each particular context is to count as the correct type of explanatory factor, and if they indeed are causes, what is to be *the* purposive cause. Emphasized parts of explanandum statements play an important role in determining what the plausible comparison situations are and thus in selecting *the* cause (cf. our above treatment).

3.2. Let us now go back to illustrate our general thesis about the nature of intentional-teleological action-explanations. We return to our simple example of Chapter 8 about the agent who escaped a thunderstorm by going to the cellar of his house. There we constructed the following kind of expanded practical syllogism to explain the agent's action:

(32)   (P1) *A* is afraid of thunderstorms.
         (P2) *A* intends to save his life.
         (P3) *A* considers that unless he goes to the cellar he will not save his life.
         _____
         (C)   *A* goes to the cellar.

The premises of this argument are supposed to answer the question: Why did *A* go to the cellar? *A* went to the cellar *because* (P1), (P2), and (P3). This answer is not, as it stands, a full ϵ-argument. But it does not have to be either. It suffices that *there be* a suitable ϵ-argument of which the premises (P1), (P2), and (P3) form a part. What is missing here, at least, is an explanatory *law*, it would seem. But, to see clearly what is involved here let us look at the matter from another angle.

The conclusion (C) describes a singular action, say *b*. If our arguments in Chapter 8 are accepted, then it is reasonable here to assume that *b* was purposively caused by the episode satisfying (P1), (P2), and (P3). Let us call this, possibly complex, episode *a*. Thus, we explain *b* under the description (C) by saying that it was purposively caused by *a* under the description (P1) & (P2) & (P3). In other words, we have $C^*(a, b)$, where (P1) & (P2) & (P3) constitute *A*'s conduct plan (say) *K* (or, rather, an essential part of it) which is needed to establish $C^*(a, b)$. The episode *a* obviously represents the episode here activating *K*.

A's effective intending or trying (*t*) to go to the cellar can be assumed to be a part of *a* (e.g. $t < a$, in the sense of the calculus of individuals). We may assume that $C^*(t, b)$, too. In fact, it seems plausible to assume that for every part *a′* of *a* (i.e. $a′ < a$), $C^*(a′, b)$ if $C^*(a, b)$. This is so because it seems reasonable to build one's semantics (i.e. the relations between event-descriptions and events) so that the claimed relation holds.

Notice that as the law backing $C^*(a, b)$ need not necessarily be couched in the vocabulary of the conduct plan *K* we have to keep track of two kinds of correspondence between events and event-descriptions.

What kind of "interesting" parts can the episode *a* have here? At least, *A*'s activated state of fear, *A*'s effective intending (*t*), and *A*'s activated state of believing are obvious candidates. Corresponding to this onto-logical side, we have on the linguistic side the analogous explanatory statements such as '(C) because of (P1)', '(C) because of (P2)', and '(C) because of (P3)'. In terms of our purposive conditional $\rhd\!\!\rightarrow$, we can thus say that the following *purposive causal* statements, which qualify as incomplete singular explanatory claims, are assertable (we use shorthand notation):

(33)   (a) (P1) $\triangleright\!\!\rightarrow$ (C)
         (b) (P2) $\triangleright\!\!\rightarrow$ (C)
         (c) (P3) $\triangleright\!\!\rightarrow$ (C).

We can continue by pointing out that there are several other incomplete singular explanatory claims which are equally well assertable. Here is a further list, given purely verbally:

(34)   (i)   (C) because A's reason was that he intended to save his life.
         (ii) (C) because A's reason was that he believed that going to the cellar was necessary.
         (iii) (C) because A considered it to be the appropriate thing to do in these circumstances.
         (iv) (C) because A believed there was a thunderstorm outside.
         (v)   (C) because A obeyed the norm: Whenever there is a thunderstorm, one ought to seek covered shelter.

Of these (iii) and (v), perhaps, contain slightly more information than originally given in our description of the example, whereas (i), (ii), and (iv) strictly express what has been said earlier. We still might want to consider other explanations, which go beyond our original description of the case:

         (vi) (C) because A was a coward (explanation by personality trait).
         (vii) (C) because A perceived sounds of thunder.

The list could still be continued. We have anyhow already shown that an emotion, an intention, a want, a belief, a reason, a norm, a personality trait, and a perception can serve as a singular explanatory factor of an intentional action. As we claimed in Chapter 8, it is ultimately the activated intention-belief pattern which carried out the explanation, in the sense that all the other mentioned factors are seen to somehow consist of that pattern when the explanation is fully spelled out.[6] Thus, an emotion "involves" an intending in the sense explicated by (32). Reasons, although explicitly concerned with wants and beliefs, were also seen earlier (Chapter 8) to depend on intendings when explanatory. Considering something as an appropriate thing to do means *having a reason* to do it (case (34) (iii)). This makes Drayian explanations fit our mould. Norm-explanations again involve an intention to *obey* the norm (cf. Chapter 8).

Personality traits also lead to intentions of a suitable sort: a coward is disposed to intend to act "cowardly". That disposition may have been activated and made specific as the cellar-going action in our example. Finally, a perception (case (34) (vii)) obviously refers to a perceptual belief; thus, it belongs to the category of normal conditions (viz. A believed that he was in the appropriate situation, cf. Chapter 7).

The above sketch of the variety of singular explanantia an IT-explanation may involve was not meant to be fully exhaustive. For instance, sensations and feelings were omitted from our list of potential explanatory factors for singular intentional actions (cf. Taylor (1970) and Nordenfelt (1974) on such explanations). Our main point was to illustrate that one and the same situation may be given a number of apparently different incomplete singular explanations. Yet, we argue, behind these singular explanations, there is, or may be, a single complete explanans, of which the mentioned explanantia are parts. In fact, we have shown what the singular premises of the explanans typically contain. In the case of our above example (in its original unexpanded form) they are simply (P1), (P2), and (P3), together with the normal conditions assumptions.

The problem we now have to face is what the explanatory law is or can be. For, we recall, every ϵ-argument is a nomological argument and thus such an explanatory law must exist. Or, when approaching explanation from the ontological angle in terms of causation, we must refer to a backing law. Let us thus turn to the problem of explanatory laws in action theory.

If the backing law is couched in psychological terminology, it must apparently be concerned with the agent's *tendencies to make practical inferences*. This is because all IT-explanations presuppose that *some* kind of a practical inference lies at the bottom of every such explanation: the explanandum-action is one in favor of which the agent's practical reasonings were concluded before his acting. Is it possible that such backing laws exist in psychology? My answer is a qualified yes. Let us now go on and discuss some reasons for such an answer.

In Chapter 7 we considered a general principle (L) about intentions, beliefs, and actions. We took (L) to back practical syllogisms of the ordinary kind, i.e. (7.13). (L) was said to be "analytic" vis-à-vis the framework of agency (i.e. (L) is invariably true, *given* the framework or "theory" of agency). However, even (L) has something nomological in it, so to speak, due to two things. First, the framework of agency, to which (L) was said to belong as a ground principle, is a *broadly factual* theory itself. It is a synthetic matter that human beings satisfy the central principles of the framework of agency. Secondly, (L) is a *causal* law in that it contains the idea that intendings, when effective, purposively cause bodily actions (see our discussion in Chapters 7 and 8 and above). This feature is due to our Jonesian theorizing. Thus (L) is broadly factual in a double sense, even if it is non-contingent relative to the framework of agency and relative to a Jonesian type of causal theory of action.

It is at least conceivable that there are action-laws in psychology in a *specific* factual sense (comparable to "All copper expands when heated"). Consider, for instance, the common sense law-sketch: A burnt child shuns the fire. There is clearly something factually informative and nomological about it. It is, however, only an incomplete sketch of a law as it stands. We do not know exactly under what initial conditions it holds true nor even what its exact semantic content is. It is, of course, conceivable that a burnt child would not always shun the fire in ordinary situations. We might, for instance, reward it with candy, or something like that, to make it occasionally falsify the law, and so on. "Self-fulfilling" and "self-defeating" predictions in general are *conceptually* possible, to be sure (cf., e.g., Scriven's (1965) argument against "the essential unpredictability of human behavior"). All such arguments purporting to show that action-laws cannot exist do, however, make rather strong rationality assumptions concerning the agent. If *freedom of the will and of decision* and *freedom of action* are analyzed along the lines sketched in Chapter 6 above, it becomes a *factual* question whether or not an agent's rational powers (to will or to decide) are subject to any law (deterministic or indeterministic). Notice, too, that we do not have to require that psychological laws be lawlike in the strong sense philosophers often think physical laws are (cf. Goodman (1955)). We may well be happy with true generalizations with a lower *degree of lawlikeness* (cf. Niiniluoto and Tuomela (1973) on this notion).

It is possible that, for instance, wants may either invariably or with some specifiable probability lead to action in suitable internal and external circumstances. Thus it is not completely implausible a priori to think that ,e.g.,

$$(L^*) \quad (y)(z)(W(A, p, y) \ \& \ B(A, q, z) \ \& \ N(A) \rightarrow (Eu)(\text{Mod}'\text{-Verbs}' \ (A, o' \ u) \ \& \ T'(o'', u)))$$

might represent a psychological law for some suitable interpretations of its predicates. (L*) closely resembles (L) (see Chapter 7). W is a predicate for wanting (cf. the I of (L)). The symbolism of (L*) has the same interpretation as that of (L) with two exceptions. Concerning q, we allow that Mod'-Verbing' merely generates Mod-Verbing (instead of being necessary for it as in (L)). The normal conditions N in (L*) are the same as in (L) except that the concept of preventing is not to be interpreted in the same broad inclusive sense as before, but in a more restrictive sense which in principle allows for the external world "not to cooperate" with the agent's tryings.

Thus interpreted (L*) cannot be regarded as flatly analytic even relative to the framework of agency (see our analysis of wanting in Chapter 4). There is the possibility that an agent does not after all form an intention to act when the antecedent of (L*) is satisfied. If no "movement of the will" takes place, after all, the Mod'-Verbing' need not occur (see our remarks in Chapter 6 about this possibility). Notice that, as intendings are wantings, (L*) could possibly be used as a backing law for practical syllogisms of kind (7.13).[7]

There may even be laws with much weaker antecedents than (L*) has. The point is simply that an agent's forming an intention may be in lawful connection to some suitable antecedent such as wanting, fearing, perceiving, etc., or even to some purely physically and non-intentionally describable external circumstances. Presently, we just simply don't know the details. The matter is a factual problem, not to be conclusively decided a priori on purely conceptual grounds.[8]

In our example (32) the needed backing law might be one which says that whenever the agent A is afraid of thunderstorms and perceives that there is one coming he concludes his practical reasonings in favor of seeking covered shelter. Notice that this law does not speak about going to the cellar at all. We would, therefore, need an additional premise saying that in the particular circumstances with which (32) is concerned, going to the cellar amounts to going to a covered shelter, or something to that effect. Redescriptions like this are allowed in our approach.

Our basic requirement is that the statement '$C^*(a, b)$' is true here. This entails only that 1) there exists a way of describing a and b such that a nomologically backs b (via a causal ϵ-argument) and that 2) a causes b according to A's conduct plan, say K. It is not required here that the law backing the causation be known. It is also not required that A's conduct plan K be known. Only existence is required in both of these cases. Furthermore, it is not required that the causal backing law be a) a psychological or b) a neurophysiological one, or c) given in the terminology of the conduct plan K. All this makes our approach flexible and helps it to escape some well known criticisms that most other versions of the causal theory of action, e.g., Davidson's, are subject to (see, e.g., Stoutland (1976a), (1976b) for such criticisms). A price to be paid for this is that we cannot say very much that is informative about backing laws presently.

There is one thing to be especially emphasized in the context of backing laws for purposive causation. We recall our earlier argument from Chapters 7 and 8 according to which the premises of a practical syllogism are in a sense causally (but also conceptually) connected to its conclusion. This thesis results from how the analogy theory construes psychological concepts. According to this construal it is possible that there is a *single* (psychological, neurophysiological, etc.) law backing *all* the instances of the practical syllogism. It is equally well possible, to go to the other extreme, that each instance is backed by a *different* law or different kind of nomic process or connection. (Intermediate cases are of course possible as well.)

Even if there were a different backing law for each instance, there is, however, a conceptual connection, between some suitable universals which these instantiations exemplify. These universals are those which the agent's conduct plan employs. Thus the agent's having *internalized* certain concepts (i.e. those his operative conduct plan employs) or his acting according to certain rules (which amounts to the previous point, cf. Chapter 3) "guides" his bodily processes to causally operate "in the right way". The seemingly very different physiological processes involved in, say, the different instantiations of making a certain promise or of building a house may indeed be genuinely different physiologically (i.e. as to the physiological universals they exemplify), but connected at the psychological level (due to the agent's internalized and operative conduct plan).

What features, if any, distinguish psychological laws from physical ones? Davidson (1970) claims that it is the "open" and holistic character of mental concepts (cf. here our above remarks on free will and free decisions). The realm of the mental is a holistic unity which is constrained by rationality assumptions and which does not have any very tight connections to the realm of the physical. The proper sources of evidence for these two realms are different and "the attribution of mental phenomena must be responsible to the background of reasons, beliefs, and intentions of the individual" (Davidson (1970), p. 98). Mental concepts and physical concepts do not fit one another in the way, for instance, physical concepts fit each other. An analogy would be to compare the fit of 'emerald' with 'grue' (the former case) and with 'green' (the latter case). As mental concepts and physical concepts do not fit together in the right way, Davidson concludes that there cannot be any true psychophysical laws. By a law is here meant something rather strong: a law must be capable of receiving genuine instantial support and of sustaining counterfactuals. It is roughly in this strong sense that typical physical laws have been regarded by philosophers (rightly or wrongly) as being lawlike.

But Davidson concludes too much here. If we concentrate on the analogue of the pair emerald-green versus the pair emerald-grue we go wrong. For physical and, say, chemical or biological concepts are equally well disparate in the sense of that analogy! Furthermore, conceptual frameworks grow and change, and therefore such a priori arguments as Davidson's do not have much force.

We may, however, concentrate on the idea that it is the *rationality* assumptions inherent in the framework of agency that give us a really distinguishing feature between psychological and physical laws. This rationality aspect, mentioned by Davidson (1970), is nothing but the *normative* feature of the mental that we have been emphasizing in this book all along. Wants, intentions, beliefs, etc. are normative entities in that human agents are rule-followers (cf. our discussion in Chapters 3 and 4). Thus psychology studies something which has inherently normative features, whereas the objects and systems, e.g., physics studies lack this feature. It follows that rule-following, norm-internalization, and other similar concepts are central for psychology, and this is also what we have been emphasizing with a causalist tone.

All this is, however, compatible with the existence of psychological laws and true generalizations with a high degree of lawlikeness. For instance, the problematic notion of intentionality, often thought to be the source of "real human freedom", seems analyzable in terms of overt behavior (potential and actual overt speech) and hence in terms of some normative and factual characteristics which do not present obstacles for the existence of psychological laws (see Chapters 3, 7, and 10). This holds both for psychophysical laws and "psycho-psycho" laws (viz. laws containing only psychological predicates). Notice that Davidson's argument above was concerned with psychophysical laws only. However, psychology can hardly manage with only "psycho-psycho" laws — the agent's subjective world has got to be connected with "objective" reality (cf. here the normal condition assumptions for a simple "proof"). But if psychology, after all, could manage with laws couched merely in psychological terminology, Davidson's argument would of course have no direct relevance.

The brief remarks above on the conceptual possibility of psychological laws (deterministic or indeterministic) will have to suffice for now. Our account does not, in fact, rely on the existence of any *specific* factual psychological laws, and hence we do not need a more elaborate defense here. We accept singular explanations such as given by (*32*) or $C(a, b)$ as perfectly good explanations. It is only that they are to be justified by adding an explanatory law to establish the required ϵ-argument. That we are not able to provide such a law at the present state of science is not a deep criticism against our position.

We have required of explanations that the ϵ-arguments involved should convey understanding concerning the explanandum. We have discussed this idea earlier and located the causal pattern of understanding and the hermeneutic pattern of understanding, which are present in each IT-explanation. In addition, notice that the hermeneutic pattern is really presupposed by each IT-explanation in that such an explanation requires that the explanandum-behavior be understood as intentional action. Furthermore, as purposive causation refers to the agent's conduct plan we get the further hermeneutic understanding provided by citing the agent's reason for his acting.

One important dimension of psychological laws that we still have to say something about is the *scope* of such laws. What kind of behavior can psychology possibly capture in its nomological nets? In our attempt to see this, we recall our earlier formalization of a singular action-explanandum:

(*23*)   Mod-Verbs (A, o, b) & $T(o', b)$.

As remarked earlier, it does not seem very likely that such idiosyncratic features (e.g. spatiotemporal ones) as the predicate T stands for would be law-governed. That is, propositional attitudes, possibly together with some environmental factors, will hardly determine such factors. Thus, action-laws are unlikely to cover more than the first conjunct of (*23*), it would seem.

What kind of action types (Mod-Verbings) could an action-law possibly "catch"? One cannot a priori say very much about this, of course. Something negative can be asserted, however. We have emphasized that all actions involve a public *result* — an end event or state logically intrinsic to the action. Consider now complex actions such as building a house, travelling around the world, taking a Ph.D. degree, playing Beethoven's violin concerto, or winning a game of chess (cf. Chapter 10). A successful performance of each of these actions requires complicated events (or states of affairs) to take place (obtain). Most of these events are external to the agent's body.

There is some reason to think that action types of the mentioned kind cannot be strictly lawfully connected to an agent's proattitudes. At best, the latter are weak probabilistic determinants of such actions, one may argue, because obviously so many factors external to the agent effect the action's result. These factors can be, e.g., purely physical, social, socio-psychological, etc. in nature. The point I am trying to make here is that complex actions of the above kind need not even in principle belong to the scope of strict *psychological* laws. Granting this, it might still be possible that such action types are deterministically connected to some "mixed", suitably complex, set of psychological, social, physical, etc. factors. If so, we would have strict action-laws, to be sure, but they would not be purely psychological any more.

This would, accordingly, mean a limitation in the scope of psychology (as it is traditionally conceived of). Indeed, some psychologists have gone so far as to suggest that psychology "ends" with something like action tendencies. That is, psychological laws could have as their consequents only predicates representing action tendencies, and psychology could thus not be concerned even with simple bodily actions. (The neo-Lewinian motivation theory of Atkinson and Birch (1970) seems to amount to this.) In the terminology we have been using, perhaps we could alternatively suggest that on a priori grounds one can at best guarantee that psychology, in the consequents of its laws, gets to an agent's effective intendings (try-ings) or, in Rylean terminology, to an agent's settings himself to do, or his embarkings on something.

Even if psychology could not capture interesting complex action types in its nomological nets and even if there were no other way of doing it, we may get some comfort from noticing that the ordinary natural sciences are in a rather analogous position. For instance, it is almost impossible for a physicist to give a detailed physical explanation, not to speak of a prediction, of the path and velocity of a falling leaf, say. This is because so many factors outside the scope of mechanics affect the situation.

Even if there were in principle no deterministic or non-deterministic laws to be found in the case of complex actions of the mentioned kind, our backing law account enables us to claim that complex actions are, nevertheless nomically produced and, hence, nomically explainable. We have been emphasizing that action-explanations are typically singular as to their explicit form. Thus, singular statements of the kind '$C^*(a, b)$' (ontic explanatory statement) or '$D_1(a) \rhd\!\!\rightarrow D_2(b)$' (linguistic explanatory statement) may give perfectly good action explanations, where a and b represent a conduct plan activator and a singular action, respectively, and '$D_1(a)$' and '$D_2(b)$' their respective descriptions. The role of nomic laws (or processes or connections) here is only implicit, although it is essential.

In the more general case, we may have singular action-explanations not only in terms of direct purposive causation but also in terms of *indirect* purposive causation $(IC^*(-,-))$ and *indirect* purposive *generation* $(IG^*(-,-))$. (Our theses concerning IT-explanation in Chapter 8 must accordingly be relaxed to include explanation by purposive generation in general and not only purposive causation; but this is a minor point.) In linguistic terms, a singular causal conditional '$D_1(t) \rhd\!\!\rightarrow D_2(r)$' would represent a singular action $\langle t, \ldots, b, \ldots, r \rangle$ (see our earlier treatment). Here b cum r would represent an overt action such as the shooting of a rabbit, viz. b = the bodily movements required for firing the gun and r = the rabbit's dying. How informative the explanation given by stating '$D_1(t) \rhd\!\!\rightarrow D_2(r)$' exactly is depends on what is contained in the agent's conduct plan, a part of whose activator t is.

The essential thing to notice here is that a nomic process (a backing theory) must be involved here in order to make the statement '$D_1(t) \rhd\!\!\rightarrow D_2(r)$' assertable. When we speak of a backing law or theory here, two things must be emphasized. First, it is naturally assumed that the backing theory correctly represents an underlying nomic process or connection *in rerum natura*. Secondly, as our definition of indirect causation (and generation) shows, we are in fact dealing with a kind of conglomeration of different backing laws (theories) in each situation. Indeed, these laws backing the different links of the nomic chain may be so different as to belong to different sciences. Furthermore, if complex action *types* are not lawfully determined at all, such conglomerations of backing laws will in principle be *different* for each instantiation of the complex action (and the relevant proattitude).

What we have just sketched seems to fit explanations in psychology quite well. Let us illustrate this by means of a couple of examples. First, the study of problem solving is an area to which our account applies beautifully. Within the *Cognitive Simulation* approach, an agent's problem solving is explained by simulating his cognitive processing starting from some elementary cognitive processes and the given "materials" of the problem solving situation (cf. Newell and Simon (1972)). The explanation of the agent's problem solving is typically given in terms of a flow diagram (or, if in fuller detail, of a program). What we get is a sequence of connected singular computational "steps" or episodes (cf. e.g., the solution of the "missionaries and cannibals" problem or the "Hanoi's tower" problem). This is a singular "software" narrative, which corresponds to the use of $\triangleright\!\!\rightarrow$, applied to mental events. The underlying "hardware" processing, which "obeys" the software description (the program) corresponds to our indirect causation $IC^*(-, -)$. This correspondence is indeed a precise one if the doctrine CSF (computational state functionalism) sketched in Chapter 5 is acceptable.

Another example of singular psychological explanations to be mentioned here is *psychoanalytic narratives*. We may explain a neurotic outburst by referring ultimately to a traumatic experience in childhood. A singular event is thus explained by telling a singular narrative which starts from the traumatic experience. However, this narrative is backed by a law, if we are to believe in Freud's theory, and thus it becomes fairly clear that and how our account in terms of singular explanatory conditionals applies here (when extended to mental episodes). (See Sherwood (1969) for a nice analysis of psychoanalytic explanations in terms of singular narratives.)

A third example illustrating the applicability of our approach to explanation is *historical narratives*. Let us consider a well-worn example: the explanation of the outbreak of the First World War. Let us suppose that it is adequate tŏ say that the assassination of the archduke of Austria was a contributing cause to the outbreak of the war. Therefore, this incident may be regarded as a partial singular explanatory event for the outbreak of the war. This singular explanation is usually made more precise by referring in more or less detail to the following mediating events. The assassination of the archduke brought about the Austrian ultimatum to Serbia. The ultimatum made Russia mobilize its troops. This again contributed to Serbia's refusal to comply with all of the conditions of the ultimatum. As a consequence the Austrian declaration of war followed.

Suppose the above outline is at least approximately correct. This singular narrative seems to fit our idea of explanation rather well by means of indirect singular causation (and generation). We may thus, perhaps, say that the assassin's action of killing the archduke generated (or contributed to generate) the Austrian declaration of war and that, accordingly, the assassin started the First World War in being a partial agent of its outbreak. It should be emphasized that there is, of course, no law in history which connects shootings of archdukes to declarations of war, or anything of the sort. Our account of explanation does not, however, require that. What is presupposed is an underlying nomic process, which in this case looks very complicated, no matter how it is conceptualized. Consider, for instance, the Russian mobilization. Behind it there may be a nomic connection roughly with the content that in situations of experienced danger (singular and collective) agents (with such and such probability) undertake measures of self-defence. In history the backing laws may be, and typically are, psychological, sociological or economic. A historian often is not interested in finding laws at all, nor need he in the first place, if our account is correct. Very likely he would not be able to find historical laws, for there may not be any. Instead he may give singular explanations which are nomically backed in different ways in different situations.

NOTES

[1]   Very briefly, the mentioned conditions (V) and (Q) are needed to further specify the substitution process that condition (5) above deals with. Roughly, (V) specifies that no new extralogical constants occur in the substituting statements. (Q) again essentially serves to preclude the introduction of new rival theories (forming a new explanans Tc_5') as disjuncts added to one or more of the members of the original set Tc.

With the above "reading specifications" for condition (5), we seem to be able to capture, by our model, those deductive arguments that according to their form seem to be acceptable explanations. A trivial example would be the explanation of a singular statement $Q(a)$ by means of $Tc = \{(x)(P(x) \rightarrow Q(x)), P(a)\}$.

[2]   If we decided to include only *singular* statements in our conduct plan descriptions we could either use $Th(\mathcal{M}) \cap S$, where $S =$ the set of singular statements of our language, or we could use the (slightly stronger) model-theoretic notion of the *diagram* of $\mathcal{M}$ instead. Both solutions would do equally well in solving our present puzzle.

[3]   It seems that the account of purposive causation developed above can be fruitfully applied to the almost completely analogous problems within a causal theory of perception, where the causal chain goes from external events to cognitions. But a detailed story of that will have to be left for another occasion.

[4]   What we are doing here can alternatively be done by means of the elegant *protosemantics* of J. Rosenberg (1974). While the Tractarian view I espouse here makes linguistic *facts* isomorphically represent non-linguistic ones, in Rosenberg's protosemantics, on the contrary, complex linguistic *objects* picture complex non-linguistic objects by suitably "designating" and "portraying" them.

We have been assuming that the "working of the mind" presupposes a representational system used for *picturing* the external world. If someone wants to dispute this plausible looking idea he has to produce an alternative account, at least.

[5]   Let us briefly consider a recent and perhaps the most sophisticated analysis of *the* cause due to Martin (1972). His "definition" for *the* cause goes as follows (cf. Martin (1972), pp. 226 and 231):

c was the cause of $e(d)$ relative to ϕ, only if

(1) c was a condition of $e(d)$, and
(2) No factor, c', was present in ϕ, such that
    (i) c is an instance of some kind D, and c' an instance of some kind C, and
    (ii) it is logically necessary that a factor of kind D occurs only if a factor of kind C occurs, and
    (iii) there were factors that were present in the effect situation and not sufficient for $e(d)$ which together with any factor of kind C occurring in the effect situation would have been sufficient for $e(d)$.
(3) c is sufficient for $e(d)$ when c is conjoined only with conditions of $e(d)$ that are factors of a kind present in ϕ.

I have used Martin's own terminology. In this definition 'c' and 'e' are "variables which range over *individual* factors"; 'd' is a variable which ranges over factor descriptions and "'d' specifies those aspects of e for which c purports to provide an explanation". 'C' and 'D' range over *kinds* of factors (such as events), and 'ϕ' is a variable which ranges over comparison situations. That c is a condition of $e(d)$ means that it is a contributory cause of $e(d)$.

First we notice that Martin's definition only gives a necessary condition for *the* cause. (I think it could, in fact, be taken as a sufficient one as well.) Secondly, Martin says very little about what indeed a comparison situation is and next to nothing about the selection of a comparison situation.

I find clauses (1) and (3) basically acceptable. However, (2) may seem more problematic at least because of its strong subclause (2) (ii). Much depends here on one's criterion of identity for events (and other causal factors). Given our account in Chapter 2, I can find no clear counterexamples to clause (2), however. Thus Martin's above conditions seem to me basically acceptable at least in the case of simple (non-composite) events.

⁶   That IT-explanations contain reference to an agent's intendings and do not settle for less helps to solve a difficulty which affects weaker forms of action-explanation. In saying this I have in mind a problem that could be called the *practical explanatory ambiguity*.

To illustrate this problem, suppose an Agent A intends something p and that he also intends something q. Suppose the intention p requires A to do X. It may now be that the intention p & q requires A to do $\sim X$. For instance, p could be the intention to be a good teacher and q the intention to be a good scientist and X would be the action of writing lectures during one's vacation, while $\sim X$ would involve doing research during one's vacation. This example could be taken to satisfy the above requirements.

Given that having the intention p & q entails having the intention p *and* having the intention q, A would be required to do both X and $\sim X$, which is impossible. For a practical reasoning with the intention p would lead to X while a practical reasoning starting from the intention p&q (or the intention p, while the intention q also is endorsed by A) would lead to $\sim X$. This is the phenomenon of practical explanatory ambiguity, as the intention p would be an explanatory intention both for X and $\sim X$. Fortunately, the whole paradox does not arise for us, as we have explicitly specified (in Chapter 6) that, on conceptual grounds, A cannot both intend p and intend q (and thus intend p & q), if they presuppose incompatible actions to be performed at a certain time. In the case of *wants simpliciter* this mentioned paradox does arise, however.

⁷   In this connection I would like to make two critical comments on Nordenfelt's covering law approach to action-explanation. Nordenfelt (1974) suggests that all action-explanations should be treated as explicit or implicit deductive-nomological explanations. His general idea is thus somewhat similar to ours. However, Nordenfelt's

analysis is unfortunately marred by serious flaws. Let me mention two grave ones here.

First, Nordenfelt thinks that deductive explanation amounts to logically inferring an explanandum from a causal law and an initial condition statement (Nordenfelt (1974), pp. 42, 71). Thus, he has to face a great number of well-known formal paradoxes of explanation; his model is much "behind" even Hempel's and Oppenheim's old formal model (see Tuomela (1972) for a list of such paradoxes).

Another great difficulty with Nordenfelt's approach is this. He proposes that the following covering law be used in all explanations of intentional actions X in terms of an end P (I state it verbally only): For all X and Y, if agent A believes that he is in circumstances Y, and if A believes that in circumstances Y he cannot do P unless he does X, and if it is possible for A to do X, then it causally follows that A does X. But we know already on the basis of our common sense that this law is flatly *false*. If A does not want to do P then he will not (probably) do X. Another difficulty with Nordenfelt's law is that the notion of cause that it employs is something like a Humean cause (and at least not anything like our purposive cause), and hence Nordenfelt's analysis is open to paradoxes due to internal and external wayward causal chains (cf. Section 2 above).

⁸ A fairly good example of an exact action-law, even if not a deterministic one vis-à-vis particular actions, is the so called *probability matching law* (cf. Chapter 12). It is concerned with choice situations of roughly the following kind. On each trial the agent is supposed to predict which one of n outcomes will occur. His performing a certain action X_i, $i = 1, \ldots , n$, marks this prediction. On each trial exactly one outcome occurs and each outcome has a certain fixed probability (π_i) of occurrence. The probability matching law now says that asymptotically each agent will perform action X_i with probability π_i.

The probability matching law seems to be true for a large variety of experimental situations (for a review and discussion see, e.g., Tuomela (1968)). It is an amazingly exact law for a psychological law. Another surprising fact about it is that in obeying it people act somewhat irrationally: to maximize the number of correct predictions in a probabilistic series of outcomes they should act differently; e.g., in a two-choice experiment they should choose the more probably rewarded alternative with probability one.

CHAPTER 10

BASIC CONCEPTS OF ACTION THEORY

1. BASIC ACTIONS AND ACTION TOKENS

1.1. It is a central feature of our causal theory of action that the basic action concepts involve causality. As we now have developed an account of purposive causation and generation we can refine our action concepts accordingly. We will define in detail such concepts as basic action, bodily action, generated action, and action token. We will also analyze arbitrary *complex* actions below and thus lay bare their structure. Everything we have said about actions and action-explanations in the earlier chapters will remain intact, for we will only add some more details into the total picture.

Before starting, there is one general remark to be made. Our causalist theory, in general, analyzes actions by reference to some mental episodes (such as effective intendings, i.e., tryings) purposively causing or generating them. This is to be taken as a broadly factual claim about actions; i.e., about what they are ontologically, i.e., in the order of being. However, as our concepts of mental episodes ultimately are built on the basis of overt, or Rylean, concepts, our action concepts are ultimately overt, too. Thus, conceptually, or in the order of conceiving, actions are overt entities, while in the order of being they extend "behind the overt surface".

In the present section and the next we shall study various action types: simple and compound basic action types as well as some complex action types. By an action type we mean a *generic* action (in some other writers' terminology). Thus, such "*X*'ings" or "Verbings" as arm raising, signalling, saying hello, promising, and building a house might be regarded as some kind of generic actions or action types. As a matter of fact, in order to get a learnable and manageable language to communicate with and to think in, we ultimately want a *finite* list of core action predicates such that all other action predicates (properties) are somehow constructible on the basis of it (cf. Chapter 2). Presently we, of course, do not know how, exactly, this finite list of core action properties will look.

Below we will characterize action types essentially in terms of two notions: 1) purposive causation (and, more generally, generation) and

2) action token. We have already discussed purposive causation and generation, and we are going to discuss action tokens in more detail later.

Here it will suffice to say that action tokens (action tokens$_1$ in our later
terminology) are (possibly complex) singular events. Since we basically
treat behavior as process-like, it is natural to think of actions (action
tokens) as complex events which are causal or generational sequences of
events. Thus an action token in general has the structure of a finite
sequence of events $\langle v, \ldots, b, \ldots, r \rangle$. In this sequence of events v
represents a "mental cause" of the overt bodily behavior event b. The
result-event r represents the *achievement-* or *task*-character of action: r
is a publicly and overtly accessible bodily-generated event which *intrinsically* belongs to the action and which serves as a criterion of whether the
action is successful or whether the task to which the action is a "response"
has been performed. (Notice that action tokens in this sense form a proper
subset of the set of exemplifications or instantiations of a property.)

We think that action tokens can be understood and treated to a great
extent independently of a systematic characterization of action types.
Action tokens are for us more basic (at least ontologically) than systematic-
ally characterized action types. However, there will be some dependence
on action types as will be shown later. But let us now proceed to a discussion of action types and leave a more detailed discussion of action tokens
until later.

In recent literature a dichotomy of actions (action types) into basic
and non-basic ones has often been made, and so have we in our earlier
chapters. Basic actions are thought to be primitive actions in some strong
generational sense: they are actions that an agent does not do *by* or
through doing something else. On the other hand, basic actions are usually
supposed to generate all the other actions and thus to play the role in
action theory "basic cognitions" or "protocol knowledge", etc., plays or
was supposed to play at least in empiricist epistemology. Thus, for instance,
Davidson summarizes his view by saying that "our primitive actions, the
ones we do not do by doing something else, mere movements of the body —
these are all the actions there are. We never do more than move our bodies:
the rest is up to nature." (Davidson (1971), p. 23).

Our quotation brings out another aspect in addition to the generational
one that basic actions are often thought to have: they are *bodily* actions,
whatever that exactly means. A third aspect is that basic actions are
something an agent has the power to do in standard (or normal) circumstances. They belong to his *action repertoire*.

In spite of the heavy criticisms directed against most of the attempts at providing a sharp criterion for classifying actions into basic and non-basic ones, we shall below make another attempt which takes into account the three features above and which is not open to the criticisms made in the literature (cf. Baier (1971)).

Our aim in this chapter is to show that, perhaps excepting psycho-kinetic actions (if there be any), in a suitable sense *all actions* are conceptually built out of simple or compound bodily actions through some kind of generation and that everything an agent *can do* is reducible to his simple and compound basic actions. Causal notions are employed essentially in carrying out this programme. The only relatively detailed attempt along these lines that I know of is that by Goldman (1970). However, his attempt is inadequate for a number of reasons: his definition of a basic action is faulty and he does not, after all, give any detailed account of the central problem of how complex actions are supposed to be built out of (simple or compound) basic or bodily actions.

Let us start our analysis by considering Goldman's definition of a basic action type:

> Property P is a *basic action type* for agent A at t if and only if
> (a) If A were in standard conditions with respect to P at t, then, if A wanted to exemplify P at t, A's exemplifying P at t would result from this want; and
> (b) the fact expressed by (a) does not depend on A's generational knowledge nor on A's cause-and-effect knowledge, except possibly the knowledge that his exemplifying P would be caused by his want. (Goldman (1970), p. 67.)

Let us consider condition (a). First, I argue that wants do not bring about actions without qualifications. Even if I accept that wanting to do X is conceptually linked at least with the disposition to do X, wanting (contrary to intending) need not even in standard or normal conditions result in any action at all. Secondly, Goldman construes the 'if-would' relation as an *ordinary* causal one. But as the many paradoxes concerning "wild" causal chains in such cases show, we need purposive rather than mere causation.

Concerning condition (b) let me just remark that it seems clear to me that in raising my arm and in shrugging my shoulder, which both are supposed to be typical examples of basic actions, I do rely on a great amount of *tacit* (at least anatomically and physiologically describable) knowledge. Of course such knowledge is very seldom conscious, except perhaps in the case of, for instance, a physiologist.

   What is the internal impetus of my raising my arm if it is not my mere wanting? I would say that it is my effective intention to do so. We recall than an intention is effective when the time for carrying it out has arrived, and it becomes a *trying*-event. Thus my trying to raise my arm is what under normal circumstances results in my arm going up and hence in my arm raising. My raising my arm, $u = \langle t, u_o \rangle$, is made up of my trying t to raise my arm and my arm rising ($= u_o$, also describable as my overt arm raising), which is a consequence of t. Thus, by using trying ($=$ effective intending) instead of wanting we can take care of the standard repertoire aspect of basic actions. When this central requirement is complemented by the two other requirements that a basic action be non-generated and that it be a (maximal) bodily event we arrive at the following causal characterization in the case of a simple (non-compound) property X.

(*1*)   Property X is a *basic action type* for A at T if and only if for any singular event u exemplifying X at T it is true that:

     (1) If u had the internal structure $\langle t, b \rangle$ and if A were in normal conditions with respect to X at T and if he truly believed that, then the causal conditionals '$D_1(t) \rhd\!\!\rightarrow D_2(b)$' and '$\sim D_1(t) \rhd\!\!\rightarrow \sim D_2(b)$' would be true in that situation. Here D_1 describes t as A's trying (by his bodily behavior) to exemplify X and D_2 expresses that b is the maximal simple overt bodily component-event of u and that b is the result event of u (*as an X*).

     (2) There is no statement D_2' which satisfies clause (1) as a substitute of D_2 and which is such that the conditional '$D'_2(u) \ \bigcirc_{\overrightarrow{2}} \ D_2(u)$' is true in this situation by virtue of the semantic principles internalized by A.

There are several clarificatory remarks to be made concerning this definition, which characterizes a basic action type X in terms of some of its exemplifications. We start with its clause (1) and leave (2) and the definition of $\bigcirc_{\overrightarrow{2}}$ until later. First, we emphasize that all action types must have some intentional tokens, i.e., tokens of the form $\langle t, \ldots, u_o \rangle$. Furthermore, we exclude the vacuity of clause (1) and assume that X has some tokens of the form $u = \langle t, \ldots, b \rangle$ (i.e. $u_o = b = r$). (We could alternatively have done this by requiring X to be a bodily action type, cf. (6) below.)

We assume now that u has the structure $\langle t, \ldots, b \rangle$ or $\langle t, b \rangle$, for short. (In our later terminology u will be an action token$_1$, in which the result event coincides with b.) The event t is describable as A's trying by his bodily behavior to exemplify X. We notice that even in this simple case of a basic action the event t activating A's conduct plan need not be only an effective intending but it may also comprise (i.e., make true a statement about) the belief that trying to exemplify or bring about X causes a bringing about of X.

What are the normal conditions in clause (1)? They involve that (a) there is an (external) opportunity for action (e.g., one can try to raise one's arm only if it is not raised); (b) the agent is not in any way prevented by external forces (e.g., one's arm should not be tied, etc.). On the other hand, in the present case normal conditions are not assumed to include internal states of the agent's body. Thus one may try to raise one's arm even if it is paralyzed, and one has some reason to believe it.

Our notions of intending and trying do involve conceptual elements (see Chapter 6). As a matter of fact, on our strong interpretation of these notions the normal conditions in the above sense must hold before one can effectively intend (try) to bring about X. Furthermore, the agent's correct belief and even knowledge (in a rather weak sense of awareness) of the obtaining of normal conditions is also entailed by our notion of trying. Thus for our present purpose an explicit standard conditions-clause can be regarded as superfluous. But we use the standard conditions-clause to leave room for other — weaker — interpretations of intending and trying and to allow for the conceptual possibility of more complex (with respect to normal conditions) basic action types than arm raisings and the like.

Our clause (1) does not only clarify the standard action repertoire-aspect but it also in effect requires that a basic action be a bodily action. For it requires b to be described (or at least indirectly expressed) as a maximal simple overt bodily component-event of u, and as the result-event of u (as an X). First we notice that b is an *overt* bodily event — it cannot thus be a mental event. (Analogously we required t above to be A's trying to exemplify X by his *bodily behavior*.) But b must also be a maximal bodily event, while not being "more" than a bodily event.

We require b to be a maximal bodily event in order to exclude from basic actions such examples as exerting oneself to raise one's arm and flexing one's arm muscles, which both cause the bodily event of the arm going up and which have been regarded as fully intentional actions (cf.

McCann (1972)). If we accept them as basic actions then either we were wrong in assuming arm raisings to be basic actions or then we were mistaken in accepting that basic actions could not be causally generated by other actions. The assumption that b is a *maximal* event excludes these cases. For I take maximality to entail that for no other event $e \neq t$ in u is it the case that $C(e, b)$, which suffices for our present purposes.

We also required b to be a simple (as opposed to a composite) event.[1] Thus the bodily event involved in my shrugging my shoulder *and* bending my left little finger would be a composite (non-simple) event in this sense, whereas both the shrugging of the shoulder and the twisting of the left little finger would separately qualify as simple events and as b's. I admit that the present simple-composite distinction may be problematic, but for the purposes of this chapter it is assumed to be clear enough.

Another term assumed to be sufficiently clear is 'bodily'. Thus, we presuppose a distinction between the body of an agent and his (physical) environment, and that is hardly a seriously problematic distinction in our present discussion.

Intentional basic action tokens u do not have any agent-external result component r. Thus, my opening the door involves the result event of the door becoming open. Hence opening the door cannot be my basic action type according to the above definition, though opening the door (let's say by pushing it) might well belong to my standard action repertoire (cf. our later definition of this). Still there is a closely related candidate for a basic action: my bodily action involved in (or required for) my opening the door. In other words, we may pick as our b the bodily movements involved in opening the door, and surely that is always possible. Thus, by means of the above kind of *functional* (indirect) characterization we can get hold of many bodily movements in our standard repertoire for which we do not have better names or descriptions.

One thing to be considered still is in what sense 'the bodily action involved in my opening the door' really is a description of an action. First we notice that the bodily movement in question may seem (in some sense) to generically differ from case to case. But even if that were the case and there were no intrinsic (non-functional) descriptions of this bodily action, in each singular case the bodily action "is there" — you may, perhaps, point at it. So it presents no great problem for us.

The bodily action $u^* = \langle t, b \rangle$, involved in the singular action $u = \langle t, b, r \rangle$, where r = the door opening and which represents my opening the door, is a proper part of the latter action (recall our characterization of

t). They differ only with respect to the presence of the result r. But r is in a sense the contribution of nature rather than by me. Thus, though u^* is obtained by means of a conceptual abstraction from u there is this much of real difference. On the other hand, u and u^* are the same as far as my own direct contribution is concerned. Whatever self-knowledge I have of t and b in the case of u I also have in the case of u^*. Similarly, if t purposively causes b in the case of u it does that in the case of u^* as well (as is also seen from my definition in Chapter 9). That b is the result of u^* should be clear from the very description of u^*. Thus, we conclude that if u satisfies clause (1) of our definition of a basic action type, u^* does so also (cf. our later definition of action token$_1$).

We have not yet commented on the causal conditional statements occurring in clause (1) of our definition. What they say is that the trying event t is both a sufficient and a necessary purposive causal condition of b. In general, causes are not necessary in this sense (cf. Chapter 9), but here we need that: otherwise an action token satisfying clause (1) might not be intentional (which we exclude). In other words, when the behavior process is carved up in our analysis, these conditions are what the essence of our causal analysis of basic action comes to. (Note: instead of using the phrase 'in that situation' we could have explicitly spoken of a Tarskian structure $\mathcal{M}$ of an appropriate type as in our definition of purposive causation.) In case u actually occurs (i.e., $(Ex)(x = u)$) we can just require the truth of '$C^*(t, b)$' instead of the causal conditionals. Thus, we see that an essential element of (purposive) causation is presupposed by the notion of action (cf. below). As we recall from Chapter 9, our notions of purposive causation $C^*(-, -)$ and $\rhd\!\!\rightarrow$ can be defined independently of action type concepts; thus no direct circularity lurks here.

Let us now consider clause (2) of (1). While clause (1) entails that a basic action is a non-generated action in the sense of R_1 (or, rather, causal generation), (2) says this for R_2, or rather $\odot_{\vec{2}}$, i.e. purposive conceptual generation.

In addition, clause (2) speaks of the semantic principles *internalized* by A. This means, roughly: the principles have to be implicitly or explicitly contained as a competence-component in A's operative conduct plan (cf. Chapter 3).

We can now see that, for instance, signalling for a left turn (with one's left arm) does not become a basic action type because of clause (2) of our definition: each of A's signallings for a left turn is (in this kind of circumstances) conceptually generated by A's left arm raising (according to A's conduct plan and some semantic rule such as: left arm raising in this situation means signalling for a left turn).

What will be the basic action repertoire of any given agent A according to our definition is a problem which cannot be solved a priori. Our definition involves references to the agent's operative conduct plan and to some causal processes, and all the relevant facts can ultimately be found out only by means of scientific research.

However, we can make "educated guesses" about what a normal agent's basic action types are. I think the reader will agree with me that the following action types will usually be included: raising one's arm, moving one's finger, blinking one's eye, the bodily action involved in tying one's shoe laces, the bodily action involved in signing one's initials. On the other hand, among actions excluded from this list would probably be: the mental act of trying to raise one's arm, flexing one's muscles for raising one's arm, flipping a switch, tying one's shoe laces, signalling for left turn, signing one's initials. (Note, I have omitted the discussion of actions of neglect, e.g., omissions, in the present context.)

Our definition of a basic action type is relative to an agent and to time. That this should be the case is clear from what we have said: the movements required for tying one's shoe laces do not belong to the standard repertoire of a two-year old child, even if they do belong to mine presently and to that child's repertoire when he is five.

A definition of a standard or typical basic action token can be given in terms of standard exemplifications of the property X. Of course, all exemplifications are not standard. Even if blinking your eye may be your basic action type it may be exemplified unintentionally as a reflex. It seems useful to distinguish what we call a standard basic action token from both other action tokens and non-action exemplifications of that basic action type (cf. Goldman (1970), p. 72):

(2)   An exemplification u of property X by A at T is a *standard basic action token* for A if and only if
        (1) property X is a basic action type for A at T;
        (2) u has the internal structure $\langle t, \ldots, b \rangle$ (where the symbols have the same meanings as earlier).

Here we assume that the occurrence of t entails the satisfaction of the normal conditions-clause of our definition of a basic action type. As we shall later see, it follows that if u is a standard basic action token then u is *intentional*.

If X is A's basic action type (at T) then obviously A has the *power* to do X intentionally (at T). Let us consider the following analysis of *power*: If A tried to do X (in normal circumstances), he would succeed. I shall here — without further discussion (cf. below) — accept this conditional (under a causal interpretation) as giving a necessary and sufficient condition for power, provided the element of causality in it is explicated in terms of purposive causation. In case X is a basic action type, this is done as in clause (1) of the definition. Then the truth of clause (1) automatically entails the truth of the causal conditional analyzing power.

We have emphasized that man moulds the world by his actions. But, perhaps to the disappointment of some, our basic actions lack this feature in that the definition of basic actions does not involve a result r which is separate from the behavior b involved in the action. But, of course, this in no way precludes basic actions from having a great variety of external factual effects, both intended and nonintended ones (cf. our notions of purposive and non-purposive causation). We just have not built them into our *concept* of basic action. To be sure, it is of interest to study actions which have *logically* built into them the feature that they have external results. Indeed we shall later put a great deal of effort into the clarification of such actions.

In the present context it may be of interest to take up one special kind of such actions. We shall briefly define a notion of a simple (non-compound) action type which comprises both some direct causal and generational intended effects and some such indirect intended effects. In other words, we define what it means to say that an elementary action belongs to one's (standard) repertoire and thus is something he can reliably do or bring about in standard circumstances. To fully capture this idea we include simple basic action types here, too.

As our definition essentially employs only previously introduced technical concepts and is easily understood we shall be satisfied with just stating the definition here for a simple or non-compound property:

(3)   Property X belongs to A's *action repertoire* (or A has the *power* to do X) at T if and only if either

(1) X is a basic action type for A at T, or

(2) for any singular event u exemplifying the property X it is true that if u had the internal structure $\langle t, \ldots, b, \ldots, r \rangle$ and if A were in standard conditions with respect to X at T and if he truly believed that, then the statement

'$IG^*(t, r)$' would be true for u in that situation. Here t is
described as A's trying (by his bodily behavior) to exemplify
X, and b cum r is described in A's operative conduct plan
K (underlying '$IG^*(t, r)$') as a maximal overt component-
event of u (*as an X*), and r is the result-event of u (*as an X*).

The above definition is thus meant to cover all the basic and generated
simple (i.e., non-compound) actions A can (in a strong epistemic sense)
perform. Thus, for instance, even if A's tying his shoe·laces is not his
basic action, it is, presumably, an element of his simple action repertoire
through its satisfying clause (2) of (*3*). Our definition essentially relies on
the notion of (indirect) purposive generation. Thus, our concept of a
simple action depends on the idea that man intervenes with the world
by producing or generating result-events in the world. As generation
typically involves causation there is accordingly a dependence on causality.
We can in fact put our interventionist idea more generally by saing that an
agent intervenes with the world by means of his bodily behavior thereby
producing certain states of affairs and preventing other states from
obtaining.

It is sometimes claimed that all human actions share this interventionistic aspect in the strong sense of being both sufficient and necessary for
some changes (or, perhaps, "unchanges") in the world. This might perhaps be taken to mean: actions are both sufficient and necessary for their
results. In our framework this could non-trivially be expressed as follows.
Let $u = \langle t, \ldots, b, \ldots, r \rangle (= \langle t, b, r \rangle)$. For the present purposes we
can think of the action minus the result r as a composite event $t + b$.
Then the above interventionistic idea would amount to requiring the truth
of $C^*(t + b, r)$ with a causal backing which guarantees not only the
sufficiency of $t + b$ for r but also its necessity. It seems appropriate,
however, not to require the necessity. This is because of possible cases of
overdetermination. Consider here, e.g., Tom's and John's simultaneously
turning on the light from their separate switches. Thus, as with causes in
general, we cannot invariably require even *ex post facto* necessity.

At this point it is suitable to briefly compare our analysis with von
Wright's non-causalist theory of action. In von Wright (1974), he claims
that a) the concept of action does not rely on the concept of causality
but rather b) the concept of causality is to be analyzed on the basis of
the concept of action. I obviously reject both a) and b) and claim their
opposites (concerning b) see Tuomela (1976a)).

According to von Wright, typically, "to act is to interfere with the course of the world, thereby making true something which would not otherwise (i.e., had it not been for this interference) come to be true at that stage of its history" (von Wright (1974), p. 39). That which is made true is a description of the *result* of the action. An action description (e.g., '*A* opened the window') makes explicit reference to the event which was the result of the action (i.e., the window's becoming open) and implicit reference to the state of affairs which would have continued to obtain had the agent not acted on that occasion (i.e., the state of affairs of the window's staying closed). This implicit reference is formulated in terms of a counterfactual conditional, and von Wright refers to it as "the counterfactual element involved in action".

It is essential to von Wright's account that this counterfactual not be causal. I can accept that, if *productive* causation is meant. However, typically *sustaining* causality is involved. For instance, the window is kept closed by sustaining causation due to its locking mechanism (and whatever else is involved).

What is more, no matter whether sustaining causality is involved here, causality gets involved in another way. It is simply that *A*'s bodily behavior (involved in his action) causally produces or generates the window's becoming open. This is a "positive" aspect in which causality is involved in action.

Von Wright also tackles the present problem in a certain indirect way. He claims that, contrary to his earlier view, a result is not necessary for action, and he concludes that the existence of a cause of the result is, in fact, *immaterial* to the characterization of something as an action (von Wright (1974), p. 132, also cf. Stoutland's (1976b) endorsing remarks).

Let us consider von Wright's argument for this conclusion in more detail, starting with an example: "Someone thinks he gets up from his chair. At the very same time some (to him invisible) agent, or perhaps the operating of a hidden mechanism, lifts him from the chair. Did *he* then not get up, *i.e.*, must we say that he did not raise himself?" (von Wright (1974), p. 130.)

Von Wright now assumes that it is possible that the agent is in good health, etc. and that he has no reason to doubt that he can get up from the chair. Then, von Wright claims, we must say that the agent acted. However, here the counterfactual statement that the result would not have occurred had the agent not acted is false. This example is, incidentally, concerned with a *basic* action in our sense (i.e. we take it to satisfy our definition (1) as $r = u_o = b$). Because our earlier example was about the overdetermined actions of turning on the light, which were simple *generated* actions, we now *seem* to have examples regarding all the simple actions in the agent's repertoire for showing that the necessity requirement is too strong.

What are we now to say concerning von Wright's conclusion that causation is irrelevant to acting? First, I simply do *not* accept the above example of the agent's getting up from the chair as an action unless the agent by his bodily behavior causally contributed to the result of the action (i.e. to his getting up). Thus, even if the agent's bodily behavior need not perhaps be necessary for the result it must yet be causally productive. If the result occurred completely independently of the agent's trying (by his bodily behavior) to get up or, more generally, independently of his operative conduct plan explicitly or implicitly involving an intention to get up, he did not perform the action of getting up. Furthermore, he did not get up by means of his *bodily behavior* (he might have, alternatively, operated a lifting mechanism) if his bodily behavior was causally (productively) independent of the result coming about. (I do not claim, of course, that the element of productive causation must be *explicitly* involved in an action description. That would be trivially false. Rather, I claim that there is an implicit conceptual dependence.)

Von Wright seems to think that all that matters here is that the agent intended by his behavior the result of the action and that he had reasons to think he was *able* to perform the action. Now, as von Wright analyzes ability in non-causal terms he seems to be warranted in concluding that causation is irrelevant to something's being a result of an action.

Does this not leave it a complete mystery that an agent ever succeeds in moulding the world? Not quite, perhaps, for an irrelevant cause, such as the invisible agent in von Wright's example above, cannot always operate when an agent acts: "our idea of acting, of being able to do certain things, others not, depends upon the not too frequent occurrence and non-occurrence respectively of such discrepancies between causes and actions" (von Wright (1974), p. 132). That this *congruence* requirement is satisfied is a contingency, says von Wright, although it is a condition which the world must satisfy if we are to entertain our present notions of action and agency.

Now one can better see the difference between von Wright's and our views. We have incorporated the idea of causality in our framework of agency, whereas for von Wright it is something external to it. To be sure, we take it to be a contingent fact (if it is) that the framework of agency applies. But if this framework does apply, causality and "moulding the world" is essentially involved. Thus a "contact" between the agent and the external world is guaranteed, so to speak. But von Wright's "conceptualism" still leaves it too much a mystery that man is ever able to change the world, as he does not build this idea of "productive contact" into his conceptual framework. He, so to speak, leaves the "software" (the conceptual) unconnected with, and sharply distinct from, the "hardware" ("the extra-conceptual"). This is basically why we oppose von Wright's account and think ours fares better.

1.2. Until now we have been discussing only simple (non-compound) properties (types). But there are compound action properties like raising one's left arm and lowering one's right arm, or, to mention more interesting examples, performing a dance, or completing a Ph.D. degree in philosophy. The last example can be analyzed as involving passing such and such examinations, writing a dissertation, etc.

Generally speaking, we accept as action properties only suitable simple ones (i.e. an X, such that X may also be "negative", an omission or a negligence) and conjunctions of such simple ones. Thus, $X = X_1$ & X_2 & ... & X_m stands for an action property in general. Here each X_i is a positive or "negated" non-compound property. There will be no disjunctive, conditional or any other truth-functional combinations qualifying as action properties. (Note: in our *language* we may of course use proper negative, disjunctive, conditional, etc. action-describing statements.)

On our way to a characterization of an arbitrary action property we start with compound basic action types. They are essentially conjunctions of simple basic action types such that the agent has the power to perform them jointly.

We again present the definition first and comment on it afterwards (cf. a somewhat related analysis in Goldman (1970), p. 202):

(4)   Property X is a *compound basic action type* for A at T if and only if

      (1) X is the conjunction of some basic action types X_1, $X_2, \ldots, X_m$; and

      (2) it is physically possible for A to be in normal conditions at a single time with respect to each of $X_1, X_2, \ldots, X_m$; and

(3) for any exemplification u of X it is true that if u had the internal structure $\langle t, \ldots, b \rangle$ and if A were in normal conditions with respect to $X_1, X_2, \ldots, X_m$ at T and if he truly believed that, then the following causal conditionals would be true in this situation for $i = 1, 2, \ldots, m$:

    (a) $D_i(t_i) \rhd\!\!\to D_i'(b_i)$

    (b) $\sim D_i(t_i) \rhd\!\!\to \sim D_i'(b_i)$

    (c) $D''(t) \rhd\!\!\to D_i(t_i)$.

Here t is described by D'' as A's trying (by his bodily behavior) to do X (and possibly as the relevant believings); each t_i, $i = 1, 2, \ldots, m$, is described by D_i as A's trying (by his bodily behavior) to do X; and D_i expresses that b_i is a maximal simple overt bodily component-event of the respective event $u_i = \langle t_i, \ldots, b_i \rangle$ tokening X_i in this situation; $b = b_1 + b_2 + \ldots + b_m$. In addition, b is assumed to be the result-event of u (*as an X*).

Conditions (1) and (2) in this definition do not require special comments except that we allow T to be a set of time points and not only a single instance of time; hence 'at a single time' in (2) will have to be understood accordingly. If T is a set of points of time then the basic actions X_i in X can be performed, e.g., in *sequence* rather than parallel in time.

Clause (3) is formulated to some extent in analogy with clause (1) of our definition of a simple basic action type. Thus (a) and (b) simply follow from the fact that each basic action type X_i, $i = 1, 2, \ldots, m$, is exemplified in the situation concerned. We assume that X has some tokens of the form $u = \langle t, \ldots, b \rangle$ and that (3) accordingly cannot be vacuously satisfied.

Condition (c) of clause (3) in effect contains the requirement that t be a *purposive cause* of each t_i. It can be assumed here that each D_i is obtained from D'' by means of a suitable intention transferral process. Thus, we may assume that '$C^*(t, t_i)$' is true for each $i = 1, 2, \ldots, m$, and as a consequence of this (c) becomes true.

We do not presently know much about the ontological relationship between t and the t_i's. Could it be that $t = t_1 + t_2 + \ldots + t_m$ (composition '+' taken in the sense of the calculus of individuals)?[2] In view of how these events are described in A's conduct plan K, I do not think we are entitled to assume that a priori. But I would think that it is correct to require that t contains each t_i as its proper part.[3]

How about the bodily components b_i then? Here I think we can safely assume that $b = b_1 + b_2 + \ldots + b_m$: the total bodily behavior of A involved in his bringing about X is just the composition or sum of the bodily behaviors b_i, $i = 1, 2, \ldots, m$, the agent performs in the situation in question. Furthermore, b will be the result of $u = \langle t, b \rangle$, and b is just the overt part u_o of u. Thus, even if results of compound actions are not in general summative, I think in the case of compound basic actions they are.

We notice that our definitions of a compound basic action type as well as of a simple basic action type are *epistemic* roughly in the sense that A is required to have at least correct beliefs about what he is doing. The difference between the epistemic notions and the corresponding non-epistemic ones would be primarily the use of purposive versus mere causation in the definitions (cf. Goldman (1970) for a discussion of a non-epistemic notion).

On the basis of what we said about the notion of power earlier it is obvious that if X is a compound basic action type for A at T then A has the power to do X (intentionally) at T (also cf. our later discussion).

Analogously with the notion of a standard basic action token, we can define the notion of a standard compound basic action token. Instead of using a full analogue of clause (2) of the definition of a standard basic action token it seems simpler here to require that the token u in question simply instantiates clause (3) of our definition of a compound basic action type.

1.3. Although we have discussed basic actions at length, the most interesting types of action are, however, compound actions with external *results* – at least if we think of man as a changer and moulder of the (external) world. The main aim of this chapter is to outline and sketch in some detail an account of the structure of everything we do (doing understood in a task-performance sense). On our way to that we have to discuss in some more detail our notion of action token, on which our characterizations of various action types relies.

As we have emphasized we think of actions as achievements with suitable causal mental antecedents. They are a kind of "response" to *task* specifications. Actions also essentially involve a *public* element – a result, by which the correctness of the achievement (action performed) can be judged. We have construed actions as sequences of singular events of the general type $\langle v, \ldots, b, \ldots, r \rangle$. On the basis of what we have just said, it should be obvious that the events b and r are intrinsically involved in an action. But according to our causal theory of action, it is also essential to logically or intrinsically include a suitable mental cause in the action. For instance, the problem of which events count as *results* of actions does not seem to me to be solvable without recourse to a suitable causal mental antecedent (cf. Chapter 8). Furthermore, if mental entities are semantically introduced according to our conceptual functionalism then there will be conceptual connections between mental concepts and action concepts. This will ensure the tightness between mental entities and behavior we seem to need while logically permitting the existence of mental states, events, and episodes as ontologically genuine entities (which is needed for them to be proper causes of behavior).

As will be seen from our characterization of complex actions below, it is essential to any action type that it be exemplifiable *at will*, i.e., it must have exemplifications which are a) action tokens and which are b) intentional. In saying this I of course presuppose that action tokens can be distinguished from mere exemplifications. For instance, the property eye-blinking can be exemplified both as a reflex and as an action. Action tokens thus form a strongly restricted subclass of all the exemplifications of a given property. A pivotal role is assigned to action tokens which are intentional exemplifications of the action property. Our characterization of intentionality always intrinsically involves the activation of the agent's conduct plan by means of a trying-event, which ultimately, through conscious or unconscious intention transferral, is about the action type the bodily event b cum the result r involved in the action token exemplify.

Let us now consider an example concerning an action token of a complex action. Assume that an agent A wants to get a cup of creamed coffee out of a slot machine. In order to get his coffee he must insert a coin in the slot, push a certain button for coffee brewing, and push another button for having the coffee creamed. Assume that A performed this complex action intentionally and got his coffee. We may now analyze this action token as consisting of the following complex event:

$$u = \langle t, \ldots, b_1, b_2, b_3, \ldots, r_1, r_2, r_3, r \rangle, \text{ where}$$
$$t = A\text{'s trying to do by his bodily behavior what is required for getting a cup of creamed coffee;}$$
$$b_1 = A\text{'s bodily movement involved in his inserting the coin in the slot;}$$

b_2 = A's bodily movement involved in his pushing the button for coffee brewing;

b_3 = A's bodily movement involved in his pushing the button for cream;

r_1 = the result-event of the coin becoming inserted in the slot;

r_2 = the result-event of the button for coffee brewing becoming pushed;

r_3 = the result-event of the button for cream becoming pushed;

r = a cup of creamed coffee coming out of the machine.

Here the trying-event may be assumed to contain as its proper parts the relevant subintentions t_1 (describable as A's trying to insert the coin), t_2 (describable as A's trying to push the button for coffee brewing), and t_3 (describable as A's trying to push the button for cream). The full action token u can, in fact, be said to contain the subactions $u_i = \langle t_i, \ldots, b_i, \ldots, r_i \rangle$, $i = 1, 2, 3$, although we do not explicitly write out the t_i's in u. Similarly, corresponding to each u_i there is a bodily action $u_i^* = \langle t_i, \ldots, b_i \rangle$ which lacks the agent-external result r_i of u_i.

What is interesting and important about our example is the occurrence of r in it. This full result-event of u is a kind of causal "interaction" effect of the r_i's. We can say here that the composite event $r_1 + r_2 + r_3$ caused r. Accordingly, A's actions of inserting the coin and pushing the two buttons causally generated his action of bringing it about that a cup of creamed coffee came out of the machine.

Notice that, in general, in an action token of a complex action there may be bodily events which belong to basic actions and thus are results at the same time.

We now go on to propose the following definition for an arbitrary action token u. (We consider u as a series of events but, normally, short enough to be itself called an event.)

(5)   Singular event
$u = \langle v, \ldots, b_1, \ldots, b_m, \ldots, r_1, \ldots, r_k, r \rangle$ is an *action token*$_1$ of A if and only if

(1) v is a singular mental event (or episode) activating a propositional attitude of A; v is a direct or indirect cause (i.e., causally contributing event) of the events $b_1, \ldots, b_m$

and an indirect cause or an indirect factual generator
(via the events $b_1, \ldots, b_m$) of the events $r_1, \ldots, r_k$, r;
(2) each b_i, $i = 1, 2, \ldots, m$, is a bodily event which does not
cause any other bodily events in u;
(3) each r_j, $j = 1, 2, \ldots, k$, is a public result-event and,
especially, r is the result-event of the total event u.

This definition should almost be understandable on the basis of what we
have said earlier in this section, but perhaps some additional remarks
are needed.

First, we shall later define action tokens in another way—hence the
subindex 1. Consider now clause (1). It is central that the propositional
attitude (e.g., want, belief) which v activates be a realistically construed
disposition (cf. Chapter 5 on dispositions and Chapter 6 on volitions).
Only then can the event v be describable as a trying, wanting, believing, etc.
as the case may be. We note that as propositional attitudes in action
contexts usually are about actions, there is in our notion of an action
token some conceptual dependence on action types. However, we do not
rely on any *systematic* notions of action types but only on an intuitive
idea of an action type.

Notice, furthermore, that if propositional attitudes and other mental
entities are introduced according to conceptual functionalism there will
be a still stronger dependence on the (presystematic) notion of an action
type. In fact, for many interpretations of v the claim that v causes a b_i
will be circular.

Causation has to be understood as excluding cases of overdetermination
(e.g., by states of shock or the like). In those cases when u consists of
subactions $u_i = \langle v_i, \ldots, b_i, \ldots, r_i \rangle$, the causal chain from v to the b_i's
must go somehow via the v_i's (which are appropriate causes of their b_i's).
In these cases at least $IG(v, v_i)$ probably will be true.

Clause (2) simply requires the bodily events b_i to be (causally) maximal
within u. All the overt bodily behavior involved in u can simply be regarded
as the composite event $b = b_1 + \ldots + b_m$.

Concerning our notion of result in the above definition, it should
be taken in a presystematic sense (which e.g., does not clearly
say whether r_j is extensional or intensional, i.e. description-relative).
Except for cases of bodily actions, the results occur "outside" A. Whereas
the r_j's, $j = 1, 2, \ldots, k$, are simple events (as are the b_i's, $i = 1, 2, \ldots, m$),
r may be a composite event.

If we do not have to display the full inner structure of a complex u, we will below sometimes write $\langle v, \ldots, b, \ldots, r \rangle$ for it; this shows that u is of the same general type as action tokens of simple actions.

In principle our definition of an action token$_1$ covers such "negative" actions as omissions and acts of negligence in which the overt events may be missing. But here we shall not discuss them in any detail.

Let us still emphasize that in our definition we are trying to give some precision to a vague notion of an action as a singular behavioral achievement event. We need the present explication in order not to make our definitions of action types directly circular. Having this much of a definition of an action token is, of course, much better than an ostensive definition or an enumerative list, etc. We shall later characterize action tokens somewhat more precisely in terms of action types.

The real test for our definition of action token$_1$ is of course what it covers and what it does not. First we notice that it obviously covers *standard basic action tokens* and actions factually and conceptually *generated* by them (see our earlier relevant definitions). The claim holds true independently of whether the basic actions concerned are simple or compound.

Secondly, examples like that about the slot-machine fit our characterization. In fact I claim that any typical *token of a complex action* (yet to be defined exactly) fits our definition. We shall not here try and go through even representative examples (see next section). Let me, however, mention two cases. As our slot-machine example shows, *parts of action tokens$_1$* can be action tokens$_1$: each u_i, $i = 1, 2, 3$ is an action token$_1$ in that example. Next, *sequences of action tokens$_1$* can be action tokens$_1$ (cf. the hammering example to be discussed later).

As a third large group, our notion of an action token$_1$ will cover lots of cases of *non-intentional actions*. Thus, standard examples of mistakes (such as misspelling one's name), which are intentional under some other descriptions, are included. But, accordingly, in our characterization, contra Davidson, not every action token is intentional under some description. For instance, a person may accidentally break a vase when doing something else in his room. His breaking the vase may satisfy our definition, for we recall that in clause (1) v (be it a wanting to flip the switch or whatever) need not, of course, purposively cause his breaking of the vase. Another type of action which is not intentional under any description would be my action of scratching my right cheek while writing this sentence. It was indirectly and nonpurposively caused by my intention (of writing) in this situation, we may assume. Still another example would be a deprived alcoholic's emptying the glass of whisky brought in front of him. He perhaps did nothing intentionally at the moment. His largely uncontrolled desire caused his drinking the whisky, which thus is seen to be an action token$_1$.

A somewhat special group of behavior, most (or many) of whose exemplifications are action tokens$_1$, are thinkings-out-loud and in general (typical) *pattern governed behaviors* in the sense characterized by Sellars (1973b) (also cf. Chapter 3).

We may also consider what our definition excludes. First, e.g., stumbling or falling down due to external factors would be excluded – these do not satisfy clause (1). Similarly, clear cases of reflexes, like the patellar reflex, fail to satisfy clause (1) and, I think, (3), too.

There may be borderline cases which our broad definition is not sharp enough to handle in its present form, but we shall not try to qualify our definition further.

Given our definition of an action token$_1$, we can characterize the notion of a *bodily action* more exactly than before, given, however, an "understanding" of the bodily events b_i. We recall that a basic action type of A is something which is in 1) the *repertoire* of A; and its tokens are (in standard circumstances) 2) *bodily actions*, 3) *intentional*, and 4) *non-generated*. Now we want to get hold of a concept of bodily action which does not at least always satisfy 1), 3), and 4), though it trivially must satisfy 2). In fact, we want to respect 4) in that only causally maximal bodily events b_i should be included. But no requirements concerning conceptual generation are needed. This latter remark is true because we are now looking for a non-epistemic notion; hence we can speak of mere causation instead of purposive causation (or generation).

But now obviously our notion of action token$_1$ will be helpful in characterizing a bodily action type: we just require each result r_j to coincide with one of the b_i's, in the general case. To start from simple actions we first get for a simple X (using our earlier symbols):

(6) X is a simple *bodily action type* for A at T if and only if every exemplification u of X which is an action token$_1$ of A is of the form $u = \langle v, \ldots, b \rangle$ such that b is the result of u.

Note that a bodily action type will include many tokens outside A's standard repertoire (e.g., various maximal or accidentally successful performances such as some complicated gymnastic movements). Thus, (simple) basic action tokens form a proper subclass of the class of bodily action tokens as defined. There is one important qualification to our definition which has to be read into it: it must be conceptually possible that X be exemplifiable at will. Thus, even if A may never *de facto* intentionally exemplify X we understand that, at least for some agent, v can occasionally be an effective intending.

Somewhat analogously with our notion of a compound basic action type we now have:

(7)   Property X is a *compound bodily action type* for A at T if and only if

    (1) X is the conjunction of some simple bodily action types $X_1, X_2, \ldots, X_m$ ($m \geq 1$); and

    (2) it is conceptually possible that $X_1, X_2, \ldots, X_m$ are jointly tokened$_1$ at a single time.

We note that (2) is weaker than the corresponding clause for compound basic action types. The present clause does not require it to be satisfiable for A, but it only requires exemplifiability in principle (in some "possible world").

An action token for a compound bodily action type will look like $u = \langle v, \ldots, b_1, b_2, \ldots, b_m \rangle$. If there were a $b \neq b_1 + b_2 + \ldots + b_m$ (corresponding to the "full" result r), it would have to be caused by some of the b_i's, it seems. But that would contradict the causal maximality requirement for the b_i's. Therefore such a b will not occur in u.

2. COMPLEX ACTIONS

2.1. Now we are ready to proceed to a characterization of complex actions. We are going to define the notion of a compound action type. By means of this notion and the notions of simple and compound bodily action type we seem to be able to exhaust all actions. That is, the result will be an exhaustive characterization of the structure of everything we do (or what is in principle capable of being done) in our task-action sense of doing.

Keeping in mind the slot machine example and our earlier discussion of complex actions we can directly proceed to the characterization of an arbitrary compound action type, which contains a non-compound or simple action type as its special case. We consider a conjunctive property

X and we assume, to simplify our exposition that all the conjuncts X_i, $i = 1, \ldots, m$, are simple, i.e., X has been analyzed as far as it goes.

(8)   Property X is a *compound action type* for A at T if and only if

(1) X is the conjunction of some simple properties $X_1, \ldots, X_m$ $(m \geq 1)$ at least one of which is not a bodily action type for A at T;

(2)   it is conceptually possible that each X_i, $i = 1, \ldots, m$, has exemplifications which are action tokens$_1$.

(3) (a) Each action token$_1$ $u = \langle v, \ldots, b, \ldots, r \rangle$ of X is (directly or indirectly) generated by some tokens of the action types X_i, $i = 1, \ldots, m$, occurring as conjuncts in X; and

(b) for each conjunct X_i such that X_i is not itself a simple bodily action type nor does X_i occur as a conjunct in any compound bodily action type consisting of conjuncts of X, it is true that every action token$_1$ of X_i is (directly or indirectly) generated by simple or composite bodily action tokens$_1$ of A.

In clause (3) (a) v is assumed to be describable as an initiating event as required in clause (1) of the definition of action token$_1$; $b = b_1 + b_2 + \ldots + b_m$; and r is the (possibly composite) full result-event of X.

Clause (1) of this definition guarantees that X is not a compound bodily action type. It is allowed that $m = 1$, in which case X is a simple action type, which is neither a basic action type nor a bodily action type.

Clause (2) is intended to guarantee that each conjunct X_i in X is an action type: it must be in principle exemplifiable by an action token$_1$. (According to our earlier reading in Section 1, X is then going to have some intentional action tokens$_1$.)

The central clause in our definition is clause (3), which tries to clarify in exactly which sense everything we do is, somehow or other, generated by our bodily actions, if not by basic actions. We shall below go through (3) in terms of a simple example which should be sufficiently complex to be immediately generalizable.

But let us first note some general aspects of the situation. Clause (3) claims that for each token $u = \langle v, \ldots, b, \ldots, r \rangle$ two generational claims are true. To see their content better let us write out the full internal structure of this token u as $\langle v, v_1, \ldots, v_m, \ldots, b_1, \ldots, b_m, \ldots, r_1, \ldots,$

r_k, $r\rangle$, where now even the v_i's (presumably parts of v) are included. Let me emphasize strongly that presently we know at best that it is conceptually possible that such structures exist as ontologically real. The factual existence of the b_i's and r_i's is hardly in doubt here, though. The question is rather about the v_i's. To be sure, we can in principle conceptually introduce as many v_i's as our linguistic resources allow. But the interesting question here is whether v in some factual sense (e.g., the part-whole sense of the calculus of individuals) ontologically contains the v_i's (as we have in our example cases claimed and assumed). On a priori grounds we cannot say much, it seems. The problem is ultimately a scientific and a factual one. (It also depends on how future neurophysiology is going to carve up its subject matter and build its concepts.) Similarly, the question whether u can always be said to contain the subtokens $u_1 = \langle v_1, \ldots, b_1, \ldots, r_1\rangle$ (cf. X_1), $u_2 = \langle v_2, \ldots, b_2, \ldots, r_2\rangle$ (cf. X_2), etc. in an ontologically genuine sense over and above mere conceptual abstracta is primarily a scientific and a posteriori problem.

We have to clarify the following two generational claims in the case of clause (3):

(i)   the u_i's ($i = 1, \ldots, m$) jointly generate u; and
(ii)   each u_i ($i = 1, \ldots, m$) is generated by some of A's (simple or compound) bodily action tokens.

We shall illustrate these generational problems by means of a new example. (We might as well have used our slot machine example.) It is presented by the diagram in Figure 10.1 representing an action token u for a complex action property $X = X_1 \ \& \ X_2 \ \& \ X_3 \ \& \ X_4 \ \& \ X_5$.

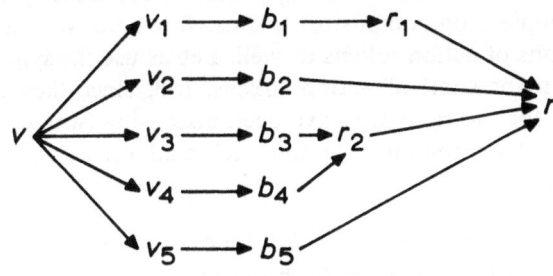

Figure 10.1

In this diagram the arrows stand for (direct or indirect) factual generation. Here we can thus use as an explicate our *IG*, which, we recall, also covers direct generation. The event v activates one or more propositional attitudes of A. For example, v might be describable as a wanting cum believing. In any case, v will somehow be about or directed towards the action X with the result r. The event v is assumed to generate each v_i, and each v_i in turn causally produces b_i. The behavioral events b_i generate the results r_1, r_2, and r as shown in the diagram.

The conceptually possible subtokens of X_1, $X_2, \ldots, X_5$ here are: $u_1 = \langle v_1, \ldots, b_1, \ldots, r_1 \rangle$, $u_2 = \langle v_2, \ldots, b_2 \rangle$, $u_3 = \langle v_3, \ldots, b_3, \ldots, r_2 \rangle$, $u_4 = \langle v_4, \ldots, b_4, \ldots, r_2 \rangle$, $u_5 = \langle v_5, \ldots, b_5 \rangle$. (Here, especially, the genuine existence of u_3 and u_4 is problematic: perhaps we only should have $u_{3,4} = \langle v_{3,4}, \ldots, b_3, b_4, \ldots, r_2 \rangle$.) In order to solve our first generation problem, we note three things. First, we can take $b = b_1 + b_2 + \ldots + b_5$. Secondly, our definitions for causation and generation in principle apply to any events, hence for any *composite* events, too. Thirdly, we are here concerned with non-epistemic notions of generation only. Thus we do not need our notion of purposive generation here but only that of mere generation (*IG*). Thus, when discussing the generation of results by behavior all the v's can be ignored for our present purposes. (They are relevant only in discussing purposive generation, intentional and voluntary actions, etc. Cf. Section 3.)

Accordingly we can accept the following principle of action token generation for two arbitrary action tokens₁ $u_1 = \langle v_1, \ldots, b_1, \ldots, r_1 \rangle$ and $u_2 = \langle v_2, \ldots, b_2, \ldots, r_2 \rangle$:

(P)   If $b_1 = b_2$, then $IG(u_1, u_2)$ if and only if $IG(r_1, r_2)$.

Principle (P) is an obvious meaning postulate for token generation in the cases of simple (non-composite) tokens. But here we want to apply it to compositions of action tokens as well. Let us use the symbol '+' for the composition (or sum) of action tokens, too, since they are events, after all. Then, for the cases we have been discussing (where we have an action token u of a compound action such that $u_1, u_2, \ldots, u_m$ are its subtokens) we accept:

(P*)   If $b = b_1 + b_2 + \ldots + b_m$, then $IG(u_1 + u_2 + \ldots + u_m, u)$ if and only if $IG(r_1 + r_2 + \ldots + r_k, r)$.

(Here we have k results r_j, some of which may be b_i's: $k \leq m$.)

Armed with (P*) we can now take a look at our example. As we see from the diagram, r_1, b_2, r_2, and b_5 generate r. As $b = b_1 + b_2 + \ldots + b_5$

we conclude on the basis of (P*) that $IG(u_1 + u_2 + \ldots + u_5, u)$ is true. This conclusion presents the answer to our first generation problem.

Let us now discuss clause (3) (b) and the second generation problem. This clause is to be regarded as a constraint on X: only those complex actions are to be called compound actions which satisfy the following "reducibility" condition: Each token $u_j = \langle v_j, \ldots, b_j, \ldots, r_j \rangle$ of a conjunct X_i of X such that X_i is not a bodily action type or conjunctive part of a compound bodily action type should be generated by (possibly compound) bodily action tokens$_1$ of A.

In our present example we have to reduce u_1, u_3, and u_4 in this way to bodily actions. We can plausibly assume that u_2 and u_5 are tokens of bodily actions. X_2 and X_5 are hence taken care of.

In the case of u_1 the generation is trivial: b_1 generates r_1, and we have $IG(b_1, r_1)$. Now the token $u_1^* = \langle t_1, \ldots, b_1 \rangle$ clearly qualifies as a bodily action as it is a proper part of the action token $u_1 = \langle v_1, \ldots, b_1, \ldots, r_1 \rangle$. The bodily action type X_1^* consisting of the bodily action tokens$_1$ involved in X_1' ing will be the bodily action type u_1^* tokens here. Although we are not requested to specify that bodily action type, we always seem to be able to do it. We can now apply our principle (P) trivially to get the result that u_1^* generates u_1.

In the case of u_3 and u_4 the generational situation is a little different. For, to make the case interesting, we assume that neither b_3 nor b_4 alone generates r_2, although jointly they do generate it. The properties X_3 and X_4 clearly must be treated together as they have this kind of "interaction" effect. Thus, instead of considering u_3 and u_4 separately, we consider $u_{3,4} = \langle v_{3,4} \ldots, b_{3,4}, \ldots, r_2 \rangle$. Now look at the bodily actions $u_3^* = \langle v_3, \ldots, b_3 \rangle$ and $u_4^* = \langle v_4, \ldots, b_4 \rangle$ involved in u_3 and u_4. As $b_{3,4} = b_3 + b_4$, then principle (P*) gives us $IG(u_3^* + u_4^*, u_{3,4})$, as required. (We could as well have said here that '$IG(u_{3,4}^*, u_{3,4})$' is true, where $u_{3,4}^* = \langle v_{3,4}, \ldots, b_{3,4} \rangle$.) We have completed our example.

Our slot-machine example (i.e. the action of getting a cup of creamed coffee out of the slot-machine) is analyzed along the above lines, but in a still simpler way. Thus, we can easily see that it satisfies clause (3)(b) as each subtoken u_i, $i = 1, 2, 3$, is generated by u_i^* as required. Clause (3)(a) becomes satisfied because r_1, r_2, and r_3 jointly generate r. For then we have $IG(u_1 + u_2 + u_3, u)$ according to (P*).

One may now ask whether action token generation could be *explicitly* defined to get a tidier account. (P*) is only a partial definition as it makes the condition $b = b_1 + b_2 + \ldots + b_m$. Now, if it were impossible that this condition *not* be satisfied, we could simply adopt as our explicit definition the following to replace (P*):

(P**) $IG(u_1 + u_2 + \ldots + u_m, u)$ if and only if $IG(r_1 + r_2 + \ldots + r_k, r)$.

I am hesitant concerning the acceptability of this principle (P**). Ultimately its status is going to depend on how science builds its ontology. Presently I do not have good a priori arguments against it. Perhaps one could think of, for instance, complicated gymnastic movements which have to be carved up into pieces so that the totality b is "more than" the sum of its parts $b_1, b_2, \ldots, b_m$. For instance, b_i might be a causal precondition of b_{i+1}, for some i, and b_i and b_{i+1} might then have some bodily "interaction effects" which make $b \neq b_1 + b_2 + \ldots + b_m$ the case. I am unable presently to invent a convincing counter-example, and, accordingly, I will tentatively accept (P**).

Does our definition of compound actions really cover all complex actions, and is our definition of action token generation sufficient? "Completeness theorems" are very hard to "prove" in cases like this.

To see some of the further problems involved, we consider an example where the agent levels a nail by striking it with a hammer. He holds the hammer in his right hand and keeps the nail straight with his left hand. Each time the nail is struck it goes down a few millimeters. After n strikes the nail has been levelled. We can analyze the situation in the following terms.

Let X_1 be the "non-bodily" action property of striking the nail (with a hammer), and let X_2 be the "non-bodily" action property of holding the nail straight. Each action token of X_1 involves a bodily component b_1 (striking-movement) plus a result component r_1 of the nail becoming struck. Each token of X_2 involves a bodily component plus a result r_2 of the nail staying straight. Now r_1 and r_2 cause r_3 – the nail's going down a few millimeters. So we call $X = X_1$ & X_2 the action type of hammering. Tokens of X have the general form $u = \langle v, \ldots, b_1, b_2, \ldots, r_1, r_2, r \rangle$.

The action Y of levelling the nail by hammering consists of n successive tokens of X. Let us write this out with the obvious symbolism: $Y = X^1$ & X^2 & $\ldots$ & X^n. (Notice that in other examples the conjuncts of Y could, of course, be non-identical properties.)

We can easily see that the tokens of Y are of almost the same general type as those of X. Their general structure is in full $u' = \langle t, \ldots, b_1^1, b_2^1, \ldots, b_1^n, b_2^n, \ldots, r_1^1, r_2^1, r_3^1, \ldots, r_1^n, r_2^n, r_3^n, r \rangle$. (The notation should be self-explanatory; r is the result of u as a Y). Nothing in our definition of action token$_1$ precludes such causal relationships between results as $C(r_1^i + r_2^i, r_3^i)$, of course.

But we seem to be able to say even more. If we substitute for X_1 & X_2 in the definition of Y we can presumably claim that the following identities hold:

$$Y = X^1 \& \ldots \& X^n$$
$$= (X_1 \& X_2)^1 \& \ldots \& (X_1 \& X_2)^n$$
$$= X_1^1 \& X_2^1 \& \ldots \& X_1^n \& X_2^n.$$

But according to the last identity Y is a compound action type. It obviously satisfies clauses (1) and (2) of our definition. It clearly satisfies (3) (a), for r will be indirectly (via the r_3^i's) generated by the r_1^i's and r_2^i's (or $r_1^i + r_2^i$'s) as is easily seen; and this essentially suffices. It is as easily seen, on the basis of our description of the example, that the r_1^i's and r_2^i's are generated by the bodily actions as required by (3) (b) (cf. our treatment of the previous examples).

One interesting feature about our present example is that the performance of each $X^{i+1}(i = 1, \ldots, n-1)$ has as its $causal$ $precondition$ the exemplification of X^i. This kind of "relative dependence" feature is very prominent in many sequential complex actions such as performing a dance, writing a book and also in various social ceremonies and rituals. (Note that the time element T will in these cases usually be a set of time points rather than a single instance.)

Our example concerning a complex sequential action proved to be analyzable as a compound action, and my general conjecture, supported by the above and other analogous examples, is that no more complex types of actions are needed (such as "conjunctive compound action types"). We can thus go on and define:

(9) X is a $complex$ $action$ $type$ (an arbitrary $action$ $type$) for A if and only if X is either 1) a simple or compound bodily action type or 2) a compound action type.

On the basis of what has been argued, we can now claim that all $actions$ are $complex$ $actions$ independently of their other special characteristics (such as their being "parallel", "hierarchical", "sequential", etc.). My thesis (or conjecture) and its earlier clarifications obviously also tells us how it indeed can be so that "we never do more than move our bodies: the rest is up to nature".

It is possible that human beings are *"serial* information processors" roughly in the sense of the Information Processing System (IPS) defined and studied in Newell and Simon (1972). If so, it should be possible to characterize any agent's activities in terms of suitable serial programs involving some feedback loops for subroutines (cf. Chapter 4). This would mean that complex actions are both serial and hierarchical, where especially the subroutines account for the hierarchical aspect. For instance, in the complex action of building a house, for instance, hammering a nail could be a subroutine, perhaps hierarchically involved in the broader subroutine of installing the windows, and so on.

It is not, however, central for our purposes to what extent human agents can operate "parallelly". (Even the IPS can perform two parallel actions, cf. Newell and Simon (1972), p. 697.) Our general system is flexible enough to take into account any parallel actions if needed.

I have defined my notions of causation and generation only for deterministic cases. If indeterminism is true, we must allow for probabilistic causation and generation as well. This addition can indeed be made, and we shall accomplish it in Chapter 12.

We can now, if our above conjecture is true, characterize action tokens somewhat more informatively. We simply say that:

(10) u is an *action token*$_2$ of A if and only if 1) u is an action token$_1$ and 2) u exemplifies either a (simple or compound) bodily action type or a compound action type.

Notice that there are action tokens$_1$ which are not action tokens$_2$. For instance, perceptual takings (notably takings-out-loud) and inferences (notably inferrings-out-loud) are not action types as they cannot be exemplified at will. Still they have action tokens$_1$ among their exemplifications (cf. Section 1).

What is conceptually possible for A to do may, of course, consist of much more than what he has the power to do (intentionally) or what belongs to his "action repertoire". This latter notion can be clarified by defining a concept of a compound action in an *epistemic* sense corresponding to our earlier notions of a (simple and compound) basic action type and "belonging to A's repertoire".

The new epistemic concept has the following features to be noticed. First, as we want to get hold of what A can do we must reduce his complex doings to basic actions rather than to mere bodily actions. Secondly, we must obviously preserve intentionality in generation for those situations where A intentionally exercises his causal powers (i.e., brings about a standard basic action token). Accordingly, we will speak about purposive generation. We thus arrive at the following characterization for an arbitrarily complex X:

(11)   Property X belongs to A's *action repertoire* at T if and only if

(1) X is the conjunction of some simple action properties $X_1, \ldots, X_m (m > 1)$;

(2)   it is physically possible for A to be in standard conditions at a single time with respect to each of X_1, $X_2, \ldots, X_m$.

(3) (a) Each action token$_1$ $u = \langle v, \ldots, b, \ldots, r \rangle$ of X is (directly or indirectly) generated by some tokens of the action types X_i, $i = 1, \ldots, m$, occurring as conjuncts in X, such that when u is intentional (direct or indirect) purposive generation is required; and

(b)   for each conjunct X_i such that X_i is not itself a simple basic action type nor does X_i occur as a conjunct in any compound basic action type consisting of conjuncts of X, it is true that each action token$_1$ of X_i is (directly or indirectly) generated by standard (simple or composite) basic action tokens$_1$ of A. In the case of intentional action tokens$_1$ of X_i (direct or indirect) purposive generation is required.

In clause (3) (a) r is assumed to be describable as an initiating event as required in clause (1) of the definition of action token$_1$; $b = b_1 + b_2 + \ldots + b_m$; and r is the (possibly composite) result event of X. In order to keep (3) readable we have not explicitly mentioned the requirement that A truly believes that he is in normal conditions with respect to $X_1, X_2, \ldots, X_m$. This is implicit in the requirement of purposive generation.

In clause (1) we require that X_i's be simple action properties in the sense of our definition of a compound action type with $m = 1$. In clause (3) we speak of intentional action tokens — we shall soon clarify intentionality in detail.

The above definition of an action repertoire is meant to be a generalization of our earlier definition of an action repertoire with simple action properties. Analogously with the earlier case, we can here claim that an agent *can* do an arbitrary X just in case X belongs to A's action repertoire.

As we analyze "can" in terms of A's causal powers we arrive at the following characterization for an arbitrary (simple or purely complex) action property X:

(12) A *can do* X intentionally at T
         if and only if
         A *has the power to do* X intentionally at T
         if and only if
         X belongs to A's action repertoire at T.

Powers and cans, as defined here, become equivalent notions. To put the content of (12) in somewhat loose terms, A has the power to do an arbitrary action X (or A can do X), if and only if, if A effectively intends to do X (and has appropriate active beliefs) then, in normal circumstances, this purposively generates A's doing X.

Which actions over and above his basic actions A has the power to do is a factual problem. In general, the result of such an investigation can be expressed in terms of generalizations relating his intendings to his actions. We recall from our earlier discussion in Chapters 7 and 9 that it depends on how the normal conditions are analyzed whether such generalizations are analytic or synthetic vis-à-vis the framework of agency.

Admittedly our above notion of power is a strong one. A perhaps has the power to do rather few actions over and above his basic actions in this sense. We shall later, in Chapter 12, define a corresponding probabilistic notion of power, which seems more realistic.

We shall not here discuss powers and cans in more detail except for one further remark. Davidson (1973) analyzes cans in terms of causal powers as we do, though in less technical detail. His basic formula is this: A can do x intentionally (under a given description D) means that if A has desires and beliefs that rationalize x (under D), then A does x. (Here x is a variable for action tokens.) Our above analysis is compatible with Davidson's analysis and can be regarded as a further elaboration of this type of analysis. Notice, however, that in my analysis a rationalizing desire-cum-belief-complex must be an event t (describable as an effective intending cum believing) described according to the agent's operative conduct plan K.

Corresponding to action sequences in the sense of those compound actions we discussed in connection with our hammering example, we can now formulate a characterization for a kind of *indirect* power, that is, power to do something given that something else is done first:

(13)   *A has the power at time instance t_0 to do X_n by t_k if and*
       only if there is a set of actions $X_1, X_2, \ldots, X_i, \ldots, X_n$
       such that

  (1) A has the power at some time instance t_1, $t_0 \le t_1 \le t_k$,
     to do X_1 at t_1; and

  (2) for each X_i either

    (a) $X_i = X_n$, or

    (b) if A does X_i at t_i, this would, together with the state of
      the world at t_i, entail that A has the power at t_{i+1} to do
      X_{i+1} at t_{i+1}, $t_0 \le t_i < t_k$.

Our hammering example may serve as an illustration of this definition. (The above non-epistemic characterization corresponds to the epistemic notion in Goldman (1970), pp. 204–206.)

The main purpose of the present chapter has been to develop a certain version of the causal theory of action to account in a detailed way for complex actions. Although the basic theory has been developed, a lot more could and should be done. For instance, we might want to have finer distinctions for different types of complex actions and we might want to extend the theory in different directions.

Two directions in which the theory might fruitfully be developed are the following. First, one may try to define suitable probability and utility measures for various types of action, for such antecedent factors as wantings and believings, and for such consequent factors as results and (non-intended) consequences of actions. Thus, instead of studying man's deterministic causal powers, we would study his causal propensities (analyzed as probabilities). Then a philosophically more adequate account could be developed than what current statistical decision theory (and game theory; cf. below) has to offer us. In Chapters 11 and 12 we will define probabilities for actions. Utilities could be defined analogously. Thus, the basic elements for a conceptually more sophisticated decision and game theory than we have had till now become available.

Secondly, and most importantly, our account can be applied to the study of various *social* actions and (social) power. The agent A might be regarded as a *collective* agent such as a government, board, or some social group. Furthermore, the action type conjuncts in a complex action $X = X_1 \& \ldots \& X_m$ could be regarded as performable by different agents in a group (cf. the carrying of a heavy table, which requires several agents).

Thus, various *cooperative* and *competitive* social actions, including social ceremonies (cf. wedding ceremony), rituals, and various games, could be profitably studied by viewing them from the angle of the theory of action. Also the concept of *work* can be profitably approached from our point of view; cf. Marx and our earlier analysis in Chapter 4. The details of this as well as an analysis of social actions must, however, be left for another occasion.

Let us finally notice that although we claim to have developed tools for conceptualizing everything it is conceptually possible to do and everything a given agent has the (epistemic) power to do, it is yet a different matter *which* of such complex structures become linguistically represented by concepts in a given society. That is a factual problem to be solved by sociological and anthropological investigation.

3. INTENTIONALITY

With all the machinery developed in the present chapter, we may now return to a sharper explication of the concept of an intentional action, which was earlier discussed in Chapter 3 and, especially, in Chapter 7.

We wish to distinguish sharply between the *semantical* and the *causal* sense of intentionality. Semantical intentionality means simply opaque or transparent referential *aboutness* as discussed earlier. The semantical concept of an intentional action is accordingly to be characterized in terms of what the agent *means* or expresses by his behavior.

On the other hand, we may ask for sufficient empirical or factual conditions for an agent's performing an intentional action. There may indeed be causal conditions sufficient for an agent's performing an intentional action. At least a causal theory assumes that. The concept of a factual (sufficient and/or necessary) condition of intentional action is the *causal* concept of intentionality.

Stoutland (1976b) has criticized causal theories of action for confusing intentionality and causality in the sense of failing to maintain the distinction between the semantical and causal senses of intentionality. He claims that causal theories of action accept the following two theses (Stoutland (1976b), p. 293):

(C1) The concept of intentional action is a causal concept in the strong sense that an agent's action is intentional if and only if it has certain kinds of Humean causes.

(C2) To explain an intentional action is to give its Humean causes; in particular, when we explain an agent's action by giving his reasons for his acting as he did, we specify sufficient causes of the act so that (having) reasons are causes of a certain kind.

(C1) is a thesis about intentionality while (C2) is a thesis about the explanation of action. What are we to say about these theses?

Stoutland seems to mean by a 'Humean cause' a notion of cause analyzed by reference to a suitable backing law. If that is enough, our concept $C(-, -)$ of singular causation would be a Humean concept in Stoutland's sense (although perhaps not in any other interesting sense, cf. Tuomela (1976a)). Still, I can accept neither (C1) nor (C2), for they speak about mere causes instead of purposive causes. Perhaps Davidson's (1963) account of action would satisfy Stoutland's criteria, but ours does not. The heavy criticisms directed by Stoutland (1976b) against the causal theories satisfying (C1) and (C2) do not affect our causal theory of action, for these criticisms mainly hang on this point.

However, if we use 'purposive cause' instead of 'Humean cause' in (C1) and (C2), then they are almost acceptable to us. That is, with the possible exception of having to speak about generation instead of causation in some cases, the thus modified theses (C1) and (C2) would be acceptable. However, it should be emphasized that (C1) then would not be an analysis of the semantical concept of intentionality but of the causal concept of intentionality. It would be a *broadly factual* hypothesis in the sense our causal theory of action is a broadly factual claim (because our causal theory simply entails (C1) in this modified and qualified sense). An agent's acting intentionally presupposes that his suitably related effective intending purposively causes or generates the action in question.

Let us now go into a more detailed analysis of the *causal* concept of intentionality. In Chapter 7 we already presented out general criterion *(7.19)* for intentionality. As said, *(7.19)* represents a broadly factual claim about agents. We may reformulate this (somewhat circular) criterion for the situation where an agent A has performed a singular action u:

(14) u is *intentional* if and only if there is a (complete) conduct plan K such that A purposively brought about u because of K.

(14) speaks about *complete* conduct plans for the reasons mentioned in Chapter 9. We also recall that the above analysis rests on the idea that

intentionality goes with the agent's having an operative conduct plan (such that it culminates in his *trying* to realize the plan). This is compatible with such things as the agent's doing intentionally something he did not specifically aim at. (Cf. also our remarks on non-intentional actions in Section 1.)

Our definition above is to be understood causalistically; that is, the *because of*-relation is to be understood to contain a causal element. However, here we prefer to speak not merely of causal connections but of other generative (both factual and conceptual) connections as well. Basically, we claim that the connections between the event activating the agent's conduct plan and his bodily behavior are causal. But the connections between behavior and various external happenings are not perhaps always causal because of the existence of non-causal nomic and conceptual connections. (Consider, for instance, Jones' trigger-pulling non-causally generating Smith's wife's becoming a widow in the case when Jones shot Smith.) When discussing intentionality we only need to consider action tokens$_1$ in which the initiating event is A's trying t, viz. a trying to bring about by his bodily behavior a b-cum-r-event satisfying his conduct plan.

With these remarks in mind, and in order to get a unified treatment of intentionality, I now make the following claim for an *arbitrarily complex* action token u having the "canonical" form $\langle t, \ldots, b_1, \ldots, b_m, \ldots, r_1, \ldots, r_k, r \rangle$:

(I) u is *intentional* if and only if t (indirectly) purposively generated r.

In other words, it is assumed here that the because of-relation can (in this context) be explicated by purposive generation, i.e. by $IG^*(t, r)$. The intentional and controlled trying-event t activates (or is a part-activator of) the agent's conduct plan, which contains his reasons for acting. (Notice that if $u(= \langle t, \ldots, b, \ldots, r \rangle)$ *as an X* was not specifically aimed at by A, then t cannot be an ontically real trying-to-do-X but it is rather to be regarded only as a more "general" activator of K.) The occurrence of t gives the beginning to intentional activity; the conduct plan K tells how the activity is directed; and the relation $IG^*(t, r)$ guarantees that the intentional direction is kept. *How* intentionality comes to be preserved is psychologically explainable by reference to adequate *feedback-controlled* bodily processes, the exact nature of which does not concern us here.

Our present task is to clarify what the exact content of (indirect) purposive generation formally comes to in the case of various action types. Let us first consider simple action properties X. Then X is either a bodily action type or a generated action type, that is, an action type at least some of whose tokens are generated. We get for a *bodily* action token $u = \langle t, \ldots, b \rangle$ (we only have to consider tokens where $v = t$):

(a)   If X is a simple action type, then u is intentional if and only if $C^*(t, b)$.

If X is a "non-bodily" or *generated* action type then we consider tokens of kind $u = \langle t, \ldots, b, \ldots, r \rangle$ (t, b, r are non-composite events):

(b)   If X is a simple action type, then u is intentional if and only if $IG^*(t, r)$.

We have to recall here that $IG^*(t, r)$ always concerns a generational process which goes via b and which includes direct purposive generation and both direct and indirect purposive causation.

Let us now go to compound actions of the type $X = X_1 \& \ldots \& X_m$. First we consider *compound bodily* actions. A typical token is now of the type $u = \langle t, \ldots, b_1, \ldots, b_m \rangle$ with the understanding that the causal process from t to the b_i's goes via the t_i's of the subtokens. Keeping this qualification in mind we get:

(c)   If X is a compound bodily action type, then u is intentional if and only if $IC^*(t, b_i)$, $i = 1, \ldots, m$.

Finally we consider an *arbitrary compound* action type having a token $u = \langle t, \ldots, b_1, \ldots, b_m, r_1, \ldots, r_k, r \rangle$. How does the causal-generational purposive process go here? Independently of whether the action is "hierarchical", "parallel", or "sequential", it is true to say that t indirectly purposively generates r. The general structure of the process is this: t causes the t_i's (assumed to factually exist here), the t_i's in turn cause the b_i's, the b_i's cause (or otherwise suitably generate) the r_j's which in some generational sense produce r. We recall that even in such simple sequential cases as our hammering example there are several phases in this generational process. To clarify such a process one can profitably employ composite events. But even in terms of simple (non-composite) events it is easy to characterize the situation. First t generates the t^i's ($i = 1, \ldots, n$) corresponding to the hammering tokens h^i. Within each h^i, t^i causes t_1^i (roughly: trying to strike) and t_2^i (roughly: trying to hold the nail); t_1^i

causes b_1, which causes r_1^i; and similarly t_2^i causes b_2, which causes r_2^i. Then r_1^i and r_2^i together cause r_3^i, which again is a causal precondition of r_3^{i+1} in a process which is such that finally r_3^{n-1} causes r_3^n (and r). It would also be true here to say that the composite event $r_3^1 + \ldots + r_3^{n-1}$ causes r_3^n, which conceptually generates r.

There are lots of different generational patterns (regarding the fine details) involved in complex actions but we cannot go through any more examples here. As far as I can see, however, all these examples *can*, in principle, be analyzed in terms of composite events so that our original formal definition of $IG^*(t, r)$ is respected (cf. the hammering example above). Our final clarifying characterization of intentionality for compound actions then simply becomes:

(d)   If X is a compound action type, then
        u is intentional if and only if $IG^*(t, r)$.

This concludes our causalist account of intentionality as we have gone through all the possible action types. Let us just notice here that our "definition" of intentionality relies on the notions of purposive causation and generation, action token$_1$, and, last but not least, on effective intending (trying). It does not, however, directly rely on the various notions of action type. But, because of its dependence on intending, the characterization of intentionality is partly circular — I do not think a causalist analysis in less circular terms can be successfully formulated.

Our analysis of intentionality has obvious connections to the notions of *power* and the having of an adequate *reason*. An agent performs intentionally what he has the power to do and what he has effective reasons for (cf. our definition for a reason in Chapter 8 and our above analysis of power).

Above we have used only deterministic causation and generation when defining intentionality. Although that seems to correspond well with our ordinary way of thinking about intentionality, we may want to relax the above explicantia. We may want to call an action token intentional even when caused or generated by an *objectively* random process, if there be such, and perhaps by means of some kind of high probability criterion. The resulting wider explicantia of intentionality are obtained by using the probabilistic counterparts for our concepts of causation and generation. They will be developed and presented in Chapter 12.

What we have said about the *explanation* of intentional actions in the previous chapters applies to complex actions as well. Thus, in explaining a complex intentional action, we typically give an account of how the agent exercised his causal powers in that situation. This we do by displaying the conduct plan he acted on and by stating that suitable external and internal circumstances, needed for the exercise of the causal powers, obtained. Technically, we sometimes may have to refer to the notion of purposive generation (IG^*) rather than to purposive causation in the strict sense. However, in such cases IG^* will nevertheless involve the purposive causation of the relevant bodily action by a suitably related effective intending.[4]

NOTES

[1] In the case of singular events we will use the dichotomy *simple* vs. *composite*, whereas in the case of action types we will speak about *simple* vs. *compound* ones.

[2] Our notion of composition or sum '$+$' can be defined in terms of the primitive notion 'o' (read: 'overlaps') of the calculus of individuals.

$$x + y =_{df} (\gamma z)(w)(w \text{ o } z \leftrightarrow (w \text{ o } x \vee w \text{ o } y))$$

Let it be remarked here that the notion of part ('$<$') we have occasionally employed can also be defined in terms of overlapping:

$$x < y =_{df} (z)(z \text{ o } x \rightarrow z \text{ o } y)$$

[3] Notice that although we claim that the "subtryings" t_i corresponding to t do exist in the case of compound basic action tokens we do not make corresponding general claim. If an agent intends to do something, then, normally through conscious or unconscious practical inference, he will intend and, ultimately, try to do by his bodily behavior whatever is required for the satisfaction of that intention. Still there does not have to be an ontologically separate trying-event corresponding to each of the required behaviors in the full action: one may effectively intend to play a melody (and play it) without intending or trying (in an ontologically "genuine" sense) to play each note.

[4] One may try to define such notions as those of a *deliberate* and a *voluntary* action analogously with how we have defined intentionality. Deliberate actions are a subclass of intentional actions preceded by deliberation or contemplation. Voluntary actions are actions flowing from wants.

Let us define voluntary actions more exactly in terms of IG as follows: u is *voluntary* if and only if $IG(w, u_o)$, where now w is to be describable as a wanting. In our analysis obviously all intentional actions are voluntary, but not conversely (cf. our action examples of "uncontrolled" action tokens$_1$). Notice, too, that there may be action tokens$_1$ which are not voluntary (e.g. my non-intentional jumping aside because of perceiving a car about to drive over me).

PROPENSITIES AND INDUCTIVE EXPLANATION

1. PROPENSITIES

1.1. One of the leading themes in our analysis of the behavior of systems (such as human agents) has been that overt behavior is to be accounted for in terms of some "underlying" covert activated processes. Thus, for instance, human actions are to be understood and explained by reference to suitable activated proattitudes. Proattitudes are to be realistically construed as dispositions (to behave) and the underlying activated bases of these dispositions are given the heaviest burden in the explanation of actions. We construed our explanatory psychological concepts according to a certain functionalist analogy account (Chapters 3 and 4), which seems to accord well with actual scientific practice.

It has to be kept in mind that our realist construal of psychological concept formation and theorizing basically amounts to a grand *factual* hypothesis about what the investigated part of the world (i.e. agents and their "inner life") is like, although we also assign an important role to social conventions and social practice in general. What the investigated aspects of the world really are like is to be (epistemologically) found out by the (ultimately) best explanatory theory.

Thus far we have construed psychological concepts and theorizing in a rather deterministic fashion, that is, as if determinism were true. But the dispute between determinism and indeterminism, no matter whether it concerns the universe as a whole or only some specific parts or aspects or conceptualizations of it, cannot be decided on a priori grounds alone.

A philosopher, in his reconstrual of concepts and frameworks, must leave open the possibility that either of these theses is true. Thus, for instance, even if our present knowledge only supports a certain kind of microphysical indeterminism, we must leave it as a logical possibility that our "manifest image" of the world with its macroentities and-processes will turn out to contain some irreducible indeterministic, presumably probabilistic, features. An agent's active and passive powers may perhaps turn out to be probabilistic in a finer analysis.

Another reason for studying probabilistic notions (e.g. probabilistic dispositions) is the epistemic one that, due to lack of knowledge, we are unable to do better than theorize probabilistically. Thus, for this reason (presumably) most of the actual scientific theorizing in psychology and the other social sciences is done in terms of random variables and the statistical correlations holding between their distributions.

In this and the next chapter, we are going to discuss some probabilistic ideas and concepts. Although this seems to lead us to theoretical problems relatively far from action theory and the foundations of psychology, we wish to ultimately show the close relevance and applicability of our notions and results to this field.

More specifically, we shall start by a study of probabilistic dispositions or *propensities* (Section 1), as we will call them, and then go on to discuss various notions of probabilistic relevance (Section 2) and inductive-probabilistic explanation (Sections 3 and 4) in terms of our realist views. In the next chapter, our results will be applied to analyze probabilistic causation and, as an application, the probabilistic causation (and generation) of complex actions.

We shall below think of propensities as a kind of generalized dispositions or as "numerical" tendencies. Furthermore, we shall construe them realistically so that they are comparable, for instance, with such physical concepts as forces. More exactly, it is *propensity-bases* (corresponding to the bases of "sure-fire" dispositions) that will be "theoretical" properties of objects in that sense.

Whatever view of propensities one accepts, one normally says that they are properties of chance set-ups (cf. Hacking (1965)). A chance set-up is a device (or part of the world) on which one or more trials, experiments or observations (cf. a dice box) might be conducted. Each trial must have a unique result which is a member of a well-defined class of possible results. A chance set-up will involve a locus object. Customary candidates for such locus objects are dies, coins, etc. and also more complex "substances" such as agents.

Consider the following statement:

(*1*)   The strength of the propensity of chance set-up CSU to produce outcome A on trial B is r.

Let us first assume that A and B are events (in the set-theoretic sense) belonging to an algebra of events (related to CSU) on which a probability measure p_{csu} has been defined. We may now think of writing (*1*) as

(*2*) $p_{csu}(A/B) = r.$

But the transition from (*1*) to (*2*) is somewhat ambiguous and also deficient. On this point there is confusion in the literature on the topic.

To clarify the issues involved, let us ask whether (*1*) is an appropriate interpretation of (*2*), and vice versa. Statement (*1*) may be taken to refer to a particular trial and to what happens in that trial (cf. Giere (1973)). In one sense this is true but in another sense it is false. Consider this. I say that the coin I am holding in my hand has a propensity of strength one half to produce a tail on the second throw in the series of throws to come. This means that I am saying, roughly, that an exemplification of the predicate 'is a second throw in such and such a series' will produce an exemplification of the predicate 'is a tail' with probability one half. This indicates what we should mean by *single case* probabilities.[1]

We do not, on the other hand, mean that this statement has to be taken to refer to some *particular* antecedently fixed instantiations of the above predicates – not in the first place at least. Let *b* name such a particular idiosyncratic instantiation of the first predicate and *a* a particular instantiation of the second one. Probability is a notion which refers to universals in the first place. The probability of *a*, given *b*, has no clear sense except when *b* and *a* are taken to be *arbitrary* exemplifications of some universals, such as our above predicates.

Single case probabilities are to be contrasted with *long run* probabilities, e.g. relative frequencies in large finite samples or limit probabilities in infinite populations. We are going to accept a view below according to which both objective single case probabilities and long run probabilities, understood in a certain way, do "make sense". As we know, limit theorems like Bernoulli's, then, give nice connections between single case probabilities and long run relative frequencies.

We will also argue that propensities can in a sense represent physical properties. For instance, Giere (1973) claims that statements of the form (*2*) represent physical propensities, although he is not prepared to give an explicit semantics for (*2*). I just don't see how one could directly interpret the probability p_{csu} as representing a categorical physical property (analogous to the concept of force). Below I will proceed in a different manner towards the same goal.

Before we go on to a more detailed discussion of propensities, a brief logical characterization of the probabilities of generic events (and properties) must be given. Below we are going to employ probabilistic laws connecting properties (or generic events). We assume, as is usual, that such properties and generic events can be represented by means of suitable open formulas of first-order predicate logic. Paradigmatically, open atomic formulas such as $A(x_1, \ldots, x_m)$, $B(y_1, \ldots, y_n)$, etc., will then occur in probabilistic laws. Thus

(3) $p(A(x_1, \ldots, x_m)/B(y_1, \ldots, y_n)) = r$

would be a kind of prototype of a probabilistic generalization (and hence partly of a probabilistic law as well). We may interpret (3) as saying that whenever $B(y_1, \ldots, y_n)$ is instantiated then $A(x_1, \ldots, x_m)$ will be instantiated with a probability of strength r. An open formula such as $A(x_1, \ldots, x_m)$ is instantiated in a given model $\mathcal{M}$ of our first-order logic whenever a sequence of individuals $a_1, \ldots, a_m$ belonging to the domain of $\mathcal{M}$ is found to satisfy $A(x_1, \ldots, x_m)$.

Note that in (3) no quantification into a probability context occurs. Formulas like

(4) $(x)(p(A(x)/B(x)) = r)$

do not, in our approach, literally represent probabilistic laws nor even make sense (cf. below).

Let us concentrate on probabilistic laws (or lawlike statements) only, whatever the class of them exactly is. We then restrict our consideration to the set of nomologically possible models S of our first-order language $\mathcal{L}$, viz. the set of models compatible with the laws of nature. (This is philosophically circular, of course, but I do not pretend to be giving a conceptual analysis or definition of nomicity but only a reconstruction of the logical aspects of the problem.)

The following two chance-elements are involved in the *semantic* representation (and, indeed, justification) of (3). First, there is the probability that our "world" or our particular experiment (or whatever is comparable) happens to be represented by a certain model $\mathcal{M} \in S$. Next, there is the question related to the chance of finding a sequence of individuals in the domain of $\mathcal{M}$ such that this sequence instantiates the generic event or property (i.e. the open formula) in question. By way of analogy this could be compared with a set-up of urns containing balls in which we ask for the probability of finding a red ball in the next draw, which is such that it first probabilistically selects an urn and then a ball from that urn.

Let us illustrate this in terms of a simple example, which clearly illustrates the situation. We use the "bags and boxes" scheme known from statistics. We imagine that we have a bag (or urn, if you like), say B, containing cubes marked with exactly one of the numbers 1, 2, 3, 4, 5. We also assume that we have five boxes V_1, V_2, V_3, V_4, and V_5. Each of these boxes contains balls marked with exactly one of the letters a, b, c, or d.

Within this chance set-up we now perform random (double) drawings of the following sort. We first draw a cube from B. Then we draw a ball from the box associated with the same number that appears on the cube. Let the probability that we draw a cube marked with number i be $\mu(i)$, for each $i = 1, 2, 3, 4, 5$, and let the probability that a ball marked with j, $j = a, b, c, d$, will be drawn from box i be $\lambda_i(j)$. These latter probabilities are obviously conditional probabilities.

Now we can compute the probability of getting a ball marked with j in our experiment. This probability is simply

$$(5) \qquad \begin{aligned} p(j) &= \mu(1)\lambda_1(j) + \mu(2)\lambda_2(j) + \mu(3)\lambda_3(j) \\ &\quad + \mu(4)\lambda_4(j) + \mu(5)\lambda_5(j) \\[1mm] &= \sum_{i=1}^{5} \mu(i)\lambda_i(j) \\[1mm] &= E\mu(\lambda_i(j)). \end{aligned}$$

Here $E\mu$ means the expected value of $j(= a, b, c, d)$ with respect to μ, given the urn probabilities $\lambda_i(j)$.

In other words, we have here computed the probability of the event of getting a ball marked with j in our experiment, and this means that we have obtained probabilities for an open formula expressing this. The j in the argument of the probability statement $p(j)$ is short for such an open formula, say $A_j(x)$, which reads 'x is a ball marked with letter j'.

In general, the probability $p(A(x))$, which is a probability for an open formula in our (first-order) language, is obtained model-theoretically as follows. We first randomly select a model $\mathcal{M} \in S$ (this corresponds to drawing a cube), and then we randomly select a sequence of individuals (in this particular case one single individual) from $\mathcal{M}$ (this corresponds to drawing a ball above). The probability in question is given by

$$(6) \qquad p(A(x)) = E\mu(\lambda_{\mathcal{M}}(A(x))).$$

As Łoś (1963a) has shown, this idea can be worked out for the general case (i.e. for full ordinary first-order languages), although the expectations then have to be calculated in terms of integrals over model spaces. The general result says, roughly, that given a set S of models (e.g. our above set of "nomologically possible" models) the probability of any (open or closed) formula f of the language $\mathscr{L}$ in question is given by

$$(7) \qquad p(f) = \int_s \lambda_{\mathscr{M}}(\sigma_{\mathscr{M}}(f))d\mu(\mathscr{M}).$$

Here $\sigma_{\mathscr{M}}(f)$ denotes the set of sequences of elements in the domain of $\mathscr{M}$ which satisfy f. The reader is referred to the appendix of this chapter for a more detailed exposition of Łoś' semantical representation theorem and the characterization of probabilities of formulas. Let us once more repeat that this important result shows that the probability of any first-order formula can be expressed as an average of probabilities in models with respect to a probability measure on the set S of models that one conceptually started with.

It must be emphasized in this connection that the basic probability measures λ and μ may and should in part be determined on *factual* rather than *a priori* grounds, even if we cannot here discuss that (cf. the discussion related to Hintikka's probability measures in Niiniluoto and Tuomela (1973), pp. 18, 186).

What Łoś' representation theorem involves for our discussion of probabilistic laws can in part be summarized as follows. The probability of an event B, given A (as expressed by our (3)), refers to the probabilistic connection between the instantiations (i.e. extensions) of the predicates A and B, not only in a given model $\mathscr{M}$ (which might represent the part of the actual world we are concerned with), but also in all the other models $\mathscr{M}' \in$ S. Thus, it says something conditional and counterfactual as well, viz. what probabilistic process might have taken place had our actual world turned out to be correctly represented by $\mathscr{M}'$ rather than by $\mathscr{M}$. Thus, this account handles the counterfactual aspects of lawhood nicely, given, of course, that S captures the set of nomically possible models.

It should be emphasized that Łoś' approach and main result apply to any first-order formula. Thus we may obviously discuss *single case* probabilities of various kinds in addition to long run probabilities (which are always involved in the case of probabilistic generalizations). We may thus consider, for instance, $p(A(x_1, \ldots, x_m)/B(b_1, \ldots, b_n)) = r$, where $A(x_1, \ldots, x_m)$ is an open formula and $B(b_1, \ldots, b_n)$ is a sentence saying that $b_1, \ldots, b_n$ satisfy the predicate B. The semantical situation remains much the same, and it again involves the consideration of the whole set S, in principle.

It has to be noticed that such typical single case probability statements as the probability that the next individual to be observed in the experiment satisfies a certain predicate A, given that the previously observed n individuals exhibited such and such complex pattern represented by B, may (at most) be of this type. Single case probabilities usually deal with some open formulas (or exchangeable individuals) and they are not strictly of the type $p(A(a_1, \ldots, a_m)/B(b_1, \ldots, b_n)) = r$, where *both* of the arguments are *genuinely* singular. These sentences receive the probability of either 1 or 0 within any model, and thus this is not what is usually meant by single case probabilities. This kind of probability statement is too idiosyncratic to have any deep methodological interest (but cf. (52)). Note here, too, that if, e.g., a monadic probability law would have the form $(x)(p(A(x)/B(x) = r)$ (i.e. (*4*) above), then the law would really involve such idiosyncratic probabilities as its instantiation values (with respect to x) — this cannot be what probability laws are understood to involve. (Note: sometimes a universal quantification occurring in the front of a probability law is to be understood to quantify *over sample spaces* or the like, e.g., for every chance set-up, $p(A/B) = r$.)

We have not yet said anything deeply philosophical about the ontology of our first-order language $\mathscr{L}$. For our present purposes it will not be very central whether the x_i's are interpreted as objects ("substances") or events (or, more generally, episodes). We recall that events (and episodes) are in any case best construed as changes in substances. Therefore, we shall not press this point here. However, when we later discuss actions and their probabilistic causes, an event-ontology will be assumed, as earlier in this book.

After all these preliminaries, we can finally go on to clarify the philosophical foundations of propensities.

We assume that a chance set-up such as a dice box may be regarded as a (normally) complex object with certain properties. This object or some part of it will be regarded as the bearer of the propensity-basis. For instance, the physical structure which makes a die symmetrical could be regarded as such a propensity-basis attributed to the die (part of the chance set-up).

We recall from our discussion of "sure-fire" dispositions (e.g. solubility) in Chapter 5 that they are connected to some "deterministic" backing laws, which account for a certain conditional overt behavior. Here the situation is somewhat different. We are dealing with the propensity (probabilistic tendency) of a chance set-up to "do something A, given B". A sure-fire disposition like solubility is manifested on every trial (a lump of sugar dissolves, when put in water). But in the case of probabilistic dispositions, I say that the disposition is *manifested* by the appropriate relative frequencies in "sufficiently large" finite (or infinite) samples (e.g. the relative frequency of tails in a statistically representative sample).

On the other hand, I say that the probabilistic disposition is objectively *displayed* in every chance-trial (or situation or (partial) model). This display presupposes (in addition to the usual overt conditions) that the covert theoretical propensity-basis is causally (or at least nomically) active, so to speak. In the case of the dice-casting experiment, we thus require that the series of outcomes becomes "what it is" (perhaps relative to a certain way of describing) in part because of, or in virtue of, the physical structure (propensity-basis) of the die. This propensity-basis plays a central explanatory role in accounting for both the single case and the long run behavior of the die.

Let an object x_1 be a chance set-up and let o be its locus object so that $o < x_1$ (i.e. o is a part of x_1). Let B^* now be the propensity basis possessed by o. Then we can trivially define another property B for our technical purposes as follows:

(8) $B(x_1) =_{df} B^*(o) \ \& \ o < x_1.$

This new B is attributed to the whole chance set-up x_1 rather than to o.

Using B as our "derived" basis predicate we may now proceed to a definition of '$P^r_{G;F}(x_1, t_1)$', i.e. 'the propensity of strength r of x_1 to do G, given F, at t_1'. We can immediately formulate the definition of this notion (in terms of the corresponding singular formula), as it is almost completely parallel to our analysis (5.1) characterizing sure-fire dispositions:

(9) $P^r_{G;F}(x_1, t_1)$ if and only if
           $(E\beta)[\beta(x_1, \ t_1)$ and $p(G(x, t)/F(x, t) \ \& \ \beta(x, t)) = r$ is a true
           probabilistic law which, given $F(x_j, t_i)$ and $\beta(x_j, t_i)$, inductively
           explains $G(x_j, t_i)$, for all i, j, such that β is indispensable for
           that explanation.]

In this definition there are two notions which will be clarified later. The first of these is inductive explanation, which will be discussed at length in

Section 4. Secondly, the notion of indispensability (for inductive explanation) will be briefly commented on later, too. It may be noticed here, however, that the general idea is that incorporated in our definition (CI_2) of Chapter 4, which defines indispensability for deductive causal explanation. In other words, if a basis predicate is to be indispensable it should not be possible to explain the conditional overt behavior of the chance set-up without reference to active "inner" properties.

A related problem is the necessity of a basis for a propensity. Here, I do not have new arguments to offer in addition to those given in Chapter 5. The problem is to be taken as a factual and scientific one, as I have emphasized. The predicates F and G in the above definition might be in need of some additional clarification for they are to be characterized so as to make x_1 a chance set-up. However, we shall not go into these details here except by referring to our earlier remarks concerning chance set-ups (cf. also Hacking (1965)).

Definition (9) requires that the explanatory basis β occurs in a (true) probabilistic law. One problem here is, of course, what a law is. Obviously, a (probabilistic) law should be essentially general and it should not be a priori (on logical or conceptual grounds) limited to only a finite number of instances. Secondly, following Hempel (1968), we may require that no full sentence formed from a predicate in a lawlike probability statement and an (undefined) individual name be a logical truth. More generally, the predicates occurring in a probabilistic lawlike statement should be *nomological*; and this requirement may be taken to cover the above two conditions. Thirdly, a law should be corroborable both observationally and theoretically (cf. Niiniluoto and Tuomela (1973), Chapter 11). Fourthly, a law should have explanatory power and, accordingly, it should sustain singular explanatory conditionals (see next chapter). A related feature is counterfactual support in general, and on this we made some remarks above. But we are not going to give a detailed analysis of lawlikeness here at all, although we must rely on the possibility of having such an account.

Consider the following related problem, which comes up when considering alleged examples of chance set-ups and probabilistic laws. Is it a law that whenever a fair die is tossed (in such and such conditions) the probability of getting a five is one sixth? Suppose that the dice-casting experiment obeys some deterministic physical laws (as it presently seems to do). Thus, in principle, it should be possible to redescribe the situation within a different conceptual framework which contains only deterministic laws. Would this preclude the suggested probabilistic generalization from being a law? I would say no, although I am not prepared to give fully adequate reasons here. Let me just say the following.

Suppose that a probabilistic generalization such as that above is proposed within the conceptual macro-framework of "the manifest image". Suppose that this probabilistic generalization – found to be very stable under normal "manifest" circumstances – is micro-explained (or approximately and "correctively" micro-explained) in terms of some deterministic laws formulated within "the scientific image". I would say that this explanation need not deprive our probabilistic generalization of its assumed lawhood at all. Rather it may strengthen our conviction to regard it as a law. (Analogous remarks can be made about laws concerning agent's wants probabilistically causing some of his actions.)

Notice, however, that I am not here speaking of *unstable* or *spurious* probabilistic correlations (cf. "the correlation between the number of stork nests and child births in Germany is high"). Such probabilistic generalizations would be excluded from being lawlike by our earlier requirements for lawlikeness.

Our (9) is clearly concerned with "generic" propensity, as its key factor is a probabilistic law. But it also covers "single case" propensity, viz. the propensity of x_1 to produce an instantiation of G, given that x_1 instantiates F. It is only that the singular probabilistic connections are analyzed in terms of nomic generic probabilistic connections (which are in part connected to long runs and relative frequencies as our earlier analysis indicates).

What single case propensities involve in our analysis is, roughly, that instantiations of β and F by x_1 (nomically) produce an instantiation of G by x_1. The presence of β in every objective single case probability assignment is central, making our approach differ from many other views which accept as meaningful and significant both single case and long run probabilities.

It should be emphasized that the basis replacing β is only a *contributing* cause of G. Consider, for instance, the case of a six turning up when throwing a die. The physical basis of this as attributed to the die only accounts for a certain symmetry in the different faces of the die turning up but not for any particular one of the faces coming up. If the dice throwing experiment is deterministic and if, thus, a six turning up in a particular trial is to be fully determined, the antecedent conditions covered by F must play an important causal role here.

2. SCREENING OFF AND SUPERSESSANCE AS EXPLANATORY NOTIONS

2.1. As we think of singular explanation, both deductive and inductive, it must be connected to some laws which in an informative way account for the expectability of the explanandum. These laws, which are probabilistic in many cases of inductive explanation, are naturally assumed to capture the nomic processes responsible (or partially responsible) for the occurrence of the event (or episode, more generally) to be explained. The explanans must, therefore, give an informative and relevant reason for the occurrence of the explanandum event (cf. Chapter 9). We shall below be especially interested in dispositional explanation, viz. explanation where a system's behavior is explained by reference to the realistically construed disposition (e.g. propensity) of the system to bring about behavior of that kind under appropriate circumstances.

Whatever inductive explanation exactly is taken to involve, it must be connected in some intimate way to probabilistic relevance and to the informativeness of the explanans with respect to the explanandum. An explanans which is probabilistically (inductively) independent of the explanandum cannot, in general, have (considerable) explanatory power. A negatively probabilistically relevant explanans normally carries some information relevant to the occurrence (or non-occurrence) of the explanandum event, but it, of course, gives no reasons to expect the occurrence of the explanandum event.

It is (nomic) *positive* probabilistic relevance that is central for inductive explanation (cf. Section 4). Under several reasonable accounts of the information transmitted by the explanans, concerning the explanandum, positive probabilistic relevance is seen to covary with such transmitted information.

One additional aspect of our relevance requirement is that one should not give more information in the explanation than is needed. This leads to a requirement that *minimal*, otherwise "inductively appropriate" (probabilistic) laws be used in inductive explanation (cf. Section 4).

Probabilistic relevance is not only central to inductive explanation, but it will turn out to be a key concept in our later analysis of probabilistic causation as well. Thus, it is appropriate to start our investigation of inductive explanation by a study of some central notions of probabilistic relevance, most of which are familiar from the literature on the topic.

However, the interconnections between these notions of relevance have not received sufficient attention.

According to our view singular inductive explanation essentially involves laws. In the case of what we shall call inductive-probabilistic explanation, the explanatory laws will be (or include) *probabilistic* ones. Our relevance requirements, to be later imposed on inductive-probabilistic explanation, will be intimately connected to the probabilistic relevance of the antecedent of a probabilistic law with respect to its consequent. In other words, a probabilistic law of the form $p(A(x_1, \ldots, x_m)/B(y_1, \ldots, y_n))$ $= r$ expresses the probabilistic relevance of the antecedent property B to the consequent property A. In a singular explanation using this law, we explain an instantiation of A by reference to an instantiation of B and to the law.

It thus seems appropriate to briefly investigate some of the types of probabilistic nomic relevance that may obtain between properties such as A, B, etc. We shall below concentrate exclusively on probabilistic generalizations which are lawlike (and hence laws, if true). Accordingly the properties A, B, C, etc. will be assumed to be *nomological*, i.e. capable of entering into lawlike connections. We shall below use '$p(A/B) = r$' as shorthand for a probabilistic law connecting two nomological predicates $A(x_1, \ldots, x_m)$ and $B(y_1, \ldots, y_n)$, assumed to express properties or generic events. (We thus assume that a corresponding open formula $A(x_1, \ldots, x_m)$, as discussed in the previous section, here indeed stands for an m-place property or generic event.)

Let us start by considering a nomic property in a given reference class. We might for instance be interested in the causes of lung cancer (B) in the class of all men (A). We may first think that emotional stress of such and such a kind (C) is relevant to B with respect to A. Then we find out that smoking (D) is indeed a much more important probabilistic determinant of lung cancer in men. We may assume that both stress and smoking are probabilistically relevant to lung cancer but that smoking is much more so. Thus, we say that smoking supersedes stress as an explanatory factor of lung cancer in the class of men. We make this idea precise by means of the following definition:

(10)   *D supersedes C* in explaining B in reference class A if and only if $p(B/A \ \& \ D) > p(B/A \ \& \ C) > p(B/A)$.

We take explanatory power here to be directly proportional to (nomic) probabilistic relevance (Cf. Niiniluoto and Tuomela (1973), Ch. 7 for a discussion of this and related notions within a somewhat different conceptual framework.) The greater $p(B/A \& D) - p(B/A)$ is, the more explanatory power D has and the more (logarithmic) information does D transmit with respect to B. Similarly, the greater this difference is the greater becomes the nomic expectability of an instantiation of B, given a joint instantiation of A and D. If, in addition, it happens to be the case that $p(B/A \& D) > \frac{1}{2}$, the explanans gives us a reason to expect B to be instantiated rather than fail to be. In other words, D would then in a certain sense explain why B rather than its negation was instantiated (or was to be expected to be instantiated).

Positive probabilistic relevance is not the only interesting relevance notion worth considering here. Another important notion is probabilistic relevance considered relative to some property or, better, probabilistic irrelevance relative to a property. In our above example we may assume that smoking is the (or a) common cause of both lung cancer and of emotional stress, and explicate this as follows. The probability of getting lung cancer, given smoking, is the same in the class of men as the probability of lung cancer, given smoking *and* stress, in the class of men. In other words, we may assume that stress is probabilistically *irrelevant* to lung cancer, given smoking. Now, whether or not we really take smoking to be a common cause of both stress and lung cancer, we might want to assume that smoking makes stress probabilistically irrelevant to lung cancer, and in this sense it *screens off* stress from lung cancer. We may thus consider the following two notions of probabilistic screening off.

(11) D *screens*$_0$ *off* C from B in reference class A if and only if
         (i) $p(B/A \& D) > p(B/A \& C)$
         (ii) $p(B/A \& D) = p(B/A \& C \& D)$.

(12) D *screens*$_1$ *off* C from B in reference class A if and only if
         $p(B/A \& C \& D) = p(B/A \& D) \neq p(B/A \& C)$.

It is obvious that screening$_0$ off is a stronger notion than screening$_1$ off. The latter notion accepts that a D may be a screening off factor even if it is not more probabilistically relevant to B than the screened off factor C.

Screening$_0$ off has been indirectly suggested by Reichenbach (1956) and screening$_1$ off by Reichenbach (1956) and Salmon (1970). As we shall see both of these notions clearly differ from our notion of supersessance. This can be seen on the basis of some observations, which immediately derive from the following application of the general multiplication theorem of probability calculus:

$$p(B/A \ \& \ C \ \& \ D) = \frac{p(B/A \ \& \ C)p(D/A \ \& \ B \ \& \ C)}{p(D/A \ \& \ C)}$$

$$= \frac{p(B/A \ \& \ D)p(C/A \ \& \ B)}{p(C/A \ \& \ D)}$$

We get (cf. the analogous results in Niiniluoto and Tuomela (1973), pp. 99–101):

(13)   If the conjunction of A and D entails C, then D screens$_1$ off C from B in A only if D is (positively or negatively) probabilistically relevant to B relative to C & A.

(14)   If D screens$_1$ (or screens$_0$) off C from B in A, then it need not be the case that D supersedes C in explaining B in A.

(15)   If D supersedes C in explaining B in A, then it need not be the case that D screens$_1$ (screens$_0$) off C from B in A.

(16)   If D screens$_1$ (screens$_0$) off C from B in A and if C is positively probabilistically relevant to B, given A, then it need not be the case that D is positively probabilistically relevant to B, given A.

(17)   If D screens$_1$ off C from B in A and if $\sim C$, B and A are jointly instantiable, then D screens$_1$ off also $\sim C$ from B in A.

In (13) the notion of entailment is to be understood simply as saying that, assuming A, D, and C to be monadic, if the generalization $(x)(A(x) \ \& \ D(x) \rightarrow C(x))$ is true by virtue of the laws of logic then A and D entail C. (In fact, we might extend the notion of entailment here to cover also entailment by virtue of laws of nature or of semantic principles.) The joint instantiation (see (17)) of some monadic predicates $\sim C$, B, and A means that $(Ex)(\sim C(x) \ \& \ B(x) \ \& \ A(x))$ can be true as far as logic is concerned. (The generalization of these notions for the polyadic case should be clear.)

We may note that (16) is not true for the stronger notion of screening$_0$ off. Furthermore, (17) applies to screening$_0$ off if it is, in addition, required in the antecedent that C not be negatively probabilistically relevant to B, given A.

Our above observations show clearly that neither screening$_0$ off nor screening$_1$ off is sufficiently closely related to supersessance nor to positive relevance to be methodologically satisfactory alone (cf., however, our notion of nomic betweenness in Chapter 12). Thus (*14*) and (*15*) amount to saying that supersessance neither entails nor is entailed by screening$_i$ off, where $i = 0, 1$. According to (*16*) a screening property D may be even negatively relevant to the explanandum.

We may illustrate our (*14*) and (*15*) by means of an example. Suppose that the property D entails the property C and hence that $p(C/D) = 1$. Assume then that $p(C/A) = 0.50$, $p(D/A) = 0.40$, $p(B/A) = 0.27$, $p(B \& C/A) = 0.15$, and $p(D \& B/A) = 0.10$. Then $p(B/A \& C) = 0.30$ and $p(B/A \& D) = 0.25$. Thus

$$p(B/A \& C) > p(B/A) > p(B/A \& D),$$

even if D screens$_1$ off C from B in A.

2.2. There is still another notion of screening off advocated by Greeno (1970). This notion concerns the relationships between *sets* (in fact, partitions) of properties rather than between single properties. (Greeno, however, uses an ordinary set-theoretic framework.) We consider the following definition:

(*18*) $\{D_1, \ldots, D_k, \ldots, D_m\}$ *screens$_2$ off*
            $\{C_1, \ldots, C_i, \ldots, C_n\}$ from $\{B_1, \ldots, B_j, \ldots, B_l\}$
            in reference class A if and only if for all i, j, k
            $p(B_j/D_k \& C_i \& A) = p(B_j/D_k \& A)$.

In this definition we must assume that all the conjunctions $D_k \& C_i$ are consistent.

In the case of two-member sets $\{D, \sim D\}$, $\{C, \sim C\}$, and $\{B, \sim B\}$ the definiens of the above definition simply amounts to

(i) $p(B/A \& C \& D) = p(B/A \& D)$
(ii) $p(B/A \& C \& \sim D) = p(B/A \& \sim D)$.

Now it is easy to see that the following results hold for screening$_2$ off ((*19*) and (*20*) easily generalize for arbitrary finite sets):

(*19*)   If D screens$_1$ off C from B in A and
            if $\sim D$ screens$_1$ off C from B in A, then
            $\{D, \sim D\}$ screens$_2$ off $\{C, \sim C\}$ from $\{B, \sim B\}$ in A.

(*20*)   If $\{D, \sim D\}$ screens$_2$ off $\{C, \sim C\}$ from $\{B, \sim B\}$
            in A and if $p(B/A \& D) \neq p(B/A \& C)$ and

$p(B/A \& \sim D) \neq p(B/A \& C)$, then D screens$_1$ off C from B in A and $\sim D$ screens$_1$ off C from B in A.

(21)   If $\{D, \sim D\}$ screens$_2$ off $\{C, \sim C\}$ from $\{B, \sim B\}$ in A and if either C or $\sim C$ is positively probabilistically relevant to B in A, then either D or $\sim D$ is positively probabilistically relevant to B in A.

Greeno's notion of screening$_2$ off basically arises from information-theoretic considerations. To see this, we consider again the sets (partitions) $D = \{D_1, \ldots, D_k, \ldots, D_m\}$, $C = \{C_1, \ldots, C_i, \ldots, C_n\}$, and $B = \{B_1, \ldots, B_j, \ldots, B_l\}$. The (logarithmic) information transmitted by the set C with respect to the set B indicates and measures the reduction in uncertainty with respect to the exemplifiability of the properties in B given some information concerning the exemplifications of the members of C. (For simplicity, here we leave out the reference class from our considerations, as it is a constant factor in the definitions below.) Technically we then define this transmitted entropy as:

$$H(C, B) = H(C) + H(B) - H(C \& B)$$

The logarithmic entropies are defined here as follows:

$$H(C) = \sum_{i=1}^{n} - p(C_i) \log p(C_i)$$

$$H(B) = \sum_{j=1}^{l} - p(B_j) \log p(B_j)$$

$$H(C \& B) = \sum_{j=1}^{l} \sum_{i=1}^{n} - p(C_i)p(B_j/C_i) \log p(C_i)p(B_j/C_i).$$

Given these definitions the following basic result can be shown to hold for the sets D, C, and B (see Greeno (1970), pp. 289–291 for a proof within a set-theoretical framework):

(22)   If D screens$_2$ off C from B, then $H(D, B) \geq H(C, B)$.

In other words, no loss in explanatory power – as measured by transmitted entropy – occurs when a set D screens$_2$ off a set C from the set B (in a certain reference class).

A few methodological comments on the notion of screening$_2$ off are now in order. Consider two-member sets **D**, **C**, and **B**. If **D** screens$_2$ off **C** from **B** then both D and $\sim D$ must make C irrelevant to B (in A). This seems to be a rather strong requirement, and it is not so easy to find methodologically interesting cases satisfying this.

Secondly, in definition (18) the notion of screening$_2$ off presupposes that all the property conjunctions C & D, $\sim C$ & D, C & $\sim D$, and $\sim C$ & $\sim D$ are consistent. But this precludes any deductive nomic connections from holding between the sets **C** and **D**. Thus, for instance, it is *not* possible that, for monadic properties, $(x)(D(x) \rightarrow C(x))$ any more than $(x)(\sim D(x) \rightarrow C(x))$ is a (nomically or analytically) true statement. This is certainly odd. Greeno's screening$_2$ off relation presupposes a peculiar atomistic and "merely" probabilistic view of the world and, when applied to properties and generic events, of our conceptual system as well.

At this point it is of interest to compare Greeno's measure of explanatory power (viz. transmitted entropy) with the measures of systematic power discussed in Niiniluoto and Tuomela (1973). These measures are also based on transmitted information. First we consider three measures based on the information transmitted from an explanatory property C on B (for simplicity we omit the reference class A; cf. Niiniluoto and Tuomela (1973), Chapters 6 and 7):

$$(23) \qquad \mathrm{syst}_1(C, B) = \frac{\log p(B) - \log p(B/C)}{\log p(B)}$$

$$\mathrm{syst}_2(C, B) = \frac{1 - p(C \vee B)}{1 - p(B)}$$

$$\mathrm{syst}_3(C, B) = \frac{p(B/C) - p(B)}{1 - p(B)}$$

The measures $\mathrm{syst}_i(C, B)$, $i = 1, 2, 3$, measure the explanatory power of property C in explaining property B, and all of these measures are based on transmitted information (syst_1 and syst_3 on the logarithmic inf-measure, $\mathrm{inf}(B) = -\log_2 p(B)$, and syst_2 on the cont-measure, $\mathrm{cont}(B) = 1 - p(B)$). All these measures obtain their maximum values ($= 1$) when C entails B. The measures syst_1 and syst_3 obtain their minimum values ($-\infty$ and $\dfrac{-p(B)}{1 - p(B)}$, respectively) when C entails $\sim B$. The measure syst_2, originally proposed by Hempel and Oppenheim, obtains its minimum value $(= 0)$ when $\sim C$ entails B. On the other hand, Greeno's measure $H(C, B)$ obtains its minimum value when the sets $\mathbf{C} = \{C, \sim C\}$ and $\mathbf{B} = \{B, \sim B\}$ are inductively independent.

Of the above syst_i measures, syst_1 and syst_3 covary with positive relevance. Hence, in their case a property D supersedes another property C in explaining B (in the sense of definition (10)) exactly when D is more positively relevant to B than C is. (The measure syst_2, in addition, is inversely proportional to the prior probabilities of C and D here.)

To say the least, different intuitions about inductive explanation seem to be involved in accounts which allow negative relevance to have explanatory power (Salmon, Greeno) and those which do not (cf. our notion of supersessance and the syst_i-measures).

3. EXPLANATORY AMBIGUITY AND MAXIMAL SPECIFICITY

3.1. It has become customary to dichotomize views of nondeductive probabilistic explanation as follows. First, there is Hempel's well known *inductive-statistical* (or I-S) model. According to this model a (singular) explanation is an argument (relative to a knowledge-situation) that renders the explanandum highly probable (see Hempel (1965), (1968)). Secondly, there is the model of *statistical relevance* (the S-R model), whose primary advocate today is Salmon (see Salmon (1971)). According to this model an explanation is an assembly of facts statistically relevant to the explanandum, regardless of the degree of probability that results. More generally, to give a scientific explanation is, according to Salmon, to show how events and statistical regularities fit into the nomic network of the world.

Even this preliminary characterization exhibits some clear points of dispute between the 1-S model and the S-R model, three of which are worth emphasizing. First, there is the problem of whether inductive or statistical explanations are arguments, and if, in what sense. Secondly, there is the problem of whether there are *objective* (knowledge-independent) inductive explanations or whether relativization to an epistemic situation is somehow inherently necessary. This problem is related to the question as to whether the possibility of giving inductive explanations involves the truth of determinism. Thirdly, there is the problem of how to explain events with low probability (both prior and posterior).

In addition to these points we must also discuss some paradoxical cases of explanation which exhibit difficulties for any account of inductive explanation (see Stegmüller (1973) for a good inventory of the paradoxes of statistical explanation).

Our basic aim in this chapter is to present a new model of inductive explanation, which in part combines features from both the I-S model and the S-R model. Before we get to that we will sketch in some detail both Salmon's and Hempel's model and present some difficulties for both of them.

In Salmon's model of explanation a key notion is that of a homogeneous reference class. It is defined as follows:

(24)   The reference class A is a *homogeneous* reference class for B if and only if every property C effects a partition of A which is statistically irrelevant to B, i.e. $p(B/A \ \& \ C) = p(B/A)$.

One problem here is that we cannot literally consider every property C in the definiens. For instance $\sim A$ must be excluded on pain of inconsistency and B because of triviality. In other words, probability statements that are true merely on the basis of the probability calculus do not count, and hence B and $\sim A$ are excluded. Salmon seems to want to restrict the allowed properties to those determined by place selection rules (in von Mises' sense). However, as is known, there are many difficulties connected with the notion of a place selection rule. Accordingly, in view of Church's definition of randomness, one might alternatively try to require that the extensions of properties C must be recursively enumerable sets (i.e. recursively enumerable subsets of the class of all sets).

But even that move (even if it were mathematically satisfactory) would not suffice, for it would not guarantee that the C's are *nomological* properties and that the probabilistic generalizations concerned are *lawlike*. This feature is what Salmon should require even on his own account of the nature of scientific explanation. As we shall below restrict ourselves to nomological properties only, we can here leave the matter at that.

It should be noticed that there is, of course, no guarantee that homogeneous reference classes exist. This is a factual problem, which is closely related to the truth of determinism, as will be seen.

Let us now consider singular explanation. According to Salmon we explain in the following way. To explain why a member x of a class A has the property B, we partition A into subclasses $A \cap C_1, A \cap C_2, \ldots,$ $A \cap C_n$, all of which are homogeneous with respect to B. We then give the probabilities $p(B/A \cap C_n)$, all of which are different, and say to which class $A \cap C_i$ the element x belongs. We put this in the form of a definition as follows (cf. Salmon (1970), pp. 76–77):

(25) $\mathscr{D}$ is an *S-R explanation* of the fact that x, which is a member of class A, has the property B, if and only if $\mathscr{D}$ is a description of the following kind:
$$p(B/A \cap C_1) = p_1$$
$$p(B/A \cap C_2) = p_2$$
$$p(B/A \cap C_n) = p_n,$$
where $A \cap C_1, A \cap C_2, \ldots, A \cap C_n$, is a homogeneous partition of A with respect to B,
$$p_i = p_j \quad \text{only if} \quad i = j,$$
$$x \in A \cap C_k.$$

Let us first notice that instead of speaking of the explanation of a fact we might have spoken about explaining an event, viz. the event consisting of x's having B. As I have formulated the S-R explanation, it is explicitly merely (set-theoretic) *description*. Note here: whatever an *argument* exactly is, it cannot be merely a description.

Definition (25) has been given in the set-theoretic terminology that Salmon employs. However, its translation into our property-framework is of course obvious.

Salmon has argued that only *maximal*, i.e. the most inclusive, homogeneous classes should be given explanatory force. This requirement blocks irrelevant partitions, he claims. To illustrate this we consider Salmon's example. Suppose we explain John Jones' recovery in two weeks from his cold on account of his taking vitamin C. This explanation, Salmon claims, refers to a too narrow class, for people with colds recover with the same high probability independently of whether they take vitamin C. Thus the explanation should be, roughly, that John Jones recovered because at least two weeks had lapsed since his catching the cold. On the basis of examples like this, Salmon concludes that an explanation should always refer to a maximal homogeneous partition of the reference class.

This maximality requirement is in fact built into Salmon's definition of statistical explanation. Suppose that the maximality requirement concerning the partition $A \cap C_i$, $i = 1, \ldots, n$ does not hold. Then we would have $p_i = p_j$ for some $i \neq j$. Consider, accordingly, a class C_k such that $A \cap C_k$ is homogeneous for B. We may assume, on the basis of the non-maximality assumption, that we can partition this C_k into subclass C_k^1 and C_k^2 such that $p(B/A \ \& \ C^1_k) = p(B/A \ \& \ C^2_k)$. But clearly, as C^1_k and C^2_k are disjoint, we here have a case where $p_i = p_j$ does not entail $i = j$:

(26)   In Salmon's model of explanation the requirement that $p_i = p_j$, only if $i = j$, entails that $A \ \& \ C_i$, $i = 1, \ldots, n$, is a maximal homogeneous partition of A with respect to B.

Obviously (26) holds when a property-framework such as ours is concerned, too.

Before going on I would like to point out that Salmon's maximality requirement seems too strong. Consider the above example, where we assumed (probably contrary to fact) that vitamin C gives an irrelevant homogeneous partition of the class of people with at least two-week old colds. Now I claim that the maximality requirement is spurious here, as (if our assumption concerning vitamin C is correct) there is no probability *law* to the effect that recovery from a cold is highly probable after two weeks from catching the cold, given that one takes vitamin C. For vitamin C does not occur as a nomic factor here at all. Neither taking nor not taking vitamin C, *ceteris paribus*, has any effect whatsoever on recovery (or respectively non-recovery). In other words, the requirement that only probabilistic laws should be employed in probabilistic explanations already suffices to block irrelevant partitions of the kind Salmon had in mind when imposing the maximality requirement. It is nomicity which should decide what kind of homogeneous partitions should be used.

Consider still another argument against Salmon's maximality principle. Assume that our above example is changed a little by assuming that a new vitamin C* is found to cure colds almost immediately with the same high probability (say r) as laying two weeks in bed does. Now Salmon requires us to lump together the classes of persons laying two weeks in bed and those cured immediately by vitamin C*, since $p_i = p_j = r$ and $i \neq j$. But this I find unsatisfactory. An explanation should give a proper amount of information only, and no reference to irrelevant information should be made. But Salmon's explanatory law here would say that if a person stays two weeks in bed *or* if he takes C* then he will recover with probability r. In its antecedent this law is too general as a more specific law (using only one of the disjuncts in its antecedent) is available.

Our first argument showed that one should not impose the maximality principle independently of considerations of which laws are at work in the situation. Our second argument was explicitly directed against the maximality principle in favor of a minimality principle (cf. our later discussion).

A few other observations that are immediate on the basis of definition (25) are the following:

(27)   In Salmon's model of explanation it is possible that
$$p(B/A) \lessgtr p(B/A \ \& \ C_k).$$

(28)   Salmon's model of explanation does not entail the screening$_1$ off principle.

(29)   Salmon's model of explanation entails that no D screens$_1$ off any C_i from B in A.

What (27) involves is that negatively relevant and even completely irrelevant factors may be used. (28) says that, oddly enough, definition (25) does not without further assumptions entail the screening$_1$ off principle. It is not the case that each C_i screens$_1$ off every property D from B in A, for it is possible that the inequality of definition (12) fails to hold for some D and C_i within a Salmonian explanation. (Presumably Salmon would like to include the screening$_1$ off principle in his model; in fact he claims that it follows from his model (Salmon (1971), p. 105).) (29) states a weaker property of screening$_1$ off: no explanatory factor C_i should be superseded in the screening$_1$ off sense.

There are still two important aspects of Salmon's model which we have not yet discussed. One is that we have not yet really motivated the need for using *homogeneous* explanantia. The second aspect is that Salmon's view of explanation is purely *descriptive*. We shall argue below that a central feature about the notion of explanation is that it puts the explanandum into "the logical space of reasons", and thereby conveys understanding concerning it.

3.2. Considerations requiring some kind of homogeneity or specificity of the explanatory reference class are familiar. Millionaires among Texans are not rare, whereas philosopher-millionaires are rare. Should we then predict that Jones, who is a Texan philosopher, is a millionaire or that he is not? Or if we know Jones is a millionaire and explain it by his being a Texan, how shall we react to the conflicting information that philosophers very unlikely are millionaires? Though ambiguities like this have been Hempel's major starting point in constructing a model of I-S explanation, I believe that S-R theoreticians like Salmon could use

CHAPTER 11

similar reasoning for backing some kind of homogeneity requirement. This is a good place for turning to Hempel's model of inductive-statistical explanation.

Consider the following monadic argument:

(30) $p(G/F) = r$
         $\dfrac{F(a)}{G(a).}$ $[r]$

Hempel explains $G(a)$ by reference to an explanans $\{p(G/F) = r,\ F(a)\}$, which is assumed to make $G(a)$ nomically expectable to a high degree (r is close to 1). The inductive probability r within brackets is taken to be the same as the statistical probability r in the explanatory law. The higher r is the better is the explanatory argument, *ceteris paribus*. We shall later return to the inductive inference represented by (30). In any case, the above is a rough statement of the naive I-S model.

As Hempel has kept emphasizing (most recently in Hempel (1968)), we can have conflicting explanations of the following sort (corresponding to our above example):

(31) $p(G/F) = r_1$
         $\dfrac{F(a)}{G(a)}$

(32) $p(\sim G/F) = r_2$
         $\dfrac{F(a)}{\sim G(a).}$

with $r_1 \neq r_2$, although both probabilities are close to one. The schemas (31) and (32) exhibit *ontic ambiguity*. One might react to this puzzle by saying that within an explanation we must assume that the explanandum statement is true. Thus (31) and (32) can not both apply at the same time. Still there is the disconcerting counterfactual consideration that, when in fact $G(a)$ is the case, had a been a $\sim G$, then (32) instead could have been used to give an explanation. Thus, no matter what happened, we seem to be able to explain what happened.

As a way of bypassing ontic ambiguity Hempel suggests that all I-S explanations be relativized to some knowledge situation, say **K**. What Hempel's real reasons are for making this move are not very clear (cf. Coffa (1974)). As Hempel himself emphasizes, we are, however, immediately confronted with an analogous problem of *epistemic ambiguity*. Suppose, for instance, that the premises of both (*31*) and (*32*) belong to the set of accepted statements **K** in a given knowledge situation. ('Accepted' means here, roughly, 'tested, and accepted to be used for explanation, further testing, etc.'.) Here remarks completely analogous to the ontic case can be made. Furthermore, even if, paralleling the above argument that $G(a)$ must be the case, we require that the statement $G(a)$ belongs to **K**, we can modify the above counterfactuality point to suit this case as well.

Both ontic and epistemic ambiguity seem to arise because of ontic or, respectively, epistemic inhomogeneity or nonspecificity. (Note that if we take the set **K** to be the set of true, rather than of merely accepted statements, we can go from the epistemic situation to the ontic one.) Hempel's first attempt to deal with epistemic ambiguity was in terms of essentially the following condition related to (*30*) (cf. Hempel (1968), p. 120):

(RMS)   For any class F_1 for which **K** contains statements to the effect that F_1 is a subclass of F and that $F_1(a)$, **K** also contains a probabilistic law to the effect that $p(G/F_1) = r_1$, where $r_1 = r$ unless the law is a theorem of probability theory.

The "unless"-clause is meant to allow **K** to contain, without prejudice to the explanatory status of (*30*), pairs of sentences such as $F(a)$ & $G(a)$ and $p(G/F \& G) = 1$, of which the latter is a theorem of the probability calculus.

(RMS) is a principle of homogeneity (not to be confused with any requirement of total evidence), and it is closely related to Salmon's principle of homogeneity. In fact, if **K** is the set of true statements, we obviously have for the above case (we may take F to be just Salmon's reference class A, for simplicity):

(*33*) F satisfies (RMS) (with **K** = the set of true statements) if and only if F is a homogeneous reference class for G (in the sense of definition (*24*)).

Obviously the quantifiers 'every property' of (*24*) and 'any class' of (RMS) must have the same domain for (*33*) to hold.

Hempel (1968) claims, however, that (RMS) is not adequate. This is basically due to the possibility of the following kind of competing explanations (invented by Grandy):

(*34*) $p(G/F \lor G) = 0.90$
$$\frac{F(b) \lor G(b)}{G(b)}$$

(35) $p(\sim G/ \sim F \vee G) = 0.90$

$$\frac{\sim F(b) \vee G(b)}{\sim G(b)}$$

Hempel claims that both of these, seemingly conflicting examples satisfy (RMS). For the only subclasses of $F \vee G$ which **K** tells us contain b are $(F \vee G)$ & $(\sim F \vee G)$, which is G; and $(F \vee G)$ & G, which is G, too. As $p(G/G) = 1$ by virtue of probability theory alone, (37) is seen to fulfil (RMS). An analogous argument goes for (35).

A modification of (RMS) seems to be called for, and Hempel does it as follows for the monadic case. We say that a predicate F_1 *entails* F_2 just in case $(x)(F_1(x) \rightarrow F_2(x))$ is logically true. F_1 is said to be an a-predicate in **K** if **K** contains $F_1(a)$; and F_1 will be said to be *probabilistically relevant* to $G(a)$ in **K** if (1) F_1 is an a-predicate that entails neither G nor $\sim G$, and (2) **K** contains a lawlike sentence $p(G/F_1) = r$. Next, a predicate M is said to be a *maximally specific predicate* related to $G(a)$ in **K** if (1) M is logically equivalent to a conjunction of predicates that are probabilistically relevant to $G(a)$ in **K**; (2) M entails neither G nor $\sim G$; (3) no predicate expression entails (but is not entailed by) M while satisfying (1) and (2).

In distinction to (RMS) the new condition (RMS*) is concerned only with probabilistically relevant predicates M, and it is formulated as follows:

(RMS*) For any predicate M, which either (a) is a maximally specific predicate related to $G(b)$ in **K** or (b) is stronger than F and probabilistically relevant to $G(b)$ in **K**, the class **K** contains a corresponding lawlike probability statement $p(G/M) = r_1$ such that $r_1 = r = p(G/F)$.

(See Hempel (1968), pp. 131–132 for examples illustrating (RMS*).) Before we discuss the adequacy of (RMS*), we formulate Hempel's model of inductive explanation (of singular statements) as follows:

(36)   An argument of the form (30) is a *Hempelian inductive-statistical explanation* of basic form relative to a class **K** of accepted statements if and only if

(1) **K** contains the explanans and the explanandum statements of the argument;
(2) r is close to 1;
(3) the probability statement in the explanans is lawlike;
(4) the requirement of maximal specificity (RMS*) is satisfied.

One thing to be noticed about the above definition is that it requires $G(b)$ to belong to **K**. It follows of course that $\sim G(b)$ cannot belong to **K**, given that inconsistent **K**'s are regarded as unacceptable. Therefore, Hempel is wrong when he claims that both (*34*) and (*35*) could qualify as explanations (for some **K**). There is no such consistent **K**. Neither can any other two explanatory arguments with contradictory explanandum statements qualify as explanations.

When Hempel discusses Grandy's examples he does not, contrary to his definition, include the explanandum statements in **K**. This suggests that in the above definition condition (1) should be replaced by:

(1′)   **K** contains the explanans statements of the argument, but it does not contain the explanandum statement.

One might defend (1′) by argúing that an explanation should be potentially convertible into a prediction. In case of (1) that would, of course, not be possible in the intended sense. If we make this liberalization, we may in fact go on and consider various other kinds of sets of accepted statements (subsets of the set of all accepted statements at a certain time). In fact, one may just think that potential inductive explanations involve certain inductive relationships between explanans statements and explanandum statements independently of whether the statements are accepted or rejected, true or false, liked or disliked, and so on.

As said, Hempel's motivation for requiring that inductive-statistical explanations be relativized to a knowledge situation are not quite clear. Perhaps he thinks that otherwise there would be no non-deterministic explanations at all in principle (see Coffa (1974) and our comments below), and he wants to give the scientist a schema for making probabilistic explanations even when deterministic processes are at work "behind the surface" (cf. e.g. coin tossing and, more generally, explanations in the social sciences in general). But even if this were Hempel's motive he partly fails. For his **K**'s become so strictly confined that within actual scientific practice such sets can hardly ever be found.

If Hempel's main motive is the avoidance of ontic ambiguity (cf. above and Hempel (1968), p. 118) the switch to the epistemic case with ambiguity problems (even if accompanied by the avoidance of epistemic ambiguity) does not alone suffice to solve the problem of ontic ambiguity.

Let us now return to Grandy's counterexample, whose exclusion seems to be Hempel's main motive for introducing RMS* to replace RMS, given, of course, that he accepts condition (1′) rather than (1). We note that, in order for (34) to qualify as an explanation, the statement $p(G/ \sim F \vee G) = 0.90$ must belong to **K** according to condition (a) of (RMS*), since $\sim F \vee G$ is a maximally specific predicate. But since it does not, (34) is not an explanation. Similarly, (35) is excluded on the basis of (a) of (RMS*). However, Stegmüller (1973) has pointed out that since

$$\vdash (F(b) \vee G(b)) \ \& \ (\sim F(b) \vee G(b)) \rightarrow G(b),$$

Grandy's examples are trivialized. That is, **K** must be taken to include $G(b)$ as that statement is entailed by its elements $F(b) \vee G(b)$ and $\sim F(b) \vee G(b)$. Therefore (34) is a superfluous argument; and the conclusion $\sim G(b)$ of (35) is again not acceptable as its inclusion in **K** would lead to inconsistency. If we accept Stegmüller's argument as I think we must in this case, the main reason for condition (a) of (RMS*) seems to be undermined.

But Stegmüller's point against (RMS*) is not perhaps as damaging as it may seem. First, I think it is too strong a requirement that **K** in general be closed under logical consequence. If it were, the notion of a set of statements accepted at time t would become so idealized as not to apply to any actual situation (i.e. any actual research community). I think, however, that in Grandy's example $G(b)$ must be regarded as included in **K**; but I have in mind the possibility of vastly more complex and more interesting cases.

Secondly, as we shall indicate, (RMS*), contrary to (RMS), helps to solve the problem of epistemic ambiguity. (We shall below concentrate on an objectivized version of (RMS*) and show that it even contributes towards solving ontic ambiguity.)

A third point to be made about condition (a) of (RMS*) is that even if Stegmüller's argument above concerning Grandy's example were accepted (for epistemically relativized explanations), there might perhaps be other examples requiring this condition. Thus, it is possible that there is some compound predicate involving F such that this compound predicate is probabilistically relevant to G although it does not entail F (see below).

Since we will be mostly interested in non-epistemic inductive explanation below, we will need an objective or non-epistemic version of the requirement of maximal specificity. In addition we shall generalize it to cover not only monadic predicates but n-place predicates in general. Furthermore, we shall make use of the notion of a nomic field.

A *nomic field* consists, roughly, of those aspects of a field of investigation which are prima facie thought to be nomically relevant to what is studied (cf. causal field mentioned in Chapter 9). A nomic field thus consists of initially possible nomically (e.g. causally) relevant factors. It leaves out of consideration fixed standing conditions, conditions assumed to satisfy *ceteris paribus*-clauses, and so on. A nomic field may also be taken to involve some general *assumptions* concerning *ceteris paribus*-conditions, boundary conditions, and the like. It is not, however, necessary to do that here nor to define a nomic field more exactly.

Let us assume that $\mathscr{F}_\lambda$ is a nomic field dealing with all the "aspects" (properties, events, etc.) represented by the predicates of a (normally finite) set λ. The members of λ generate $\mathscr{F}_\lambda$, we say. In the general case λ will include any kind of polyadic predicates. Now we may define the notion of a maximally strong predicate in $\mathscr{F}_\lambda$.

If λ includes only primitive monadic predicates $P_1, \ldots, P_n$, then of course a maximally strong (nomic) predicate is of the form:

$$(37) \qquad Ct_i(x) = \pm\, P_1(x) \ \& \ \pm\, P_2(x) \ \& \ldots \& \ \pm\, P_n(x),$$

where the pluses and minuses mean positive or negative occurrence of the predicate in question, and that index i obviously varies from 1 to 2^n.

In the general polyadic case maximally strong (nomic) predicates can be defined by means of Hintikka's theory of distributive normal forms (see Hintikka (1965)). They are defined with respect to the notion of *depth* (cf. Chapter 4). A maximally strong nomic predicate will now be of the form

$$(38) \qquad Ct_i^d(x_1, x_2, \ldots, x_{j-1}, x_j).$$

Predicates of this form describe, varying the values of j, all the possible ways in which the value of the variable x_j can be related to the individuals specified by $x_1, x_2, \ldots, x_{j-1}$. Here "all the possible ways" means all the ways which can be specified by using only the predicates in λ, propositional connections and at most d layers of quantifiers. We can also say that (38) determines a list of all kinds of individuals x_j that can be specified by these means plus the "reference point" individuals represented by $x_1, \ldots, x_{j-1}$. (See Hintikka (1965) for an exact recursive definition over d of such polyadic Ct_i^d-predicates.)

Now, as our counterparts to Hempel's (1968) maximally specific predicates, we have suitable Ct_i^d-predicates or, rather, suitable disjunctions of such predicates. Thus consider (within $\mathscr{F}_\lambda$) a probabilistic law of the form $p(G(x_1, \ldots, x_m)/F(y_1, \ldots, y_n)) = r$. We say that a predicate $M(x_1, \ldots, x_k)$ is a *maximally specific predicate* related to $G(a_1, \ldots, a_m)$

in $\mathscr{F}_\lambda$ if and only if (1) M is logically equivalent to a disjunction of Ct_i^d-predicates (for some d) which is probabilistically relevant to $G(a_1, \ldots, a_m)$; (2) M entails neither G nor $\sim G$; (3) no predicate expression stronger than M satisfies (1) and (2).

We say of two predicates $F_1(x_1, \ldots, x_m)$ and $F_2(y_1, \ldots, y_n)$ that F_1 entails F_2 just in case the statement

$(x_1) \ldots (x_m)(F_1(x_1, \ldots, x_m) \rightarrow (Ey_1) \ldots (Ey_n)F_2(y_1, \ldots, y_n))$ is true in virtue of (a) principles of logic, (b) semantical postulates, or (c) laws of nature. Logical equivalence of course means mutual logical entailment. We have broadened Hempel's notion of entailment so as to include meaning-entailment and nomological entailment. (This move guards against some criticisms directed against Hempel's analysis, such as the ball-drawing example in Salmon (1970), p. 50.)

Now we can formulate our generalized and non-epistemic counterpart to Hempel's (RMS*):

(RMS**)   For any predicate M in λ such that either (a) M is a maximally specific predicate related to $G(a_1, \ldots, a_m)$ in $\mathscr{F}_\lambda$ or (b) M is stronger than F and probabilistically relevant to $G(a_1, \ldots, a_m)$ in $\mathscr{F}_\lambda$, if $p(G/M) = r_1$ is a lawlike statement then $r_1 = r = p(G/F)$.

Even if our (RMS**) is not relativized to any knowledge situation $\mathbf{K}$, it is relativized to a nomic field, which here formally amounts to a relativization to a set of predicates λ. As is the case with (RMS*), (RMS**) requires (condition (a)) the consideration of *any* maximally specific predicates M which are probabilistically relevant to G.[2] (Such an M will never be a Ct_j-predicate in the case of non-compound G's as every Ct_j-predicate entails either G or $\sim G$.) Clearly, RMS** is stronger (relative to $\mathscr{F}_\lambda$) than Salmon's maximal homogeneity principle relativized to $\mathscr{F}_\lambda$ because of clause (a). However, Salmon's condition is meant to concern any nomic field, and hence any set λ.

Let us consider a nomic field with $\lambda = \{F, G, H\}$ and an explanatory law $p(G/F) = r$. According to condition (b) of (RMS**), for any predicate M stronger than F, $p(G/M) = r$. Here M could be $F \& H$, where H could represent a propensity basis. Notice that one function of (b) now is to allow the use of $p(G/F) = r$ as an explanatory law instead of our *having to* use $p(G/F \& H) = r$, *ceteris paribus*.

Condition (a) of (RMS**) concerns only the maximally specific predicates. Note here that (b) already covers those maximally specific predicates which are stronger than F. As we have seen, the question is whether there could be maximally specific predicates which do not entail F but which still might make G probable by a different amount than F does. According to our definition of a maximally specific predicate such a predicate would have to be an a-predicate as well. For instance, if the predicates F, G, and H are assumed to be monadic we have eight possible Ct_j-predicates. Of these we can form 2^8 disjunctions. However, there are only seven possible candidates for maximally specific predicates stronger than F. The number of maximally specific predicates which do not entail F is larger, though.

In general, we can write those disjunctions which entail F in the form $F \& D_i$, where D_i is a suitable disjunction formulated in terms of G and H. Those disjunctions which entail $\sim F$ can be written correspondingly in the form $\sim F \& D'_j$. The question now is whether it is possible that, for some i, j,

$$(39) \qquad p(G/(F \& D_i) \vee (\sim F \& D'_j)) > p(G/F \& D_i).$$

Only if (39) may be the case are maximally specific predicates which do not entail F needed.

As far as I can see it is at least compatible with Kolmogorov's axioms that (39) is true. We can also say that (39) can be the case because we do not in our (RMS**) relativize the extensions of predicates initially to a fixed reference class A as do Salmon's homogeneity principle and Hempel's (RMS) (the latter only implicitly; cf. our (33)). For then we would just be dealing with sub-classes of A such as $A \cap F$, $A \cap F \cap H$. Here, on the contrary, all the extensions of all the λ-predicates may have to be considered. Thus we cannot drop condition (a) of (RMS**).

Our (RMS**) as well as Hempel's (RMS*) is concerned with *minimal* explanations in the sense that the explanandum event should be referred to the narrowest nomically relevant class. Compared with the deductive case, this means that one should use an explanatory law such as $(x)(F(x) \& H(x) \to G(x))$ rather than $(x)(F(x) \to G(x))$, when it makes a difference. We explain, in principle, by reference to a maximally specific property such as $F \& H$ and not F, when it makes a difference.

Salmon (1970) and Stegmüller (1973) argue that one should refer to maximal rather than minimal classes of maximal specificity.[3] First we notice that condition (b) of (RMS**) allows the use of non-minimal maximal specific classes (or properties, rather), although it does not require the use of maximal classes of maximal specificity. Secondly, and most centrally, Salmon's and Stegmüller's arguments (when they are acceptable) are already taken care of by our requirement that only probabilistic *laws* (or lawlike statements) be used in explanation (see our earlier discussion of this).

Hempel (1968) shows that his (RMS*) makes it possible to get rid of epistemically ambiguous explanations. (His (RMS) is not sufficient for that, see Stegmüller (1969), p. 696.) Analogously, we can show that our (RMS**) helps to take care of ontic ambiguity (relative to a nomic field $\mathscr{F}_\lambda$). To see this, consider the following two proposed candidates for explaining $G(a)$:

(40) $p(G/F_1) = r_1$
         $F_1(a)$
         —————
         $G(a)$

(41) $p(G/F_2) = r_2$
         $F_2(a)$
         —————
         $G(a)$

Let us assume that F is one of the maximally specific predicates in our nomic field $\mathscr{F}_\lambda$ assumed to contain F, F_1, F_2, and G. Now, condition (a) of (RMS**) guarantees that $p(G/F_1) = p(G/F)$ and that $p(G/F_2) = p(G/F)$. Thus $r_1 = r_2$, and no ontic ambiguity can arise. In the epistemic case we would in addition consider a set **K** to which the premises of (40) and (41) belong. Otherwise, the proof blocking the existence of epistemic ambiguity (relative to $\mathscr{F}_\lambda$ and **K**) would be the same.

Our relativization of the above proof to a *finite* nomic field $\mathscr{F}_\lambda$ of course guarantees that a maximally specific predicate required in the proof exists. One problem with Hempel's (1968) proof related to his (RMS*) is that there may be deductively closed **K**'s for which no such maximally specific predicates exist (cf. Coffa (1974)).

Can one exclude ontic ambiguity even when no relativization to a nomic field is made? To answer this question we must obviously somehow go through *all* the possible nomic fields, i.e. all the sets $\lambda_i (i = 1, 2, \ldots)$. In other words, starting with a certain set λ_1 of predicates, we may assume that corresponding to λ_1 there is a maximally specific (and property-expressing) predicate M_1 which determines the value of the *lawlike* statement $p(G/M) = r_1$. Now we form an infinite increasing sequence $\lambda_1 \subseteq \lambda_2 \subseteq \ldots$. We may assume that the probabilities r_i associated with the λ_i's have a limit, say r_w. But we must make this limit assumption *separately* here. What is more, there may be infinite sets which do not even have an associated maximally specific predicate and hence a probability value associated with them. Thus, we need two additional ontological assumptions before we get to such a limit probability r_w. In other words, in the general case we can get rid of ontic ambiguity *absolutely* only at the expense of the above two non-trivial assumptions, which are in part *factual* as to their nature.

If the limit r_w exists it need not have the value 1. But if it does, then we are led to determinism in the sense of always having available, in principle, probability laws with values of 0 or 1 only.

4. AN ANALYSIS OF INDUCTIVE EXPLANATION

4.1. Scientific explanations consist of what is explained – the explanandum – and of what explains it – the explanans. Both are construed here as statements (or sets of statements, if you like). When giving an explanation the explainer states the explanans to the explainee as an answer to the latter's explanation-query concerning the explanandum. As earlier, we shall think of the explainer and the explainee as idealized persons who can be regarded as being part of the fixed pragmatic boundary conditions of the situation (cf. Chapter 9). Accordingly, we shall concentrate our attention on the explanans, the explanandum and some of the relationships obtaining between them. We are going to assume that one can capture some central elements of explanation into a relation, say $E(L, T)$, which holds between two statements (or sets of statements) when the former is explained by (or contributes to the explanation of) the latter. Thus if L is an explanandum statement and T an explanans statement there is such a relation which in standard cases either applies or does not apply (i.e. $E(L, T)$ or not-$E(L, T)$ applies).

The relation E is *deductive* when $T \vdash L$, otherwise we liberally call it *inductive*. Earlier we used the symbol ϵ for deductive explanation. Below we shall use ρ for inductive explanation. When T contains a probabilistic law as its element we speak of *inductive-probabilistic* explanation. We will concentrate below on this kind of explanations and only on cases where L is a singular statement. (See Niiniluoto and Tuomela (1973) for a discussion of inductive explanation in general and the inductive explanation of generalizations in particular.)

In Chapter 9 we defined deductive explanations in terms of answers to why-questions. More explicitly, we took deductive explanations to be ϵ-arguments, or part of ϵ-arguments, conveying understanding concerning why the explanandum is the case (see *(9.5)*). This amounts to saying that deductive explanations are complete conclusive answers or parts of complete conclusive answers to appropriate why-questions. We can obviously extend this line of thought to the inductive case, too. Thus we may regard inductive explanations as ρ-arguments or parts of ρ-arguments which convey understanding (typically) concerning why the explanandum is the case. In other words, inductive explanations can be regarded as complete and *inconclusive* answers or parts of such inconclusive complete answers to suitable why-questions, we claim. Now, however, the why-questions are to be taken very broadly so as to include, e.g., the citing of probabilistic causes as well as giving other nomic probabilistic reasons among their answers (rather than restricting them to nomic deterministic necessitation). Sometimes how possible-questions will be asked, too, but it seems that whenever we are in a position to give a nomic how possible-explanation we can also give a why-explanation in our present wide probabilistic reason-giving sense. We shall later return to give a more explicit characterization of inconclusive answerhood. Presently, it is more urgent for us to discuss the form of inductive-probabilistic explanations.

We have been claiming that all explanations are intimately connected to giving *reasons* for the (potential or actual) *nomic expectability* (of some degree or other) of the explanandum. To explain is not merely to describe, as, for instance, Salmon thinks. It is, rather, intimately connected to the logical space of reasons and justifications. We may speak here of the *descriptive* and the *prescriptive* or *reason-giving* dimensions of explanation, for both are centrally involved.

On the descriptive side we show that our singular explanandum event (or state) occurred (or obtained) as a consequence of nomic processes or connections (expressed respectively, e.g., by causal laws or by laws of nomic "coexistence"). More specifically, we are concerned with *direct* explanations which refer to a *minimal* law and hence to the most specific factor nomically connected to the explanandum. This is a feature which both the DE-model of explanation (described in Chapter 9) and the model of inductive explanation to be characterized below, exhibit.

On the reason-giving side we have the connected features of *nomic expectability* and *informativeness* as displayed by the inductive argument. An explanans should make the explanandum nomically expectable relative to the explanans, and it should convey relevant information concerning the explanandum. These features become closely related when our requirement of (nomic) *positive probabilistic relevance* is adopted. What this requirement amounts to in the case of inductive explanation is this (where P is an inductive, or epistemic, probability):

$$(42) \qquad P(L/T) - P(L) > 0.$$

The difference $P(L/T) - P(L)$ measures the information transmitted from T to L (see Niiniluoto and Tuomela (1973) Chapters 6 and 7 on this). The higher the likelihood $P(L/T)$ is and the lower the prior probability $P(L)$ is the greater is the information transmitted from T to L. On the other hand, $P(L/T)$ clearly measures the nomic expectability of L on the basis of T, while the difference $P(L/T) - P(L)$ measures the nomic expectability due to T alone.

Very often we are looking for the grounds for the fact that L rather than $\sim L$ is the case: Why L rather than $\sim L$? What were the reasons for expecting L rather than its negation to be true? How was it possible that L rather than $\sim L$ was the case? If we have this in mind we obviously must impose the following so-called "Leibniz-condition":

$$(43) \qquad \text{If} \quad \rho(L, T), \quad \text{then} \quad P(L/T) > \tfrac{1}{2}.$$

This condition seems very reasonable if our notion of inductive explanation somehow involves the idea that an explanation should be potentially convertible into a prediction. Also if the reasons for *accepting L* (rather than $\sim L$) are in question, (43) must be satisfied. On the other hand, imposing (43) on inductive explanations means that not all "chance-events" are explainable: the event described by L is, but not (the) one described by $\sim L$.

One of our central ideas concerning explanation is that the behavior of a system is to be explained by reference to its (activated) stable behavior dispositions, realistically construed. In the probabilistic case, we obviously deal with propensities — or, rather, propensity bases — as explanatory properties. This may involve that even *minimal* inductive-probabilistic explanations must refer to *theoretical* properties like physical microstructures or, in the case of agents, wants, beliefs, subconscious wishes, and so on. In our (RMS**) this is seen in that the maximally specific predicates M *indispensably* contain such theoretical predicates. We might, in fact, define a propensity basis to be nomically indispensable for inductive probabilistic explanation if (RMS**) is not satisfiable without the basis-predicate occurring essentially in the minimal probability law $p(G/M) = r$ of (RMS**). It is easy to see how this characterization could be made precisely analogous to our definition of a causally indispensable predicate given in Chapter 4. Let it be emphasized here again that the problem, whether or not a propensity basis is nomically indispensable for inductive-probabilistic explanation, is a factual problem, not to be decided on the basis of arm-chair philosophizing.

According to Hempel's idea of inductive-probabilistic explanation (see (36)), an explanation is an argument requiring that the probability value be close to 1. On this view events with low or medium probability cannot be explained. Although we accept that an explanation is an argument, we do not accept Hempel's high probability-requirement nor any idea requiring the direct inductive detachment (corresponding to the deductive case) of the explanandum, such as $G(a)$ in (30), from the premises. According to Hempel, furthermore, the "goodness" of explanation is directly measured by an inductive probability having the same value r as the statistical probability in the law. We do not accept this. Although the degree of nomic expectability as measured by $p(L/T)$ covaries with r, that is not all there is to explanation, as we have seen (also see the discussion in Niiniluoto and Tuomela (1973), Ch. 7). Furthermore, there is no compelling need to measure the goodness of explanation in quantitative terms at all.

We shall later return to the problem as to how one might be able to detach a conclusion in an inductive-probabilistic argument. We shall also in Chapter 12 return to the explanation of low-probability events (cf. radioactive decay processes). Presently, our Leibniz-condition precludes such an explanation in the sense of explanation that we focus on here.

Now we can go on to present our model of (non-epistemic) inductive explanation. This model, i.e. the notion of a p-argument, is solely concerned with the *logical* aspects of (potential) inductive explanation. Thus, it omits (almost) all the epistemic and pragmatic features of explanation. Our model relies strongly on the notion of *lawlikeness* and hence on the notion of a *nomological* predicate.

A central, formal feature of the model to be presented is that it connects deductive and inductive explanation in some formal respects. Many of the conditions from our DE-model (of deductive explanation) carry over intact. Some modifications are naturally needed and one entirely new kind of condition – (RMS**) – is added. Notice that an inductive explanation does not have·to be concerned with any probabilistic laws at all. However, when it does, (RMS**) will take care of them.

Suppose, thus, that we are given a putative explanandum L and a putative explanans T. As in Chapter 9, we form a set of ultimate sentential conjuncts Tc of T. A probability statement $p(G/F) = r$ may either occur directly as a member of Tc, or as a member of the largest set of truth functional components of T (if, e.g., a disjunction $(p(G/F) = r) \lor S$ is a member of Tc).

We shall be concerned below only with *direct* and *single-aspect* inductive explanations of *singular* statements. In the case of inductive-probabilistic explanations, this means that only one covering probabilistic law of the form $p(G/F) = r$ will occur in the explanans as a member of Tc. We will restrict ourselves here to this kind of probabilistic laws for reasons of simplicity. In principle, it seems possible to extend our approach to cover probabilistic laws, e.g., of the kinds $p(G/F) \geq r$ and $r_1 \leq p(G/F) \leq r_2$ as well. As earlier, G and F will be atomic or compound nomological predicates.

If the original explanans T contains several probabilistic laws relevant to the explanandum, one may be able to "combine" them into one law, and thus into a *single-aspect* explanation, by means of the probability calculus; the resulting law can be used in Tc. (If the probability laws are not at all connected as to their vocabulary, this is of course not possible.)[4]

Let L be a singular non-probabilistic statement and let Tc be a set of ultimate sentential conjuncts. In the probabilistic case we let L be $G(a_1, \ldots, a_m)$, i.e. a singular statement with m singular names. ($G(a_1, \ldots, a_m)$ may, but need not necessarily, represent an atomic statement corresponding to an n-place predicate G.) We now define a relation $\rho(L, Tc)$ which can be read 'Tc (potentially) inductively explains L'. (Here "explaining" must be understood widely enough to cover "contribution to explanation".)

As earlier, the probabilistic case is relativized to a nomic field $\mathscr{F}_\lambda$. Our (RMS**) will appear below in one of our conditions ((v)).

We define our model (call it the *IE-model*) by this:

(44) $\rho(L, Tc)$ satisfies the logical requirements for (direct and single-aspect) potential singular *inductive explanation* if and only if

    (i) $\{Tc, L\}$ is consistent.

    (ii) $I(Tc, L)$ but not $Tc \vdash L$.

    (iii) Tc contains among its elements at least one lawlike statement and at most one lawlike probability statement. (Tc is not allowed to contain any generalizations which are not lawlike.)

(iv) For any Tc_i in the largest set of truth functional components of T, Tc_i is noncomparable with L.

(v) If Tc contains a lawlike probabilistic statement, it must be of the form $p(G/F) = r$, where F and G may be complex predicates, and Tc must contain a singular statement of the form $F(a_1, \ldots, a_n)$ such that the following requirement is satisfied:

For any predicate M in λ such that either

(a) M is a maximally specific predicate related to $G(a_1, \ldots, a_m)$, i.e. L, in $\mathscr{F}_\lambda$, or

(b) M is stronger than F and probabilistically relevant to $G(a_1, \ldots, a_m)$ in $\mathscr{F}_\lambda$,

if $p(G/M) = r_1$ is a lawlike statement then $r_1 = r = p(G/F)$.

(vi) It is not possible, without contradicting any of the previous conditions, to find sentences $S_i, \ldots, S_r (r \geq 1)$ at least some of which are lawlike such that for some $Tc_j, \ldots, Tc_n$ $(n \geq 1)$:

$Tc_j \& \ldots \& Tc_n \vdash_p S_i \& \ldots \& S_r$

not $S_i \& \ldots \& S_r \vdash Tc_j \& \ldots \& Tc_n$

$I(Tc_s, L)$,

where Tc_s is the result of replacing $Tc_j, \ldots, Tc_n$ by $S_i, \ldots, S_r$ in Tc and '$\vdash_p$' means 'derivable by means of predicate logic without increase in quantificational depth such that the derivation does not merely consist of universal or existential instantiation'.

In this definition we primarily think of the following interpretation for the relation of *inducibility* $I(Tc, L)$, i.e. 'Tc inductively supports L':

(ii′) $I(Tc, L) =_{df} P(L/Tc) > P(L)$ and $P(L/Tc) > \frac{1}{2}$.

In other words, I is taken to stand for the positive probabilistic relevance requirement together with the Leibniz-condition. In (ii′) the *inductive* (or epistemic) probability measure P is defined for sentences, which we can do for first-order statements. We have not, however, shown how to define P for probability statements such as $p(G/F) = r$ (but see (52) below and the discussion preceding it). That is a complicated task which we need not nor cannot undertake here. For we can in principle get along with comparative and even qualitative inductive probabilities as well. Thus, in the comparative case, (ii′) would just amount to the conditions that L

be less probable than L given Tc, and that L given Tc would be more probable than $\sim L$ given Tc (as for the qualitative case see below).

What definition (44) characterizes is the direct single-aspect inductive explanation of singular statements. As we have seen earlier, both conditions (v) and (vi) are a kind of *minimality* conditions. Condition (v) makes explanation single-aspect and it requires the use of maximally specific laws (relative to λ), whereas condition (vi) makes inductive explanation direct. Note that (vi) contains condition (5) of the DE-model of Chapter 9 as its special case. But in (vi) the new replacing set Tc_s (which is required to satisfy (i)–(v)) is supposed to preserve inducibility and not only deducibility, as compared with condition (5) of the DE-model. Conditions (i) and (iv) coincide with respect to content with their deductive counterparts (cf. Chapter 9). Condition (ii) substitutes inducibility for deducibility and excludes deducibility.

Let us now consider these conditions more closely. Are all of them *necessary*? Conditions (i) and (ii) are obviously necessary. Condition (iii) is tailored for single-aspected probabilistic explanation, such that it also covers non-probabilistic inductive explanation (see our earlier comments).

The special need for the noncomparability requirement in the inductive case can be demonstrated by means of examples such as the following:

(45) $(p(G/F) = r) \vee (x)G(x)$
$$\frac{F(a)}{G(a)} \quad .$$

(45) is an objectionable form of a singular inductive explanation, and it is excluded by condition (iv). (The same holds if we change this example by using $(x)(G(x) \rightarrow F(x))$ instead of $p(G/F) = r$ in it.)

Condition (v), the requirement of maximal specificity, was discussed at length earlier. Notice that when we are concerned with single-aspected inductive-probabilistic explanations we have, according to (iii), only one probabilistic law to deal with; (v) tells us it must be maximally specific and in that sense *minimal*. This excludes, e.g., such disjunctive laws as $(p(G/F) = r) \vee (p(G/F') = s)$, allowed by (iii), from occurring. Condition (v) is the only one which deals with the inner structure of probability statements. The other conditions treat probability laws only as sentential components.

Condition (vi) is needed because it excludes the occurrence of irrelevant and redundant elements in Tc and because it makes explanations *direct* (see the discussion in Tuomela (1972)). If $G(a)$ can be explained by means of $p(G/F) = r$ and $F(a)$ then we do not want Tc to contain Newton's laws or other irrelevancies. Condition (vi) needs essentially the same qualifying conditions as in the deductive case (see Chapter 9 and Tuomela (1976b)). (Notice, however, that (Q) will have to include inducibility, thus becoming

(Q')   not $I(Tc'_s, L)$.

This condition excludes the addition of alternative, rival (strict or probabilistic) laws as disjuncts weakening the original members of Tc.)

Let us consider some cases that (vi) serves to exclude. Suppose that our explanatory law, as specified by (v), is $p(G/F) = r$. Let us see whether the weaker statement $p(G/F) \geq r$ could be used in a Tc_s to eliminate the original law. It cannot, because $p(G/F) \geq r$ follows from $p(G/F) = r$ only on the basis of the calculus of probability and not on the axioms of logic alone. This directly blocks the use of $p(G/F) \geq r$ (and, for that matter, other "non-sharp" probability laws). Furthermore, (v) also excludes such non-sharp laws. Disjunctive weakenings of the form $(p(G/F) = r) \vee (p(G/F') = s)$, with F entailing F', and $(p(G/F) = r) \vee (x)(G(x) \rightarrow F(x))$ become excluded because of (v), if not otherwise. (There seem to be ways of handling the above types of cases, even if the single-aspectedness requirement of (v) is lifted, but here we will not discuss them.)

In our IE-model of inductive explanation very much hangs on how the relation of inducibility I is understood. If it is measured in terms of quantitative inductive probabilities (as in (ii')) then the account of inductive inference developed in Niiniluoto and Tuomela (1973) applies. That theory of inductive inference is concerned only with monadic languages as far as the details have been worked out, and it excludes the treatment of probabilistic laws. However, in other respects the results of that work can be used here.

Whether or not our conditions (i)–(vi) really are *jointly sufficient* to capture the logical aspects of inductive explanation seems to depend on the details of the interpretation of I. I conjecture, however, that the interpretation (ii') will prove satisfactory, even if that interpretation concerns a rather weak notion of inducibility. As we emphasized, only (and at most) a *comparative* notion of inducibility seems needed. We may even try to get along with a *qualitative* notion of inducibility. Rescher has, in fact, studied the formal properties of the qualitative notion of positive relevance jointly with the Leibniz-condition (see Rescher (1970), pp. 85–87; also cf. Niiniluoto (1972)).

We give a brief summary of some of the formal properties of the qualitative inductive relation I. The properties to be given to this relation satisfy our (ii'), i.e. the positive relevance and the Leibniz-requirements. We state them as follows (cf. Niiniluoto (1972) and Rescher (1970)):

(46)　　(a) $I(p, q)$ is invariant with respect to the substitution of logically equivalent statements to replace p and q in its arguments.

(b) If $p \vdash q$, then $I(p, q)$, given that q is not a logical truth.

(c) If p is consistent and if $p \vdash \sim q$,
then not $I(p, q)$.

(d) If $I(p, q)$, $I(r, q)$, and $p \vdash \sim r$, then
$I(p \lor r, q)$.

(e) If $I(p, q)$, $I(p, r)$ and not $I(p, r \& q)$, then
$I(p, r \lor q)$.

(f) If $I(p, q)$, $I(p, r)$ and $q \vdash \sim r$, then
$I(p, q \lor r)$.

The relation I does not satisfy any interesting transitivity principles. For instance, it does not satisfy these:

(1)　　If $I(p, q)$ and $I(q, r)$, then $I(p, r)$　(Transitivity).

(2)　　If $I(p, q)$ and $q \vdash r$, then $I(p, r)$　(Special Consequence).

(3)　　If $I(p, q)$ and $r \vdash q$, then $I(p, r)$　(Converse Consequence).

(Furthermore I does not, without qualifications, satisfy the modified counterparts of (2) and (3) as discussed in Tuomela (1976a).) Furthermore, I is not symmetric. It also fails to satisfy the following Conjunction principle:

(4)　　If $I(p, q)$ and $I(p, r)$, then $I(p, r \& q)$.

The next question we have to consider concerns which properties our inductive explanation relation $\rho(L, Tc)$ satisfies. Clearly, it must be stronger than I because of conditions (iv), (v), and (vi). Let us start by first noting that ρ fails to be reflexive (it is indeed irreflexive), symmetric and transitive. (Cf. the relevant discussion in Tuomela (1972), (1973a).)

Of the properties mentioned in (46) we first consider (a) and use $\rho(L, Tc)$ instead of $I(Tc, L)$. ρ does not satisfy (a) for at least two reasons. First our notion of lawhood is not invariant with respect to logical equivalence. Thus logical equivalence-transformations of the explanans may violate (iii). Secondly, such equivalence-transformations clearly may violate the noncomparability condition (iv) both in the case of the explanans and the explanandum. Our ρ does not nonvacuously satisfy (b) because of its explicit denial in (ii). But (c) is true as it follows from (i).

Concerning (d), it is false in general, for (vi) of (44) requires that an explanans, such as p or r, cannot be weakened. Finally, ρ fails to satisfy (e) and (f) because of (vi).

We have come to see that there is very little that can be positively stated regarding the general formal properties of ρ at least if only a *qualitative* I, such as above, is relied on. It is a rather complex relation and, in addition, it depends on how I exactly is characterized. Therefore, it is understandable that much is left to depend on the particular cases of explanation under consideration.

One remark, which connects our model to Salmon's ideas and our earlier discussion, can be made here. We recall that our (RMS**), and hence (v), is stronger than Salmon's homogeneity principle, given a relativization to $\mathscr{F}_\lambda$. According to (b) of (vi), for all predicates M of the form $F \& D$ and hence for all D, $p(G/F) = p(G/D \& F) = r$. Thus, we get the following result for a simple paradigmatic inductive-probabilistic explanation:

(47)   If $Tc = \{p(G/F) = r, F(a)\}$, $L = \{G(a)\}$
         and $\rho(L, Tc)$, then there is no property D in λ
         which screens$_0$ off (nor screens$_1$ off) F from G.

In our model of inductive explanation the goodness of explanation is not (explicitly at least) measured in terms of an inductive probability as in Hempel's model. The explanans does not have to make the explanandum inductively probable to a high degree; positive relevance and the fulfilment of the Leibniz-condition is all that has been required about the matter.

In what sense is an inductive explanation an argument and an answer to a why-question? Let us first consider the following paradigmatic inductive-probabilistic explanation:

(48) $p(G/F) = r$
         $\dfrac{F(a)}{G(a)}$.

We may read this as: $G(a)$ was the case *because* $F(a)$ and $p(G/F) = r$. (Of course this 'because' is a weak one in a double sense: it is not always a causal phrase but rather a reason-giving phrase in a broad sense and it does not completely guarantee the explanandum.) We must assume here that conditions (i)–(vi) of our model of inductive explanation are satisfied.

That (i)–(iv) are satisfied is trivial. Condition (v) may be assumed to be satisfied, for, when speaking of merely potential explanation, we may so choose our nomic field and laws that (v) becomes true. In the case of *actual* explanation, the premises must be (approximately) true; it is then partly up to the world as to whether (v) is satisfied. Condition (vi) seems to be satisfied for any reasonable account of inducibility, although I do not presently have a strict proof to offer because of the lack of a detailed theory of the measure P. (Note that the law $p(G/F) = r$ cannot be weakened because of (v). The problem is whether the singular premise $F(a)$ can be weakened.)

Now we can summarize part of our discussion by saying that an inductive-probabilistic explanation of a singular event described by L is a complete inconclusive answer, or part of a complete inconclusive answer to the question 'Why L?'. This needs some qualifications, but let us first define inconclusive answerhood on analogy with conclusive answerhood (cf. (9.5)):

(49)   A statement T is a potential *inconclusive explanatory answer* to the question 'Why L?' (relative to fixed E, E' and B) if and only if
(1) T constitutes, or is a part of, the premises of some potential p-argument for L;
(2) T conveys understanding concerning why L is the case;
(3) (2) is (at least in part) the case because of (1).

Here we of course mean by a p-argument an argument satisfying (i)–(vi) of (44). The inductive relation I will be required to satisfy only the positive relevance requirement. Clauses (2) and (3) are understood on analogy with their counterparts in (9.5).

Can we now identify an inductive-probabilistic explanation with an inconclusive answer of the above kind? We cannot. The above notion is too wide, even if it can be taken to include all inductive-probabilistic explanations. For all explanatory p-arguments give inconclusive explanatory answers in the above sense, but such a p-argument need not contain *probabilistic* laws at all.

It should be noted that there may be inductive-probabilistic explanations which are answers to how possible-questions, but not in the first place, or as plausibly, to why-questions (or to why-questions about why the explanandum rather than its negation). Consider the following artificial example. We ask: How is it possible that the philosopher Jones is a

millionaire? Our explanatory answer may consist of stating that, although
it is very unlikely that a philosopher be a millionaire, still the fact that
Jones is a Texan and that Texans with some lawful probability $r(<\frac{1}{2}$, say)
are millionaires may serve as a reason-giving explanans here. Or suppose
a new bridge collapses. How was it possible? We point to a slight structural defect and/or to a storm and show that with some small probability
the collapse could be nomically expected. Under our wide interpretation of
'why', we do in these cases also give a reason-answer to a why-question,
and not only to a how possible-question, however.

Although I am willing to claim that all inductive-probabilistic explanations *can* be regarded as answers to suitable why-questions (understood
in a broad sense), we may often be more interested merely in how possible-
questions (see above). I will not discuss this further here. The whole matter
is not so central for the aims of this book, since we in our applications
will concentrate on intentional-teleological action-explanations and hence
only on why-questions, anyway (see Chapter 12).

4.2. We have noted that an explanandum of an inductive-probabilistic
explanation, i.e. of an explanatory p-argument, cannot be directly *detached*
from the premises. There is, however, more to be said on this topic;
these additional remarks will also serve to clarify a sense in which an
explanatory p-argument conveys (pragmatic) understanding concerning
the explanandum.

Suppes (1966) has argued that it is misleading to construe the "statistical
syllogism" as (*48*). Instead we should consider

$$(50) \qquad \begin{array}{l} \mathrm{p}(G/F) = r \\ \underline{\mathrm{p}(F(a)) = 1} \\ \mathrm{p}(G(a)) = r. \end{array}$$

(*50*) represents a formally valid inference in the probability calculus, and
here the conclusion is detachable from the premises (and the axioms of
probability). So far so good. What is awkward here is that the second
premise must be represented by $\mathrm{p}(F(a)) = 1$. Suppes adopts the *extra*
axiom that if an event occurs it gets the (subjective) probability 1, and
the whole probability field is thus conditionalized to this event.

But if we consider our framework we see that Suppes' suggestion is
not very satisfactory. We recall that the probability of a closed formula
within each model is 1 or 0. What the suggestion $\mathrm{p}(F(a)) = 1$ amounts
to is that *one* model – that representing the actual world – gets all the

probability mass. I don't find this a good way of connecting the actual and the possible (probable). However, Suppes' conditionalization approach to the formulation of the statistical syllogism represents a technical solution, which is formally compatible with our model. We can still say here that $G(a)$ is our explanandum and that it is explained by means of $p(G/F) = r$ and $F(a)$. Thus, while argument (48) is not an argument with a detachable conclusion, there is, at the back, so to speak, a *deductive* probabilistic argument (with a detachable conclusion).

What is still somewhat disconcerting about Suppes' formulation is that even if we must assume $G(a)$ to be the case it still only gets the probability r. How is this to be reconciled with the fact that we gave $F(a)$ the probability 1? The answer to this problem must be that $p(G(a)) = r$ here is to be taken *relative* to the occurrence or obtaining of $F(a)$ (and hence $p(F(a)) = 1$) and the probability law. The occurrence of G thus plays no part here. What (50) then amounts to is a most trivial probabilistic conditionalization.

There is still another and, I believe, much better way of showing that at the back of a probabilistic syllogism there lies a deductive argument. This is due to the fact that explanation is closely connected to the pragmatic and, so to speak, *prescriptive* dimension of science. My claim now is that to each probabilistic explanation of form (48) there is a suitable corresponding deductive *practical syllogism* (and hence a practical argument). Let us see what is meant by this.

All laws, whether strict or probabilistic, may be taken to play the role of material rules of inference (or partial rules of inference); cf. Sellars (1969c) and Tuomela (1976a). Consider the following schema:

$$(51) \qquad \frac{F(a)}{G(a)} \qquad (R)$$

Let '$F(a)$' stand for 'a is a piece of copper which has been heated' and '$G(a)$' for 'a expands'. Now, rule (R) says: "Given $F(a)$ one is entitled, in normal conditions, to infer $G(a)$". The descriptive counterpart of the prescriptive rule (R) is (or at least, contains) a law with the form $(x)(F(x) \rightarrow G(x))$. I have omitted considerations related to normal conditions, etc., and have also otherwise simplified the situation, but I assume that my present analysis can in principle be accepted for the non-probabilistic case.

Let us now reinterpret (51). '$F(a)$' will stand for 'a is a sequence of two tosses of a fair die in a dice-throwing experiment' and '$G(a)$' will stand for 'a exhibits at least once one of the numbers 1, 2, 3, 4'. Here the rule (R) is a derived probabilistic rule roughly of the form: "If $F(a)$ then one is entitled to infer $G(a)$ with (inductive) probability $\frac{8}{9}$". This is the prescriptive counterpart of the descriptive probability law $p(G/F) = r$, analyzed in Section 1. To infer something with (inductive) probability $\frac{8}{9}$ analogically refers to one's chance of "selecting" or "finding" a model *and* an "exchangeable" individual (a) in that model so that '$G(a)$' is true, given that '$F(a)$' is true in it. Let us accept this simplified account so far.

Two remarks may be made now. First, concerning the probabilistic inference, as we said, the *inductive strength* or certainty with which each particular probabilistic inference is made is $\frac{8}{9}$. In other words, it seems compelling here to accept a principle according to which the inductive probability in question is, in general,

$$(52) \qquad P(G(a)/F(a) \ \& \ p(G/F) = r) = r.^5$$

If you like, we may also write this as:

$$P(G(a)/F(a)) = r, \quad \text{given that} \quad p(G/F) = r.$$

Secondly, our rule (R) says, both in the probabilistic and the non-probabilistic case, that one *may* infer the conclusion but not that one *ought* to. This suggests that (51) still is enthymematic in some respects. I shall now briefly indicate how one may come to think that there is a deductive argument behind (51) such that this argument, indeed, leads to a practical acceptance of the conclusion.

In order to do this, we must consider the scientist's objectives and his means for achieving them. Let us be simple-minded and think that our scientist, call him A, only strives for *informative truth*. That is, we assume that in his work he intends to maximize his expected utilities, where his sole source of utility is the information content of the hypotheses (and other statements) considered.

Let us now consider the following practical syllogism which, for simplicity, is only concerned with the probabilistic interpretation of (51):

(PS) A intends to maximize his expected utilities.

   A considers that unless he accepts all the singular conclusions of schema (51), given an appropriate singular premise of it, he will not maximize his expected utilities.

   $F(a)$ is such a singular premise and $G(a)$ is such a singular conclusion.

   ⎯⎯⎯⎯⎯⎯⎯⎯⎯⎯⎯⎯⎯⎯⎯⎯⎯⎯⎯⎯⎯⎯⎯⎯⎯⎯⎯

   A accepts $G(a)$.

This practical syllogism is of the general type we have been discussing earlier in this book (cf. (7.9), (7.13) and also the normative version (7.16)). We have omitted considerations pertaining to normal conditions from (PS). In order to be sure that (PS) represents *logically conclusive* practical inferences, we assume that it is applied *ex post actu*, i.e. after A has accepted $G(a)$. Then it also can be used to explain A's action of accepting $G(a)$, as we have in fact seen in earlier chapters.

Now, the knowledge that $p(G/F) = r$ (and through it the rule (R)) figures at the back of A's belief that he *must* accept $G(a)$, given $F(a)$, in order to maximize his expected utilities (cf. (7.16)). For this belief is intimately connected to his degree of belief that an exemplification of F will be followed by an exemplification of G. This degree of belief is just the subjective counterpart of the inductive probability P in (52). To see a little better how these two beliefs may be connected, we explicate (PS) in more technical terms.

We assume that the informative content of a statement is all that A strives for. He wants to maximize the expected information content of the statements he (provisionally) accepts as his knowledge or at least as his corroborated beliefs. Consider some arbitrary instantiating statements $F(a)$ and $G(a)$ in the case of (51). We let $h = G(a)$ and $e = F(a)$ & $p(G/F) = r$. (Alternatively, we could take $e = F(a)$ and use $p(G/F) = r$ as a backing material rule of inference.) Then we define the relevant epistemic utilities as follows:

$$u(h, t, e) = \text{the utility of accepting } h \text{ on } e \text{ when } h \text{ is true;}$$
$$u(h, f, e) = \text{the utility of accepting } h \text{ on } e \text{ when } h \text{ is false.}$$

The expected utility of accepting h on the basis of e now becomes

$$(53) \qquad E(h, e) = P(h/e)u(h, t, e) + P(\sim h/e)u(h, f, e),$$

where P is a suitably defined inductive probability. We may now define the utilities in terms of informative content as follows (cf. Hintikka and Pietarinen (1966), Niiniluoto and Tuomela (1973), Chapter 6):

$$(54) \qquad (a_1) \; u(h, t, e) = \text{cont}(h)(= 1 - p(h))$$
$$(b_1) \; u(h, f, e) = - \text{cont}(\sim h).$$

Alternatively, we may consider:

$$(55) \qquad (a_2) \; u(h, t, e) = \text{cont}(h \vee e)$$
$$(b_2) \; u(h, f, e) = \text{cont}(\sim h \vee e).$$

Both ways of defining utilities (i.e. both combinations a_1 & b_1 and a_2 & b_2) lead to this:

(56) $E(h, e) = P(h/e) - P(h)$
          $= P(G(a)/F(a)$ & $p(G/F) = r) - P(G(a)).$

Now let $P(G(a)) = r_0$. Then on the basis of (52) we get:

(57) $E(h, e) = r - r_0.$

But as we have seen, the difference (57) just represents the quantitative counterpart of our requirement of positive relevance. The positive relevance requirement entails that $E(h, e) > 0$. Now the fact that $E(h, e) > 0$ guarantees that $\sim h$ has an expected utility $E(\sim h, e) < 0$. However, that does not suffice to make h acceptable yet, for $P(h/e)$ may still remain low (if $P(\sim h)$ is high). A "minimal" requirement for the acceptability of h on the basis of e is that $P(h/e) > P(\sim h/e)$, viz. the Leibniz-requirement.

If we ignore the unimportant fact that instead of using the difference $r - r_0$ we simply use the qualitative positive relevance requirement $r > r_0$ in our model of inductive explanation, we can now say the following. Our above practical syllogism (PS), under a natural interpretation, requires for its validity the truth of $I(Tc, L)$, i.e. of $I(F(a)$ & $p(G/F) = r$, $G(a))$. This close connection may be taken to show that (PS) really amounts to giving at least a (partial) pragmatic explanation for the explanatory relation $\rho(L, Tc)$ to hold. (PS) explains and justifies our definition of $\rho(L, Tc)$ (to a great extent at least), for: 1) it may be taken to show that $I(Tc, L)$ has been "philosophically" correctly, or approximately correctly defined and, in particular, it gives a pragmatic explanation of the material probabilistic rule of inference (R); 2) it makes inductive-probabilistic explanation respectable in telling us that each inductive-probabilistic argument is backed by a related practical argument with a deductively detachable conclusion; 3) it shows that an explanatory ρ-argument conveys (pragmatic) understanding concerning the explanandum in the sense of guiding the scientist's actions ("acceptings").

What has been said about the relationship between inductive-probabilistic explanations and practical inferences seems to be generalizable to also cover other kinds of inductive explanations as well as "multigoal enterprises" with several sources of epistemic utilities. Here we shall not do that, however.

Suppose now our scientist A makes inductive-probabilistic explanations according to our model. He then comes to accept $G(a_i)$ for a great number of values of i given $F(a_i)$. The rule of inference he uses is that incorporated in the practical syllogism (PS). This rule of inference involves as its elements both the law $p(G/F) = r$ and the inductive relation I (positive relevance and the Leibniz-condition) as we have just explicated. The probability law guarantees that in the long run (whatever that is taken to be) approximately $r\%$ of F's will be G's. The applicability of the inductive relation I to characterize A's practical inferences guarantees that the F's are positively relevant to the G's and probabilify them more than the $\sim G$'s.

The problem that A will accept false G's in approximately $(1 - r)\%$ of the cases must be considered here. Now, if A would include the law $p(G/F) = r$ in the same set as all the accepted statement pairs $F(a_1)$, $G(a_1)$, $F(a_2)$, $G(a_2)$, . . ., then obviously the total (infinite) set of accepted statements would be *inconsistent*, for *all* the singular pairs of the above kind (rather than only $r\%$ of them) would be accepted. But I think this objection is not lethal. Accepted sets of statements should be regarded as more "local" than the above one, and they should be taken to be (desirably) consistent *pragmatically* restricted finite sets of statements accepted on the basis of (PS). If we were omnipotent and if we thus were using *total* infinite sets of the indicated kind, we would not need induction.

The problem discussed is related to the notorious lottery paradox. Suppose we take seriously the idea of forming a set of accepted statements. If we then accept, on the basis of the evidence and $p(G/F) = r$, $G(a_1)$ and $G(a_2)$ should we not then accept their conjunction (and so on, for an infinite conjunction of thus accepted statements)? This would in fact be an acceptance mechanism leading just to the above seemingly problematic situation of inconsistency.

But we do not have to accept an unbounded acceptance rule for conjunctions. In fact, our inductive relation I (interpreted as (ii')) *prohibits* it. For we recall that I does *not* satisfy the following conjunction principle:

If $I(e, h)$ and $I(e,k)$, then $I(e, h \& k)$.

This nicely supports our claim above that we should be satisfied with more "local" sets of accepted statements. It also blocks the application of the main mechanism, which could be thought to lead to the inconsistent total set discussed above.

APPENDIX TO CHAPTER XI

Below we shall briefly sketch the main content of Łoś' approach to the semantical representation of probabilities of formulas (cf. Łoś (1963a), (1963b), and Fenstad (1968)). We will not present Łoś' important results in their full generality but confine ourselves to a somewhat restricted case.

Let $\mathscr{L}$ be a first-order language and S the set of models of $\mathscr{L}$. We will now consider how the probabilities of formulas of $\mathscr{L}$ can be represented in terms of the models in S. In a more general treatment we would consider a theory (deductively closed set) of $\mathscr{L}$ and we would relativize the probabilities to that theory. We would also choose our model space S so that, for all $\mathscr{M} \in$ S, the intersection of the sets of all statements true in $\mathscr{M}$ becomes equal to that chosen theory. Thus we have the possibility to alternatively choose S as, for instance, the set of *nomically possible* models (related to $\mathscr{L}$).

Here we consider, for simplicity, only the set S of *all* models of $\mathscr{L}$. Let S(f) denote the set of models in which the formula f of $\mathscr{L}$ is true. The completeness theorem of first-order logic implies that S(f_1) = S(f_2) entails that f_1 and f_2 are interderivable. Hence we may define without ambiguity a set function μ by

$$\mu(S(f)) = p(f),$$

where p is a probability measure in $\mathscr{L}$.

The collection $\mathscr{A}_1 = \{S(f): f \text{ is a sentence of } \mathscr{L}\}$ is an algebra of sets in S and μ is continuous in $\mathscr{A}_1$. This is an immediate consequence of the compactness theorem of first-order logic. It follows that μ may be extended to a continuous probability measure on $\langle S, \mathscr{A} \rangle$ where $\mathscr{A}$ is the σ-algebra generated by $\mathscr{A}_1$. If **1** denotes (the equivalence class of) a provable formula of $\mathscr{L}$, then p(**1**) = 1 and thus $\mu(S)$ = 1. Hence any probability p in $\mathscr{L}$ induces a σ-additive probability measure on S.

Let f now be an arbitrary formula of $\mathscr{L}$. We define a new measure μ_f on $\langle S, \mathscr{A} \rangle$ by

$$\mu_f(S(g)) = p(f \& g), g \in \mathscr{L}.$$

It can be seen that μ_f is absolutely continuous with respect to μ. Hence, by the Radon–Nikodym theorem of measure theory it can be represented as

$$\mu_f(B) = \int_B a_f(\mathscr{M}) \, d\mu(\mathscr{M})$$

for every $B \in \mathscr{A}$. The following computation then represents p(f), for an arbitrary formula f of $\mathscr{L}$, in terms of μ:

$$p(f) = p(f \& \mathbf{1}) = \mu_f(S) = \int_S a_f(\mathscr{M}) \, d\mu(\mathscr{M}).$$

Now, the function a_f can be taken to be a probability measure. In fact we may define:

$$\lambda_{\mathscr{M}}(\sigma_{\mathscr{M}}(f)) = a_f(\mathscr{M}),$$

where $\sigma_{\mathscr{M}}(f)$ represents the set of sequences of elements from the domain of $\mathscr{M}$ which satisfy f. The probability function $\lambda_{\mathscr{M}}$ represents "probability within a model".

We may now state Łoś' representation theorem as follows:

Łoś' Theorem: Every continuous probability on an algebra of formulas may be represented in the following form:

$$p(f) = E\mu(\lambda_{\mathscr{M}}(\sigma_{\mathscr{M}}(f))) =$$
$$\int_S \lambda_{\mathscr{M}}(\sigma_{\mathscr{M}}(f)) \, d\mu(\mathscr{M})$$

Thus p(f), the probability of f, is always expressible as an average value of probabilities in models with respect to a measure on the set of all models. This important result connects, on a fairly general level, probabilities defined for full first-order languages with the set-theoretic Kolmogorovian scheme. *A fortiori*, such linguistic approaches to inductive logic as Carnap's (1950) and Hintikka's (1966) are shown to be semantically representable in principle.

If we assume that in sampling individuals every sample of individuals counts equally in determining sample probability, this exchangeability assumption entails that, for finite models $\mathcal{M}$ and *open* formulas f

$$\lambda_{\mathcal{M}}(\sigma_{\mathcal{M}}(f)) = \text{the relative frequency of individuals satisfying } f \text{ in } \mathcal{M}.$$

Thus, in *finite* models of universes of size N, p(f) equals the average value of relative frequencies in models:

$$\text{p}(f) = \sum_{\mathcal{M}} \frac{r}{N} (f, \mathcal{M}) \cdot \mu_{(r)}^{(N)} (\mathcal{M})$$

Here $\frac{r}{N}$ gives the relative frequency of f in $\mathcal{M}$ in the obvious fashion, and $\mu_{(r)}^{(N)}$ ($r = 1$, ..., N) expresses the prior probability of the set of models in which exactly r of N individuals satisfy f.

As Fenstad (1968) has shown, these probability distributions for formulas in finite universes converge with increasing N (under the exchangeability assumption) so that in an infinite universe, p(f) becomes the expected value of the limit distribution, given that a certain finiteness assumption is assumed (condition (II) of Fenstad (1968), p. 8). (However the finiteness assumption is disputable, but we cannot here consider it.)

So far no mention has been made of quantified statements occurring as arguments of the probability measure p. They can be handled in terms of Gaifman's condition, which says that the probability of an existentially quantified formula, e.g. $(Ex)A(x)$, is to equal the *supremum* of the probability of the disjunction $A(a_0) \vee \ldots \vee A(a_n)$ in the domain under discussion (see e.g. Scott and Krauss (1966) for this). However, our theory of probabilistic explanation does not deal with such probability statements. I especially want to emphasize that the covering laws of probabilistic explanations are *not* of the form $\text{p}((x)(F(x) \rightarrow G(x)) = r$.

NOTES

[1] The importance of single case probabilities is still more clearly illustrated by the following game. We are coin-tossing. If heads are in the lead over tails after the nth toss then we say that the event Red has occurred, otherwise the event Blue. The events Red and Blue are clearly not independent. For instance ⟨Red, Blue, Blue, Red⟩ is an impossible quadruple. If the coin is fair, it is clear that the probability of obtaining Red is sometimes 0, sometimes $\frac{1}{2}$, sometimes 1. It depends on everthing that has happened before, we may say, and hence it is to be regarded as an appropriate kind of relative probability.

It is in the above sense that single case probabilities are of unique events, although we must speak about these events in terms of their being suitable *exemplifications of universals*.

[2] It should be noticed that the maximally specific predicates of (RMS**) are *disjunctions* of Ct_i-predicates. In Hempel's (RMS*) a maximally specific predicate is equivalent to a *conjunction* of predicates. Whether or not Hempel's and my notions of maximally specific predicate are equal in the monadic case obviously depends essentially

on whether Hempel would accept that his maximally specific predicates could be equivalent to *negations* of conjunctions of predicates.

In any case, conditions (a) of (RMS*) and (RMS**) work identically in excluding Grandy's counterexamples. Consider (*34*), for instance. The predicate $\sim F \vee G$, which is equivalent to the disjunction $F \& G \vee \sim F \& G \vee \sim F \& \sim G$ of conjunctions, is a maximally specific predicate on both accounts. As the statement $p(G/ \sim F \vee G) = 0.90$ is not true (we take **K** to be the set of true statements), (*34*) is disqualified.

[3]   Stegmüller accepts the following condition to replace Hempel's (RMS*) (Stegmüller (1973), p. 324): For every nomological predicate F' and every r the following holds:

(a)   if $\{p(G/F') = q, F'(a), F' \subseteq F\} \subseteq \mathbf{K}$,   then $q \geq r$;
(b)   if $\{p(G/F) = q, F'(a), F \subseteq F' \cap \sim (F = F')\} \subseteq \mathbf{K}$,   then $q \neq r$.

In Stegmüller's requirement clause (b) is adopted to guarantee that *maximal* homogeneous classes be referred to in an explanation, the requirement that we have criticized above. Furthermore, basically because clause (a) only requires that $q \geq r$, Stegmüller's condition does not escape the problem of epistemic ambiguity (in the sense defined earlier).

[4]   As an example of a multiple-aspect explanation, we might consider the explanation of $G(a)$ by means of the laws $p(G/F) = r$ and $p(G/H) = s$ as well as the initial conditions $F(a)$ and $H(a)$. Given that F and H are not probabilistically connected in our system, we must analyze this situation simply as follows. We have two single-aspect explanations (one corresponding to each probabilistic law); the fact that these explanations are in the obvious way *connected* is just an extra feature that our model does not directly speak about, although clearly both explanations together convey more information concerning $G(a)$ than either one alone does. (Cf. problems of explaining and diagnosing illnesses, such as schizophrenia, where the above kind of multiple causes and/or symptoms occur.)

[5]   Here, as elsewhere, the symbol 'r' in the probability law is a *numeral*. Thus, we escape David Miller's paradox, which relies on the possibility of using arbitrary singular terms to refer to numbers.

CHAPTER 12

PROBABILISTIC CAUSATION AND HUMAN ACTION

1. PROBABILISTIC CAUSES

Let us consider the following example, which is supposed to illustrate probabilistic causation. We conceive of a slot machine with 1000 different outputs. If a coin is inserted in the machine, it produces one of the outputs, 1, 2, . . . , 1000, according to some (objective) probabilistic process. We may, for simplicity, assume that each output is realized with an equal probability of $\frac{1}{1000}$, when a coin is inserted. In this example, we are then prepared to assert that inserting the coin (probabilistically) *causes* one or the other outputs to be realized. Furthermore, let us suppose that in fact output no. 617 (a chocolate bar) was realized on a particular trial. Then we say that inserting the coin on that occasion (probabilistically) caused the chocolate bar to come out. An explanation of the chocolate bar's emerging (a "low-probability" event) can be given by saying that it was causally produced by my inserting a coin. (It did not come about because of the dysfunctioning of the machine, for example.)

In discussing singular causation it seems, then, that the possibility of causation under indeterminism must be taken into account. According to our present knowledge, at most some microphysical processes, such as radioactive decay, seem to be objectively probabilistic. Thus, the realm of genuinely probabilistic causation seems to be beyond the scope of our present discussion, unless perhaps some brain processes or thought processes can be shown to be objectively probabilistic.

How about dice-throwing? Does it not represent an objectively probabilistic process? There is reason to believe that it does not, for a deterministic physical description seems to be available in principle for each singular dice-tossing. Still the process, conceptualized in terms of the "manifest image", may be regarded as stable enough to be considered a lawful probabilistic one. Thus, we could speak of two kinds of objective probabilistic processes: 1) processes which cannot be "carved up" or reconceptualized so as to make them deterministic at any other "ontic level", and 2) processes which are objectively probabilistic at some ontic level but which nevertheless can be reconceptualized in terms of another ontic level and shown to be deterministic. What "gambling device laws" (e.g. those governing dice throwing or roulette wheels) capture seem to be processes of the second type.

In this connection we must not forget those probabilistic processes, which, though perhaps stable, get their probabilistic character merely because of our lack of knowledge. Such "epistemic" probabilistic connections and processes are what the social scientists investigate in practice. Whatever formal properties we are below going to give to probabilistic causes also apply to this "epistemic" situation, which, although perhaps of less interest to a philosopher, is of central importance to the social scientist.

We have thus distinguished three cases in which probabilistic causation may be involved; of these, only the first presupposes the falsity of "ontic determinism", understood in a strong sense. Let us now proceed to a closer discussion of singular probabilistic causation. We will in the first place be interested in *direct* (productive) probabilistic causation. It will be shown that the account of direct and productive deterministic singular causes (developed in Tuomela (1976a) and discussed above in Chapter 9) can be (in a sense) generalized to also cover probabilistic causation. Below we shall mostly concentrate on some formal aspects of probabilistic causation; we have to leave out many important relevant philosophical problems.

We start by defining a probabilistic conditional $\triangleright_{\overrightarrow{r}}$, which generalizes the "strict" conditional $\triangleright\rightarrow$ defined and discussed in Chapter 9. We do it for two singular (event) descriptions D_1 and D_2 as follows:

(1)   '$D_1 \triangleright_{\overrightarrow{r}} D_2$' is true if and only if there is a true probabilistic theory T which, together with D_1, inductively explains D_2, viz. $\rho(D_2, T \& D_1)$.

In this definition D_1 and D_2 are singular statements, e.g. $D_1 = F(a)$ and $D_2 = G(b)$ (where F and G can be complex predicates). In this case the probabilistic backing law would be $p(G/F) = r$. The names 'a' and 'b' can be names of some events which have occurred. (We shall below use event-ontology; also substance-ontology might be used.) Then we can read '$F(a) \triangleright_{\overrightarrow{r}} G(b)$' as 'the occurrence of a (as an F) inductively explains the occurrence of b (as a G)' or 'given that the F-event a occurred it was reasonable to expect that a G-event (which happened to be b) occurred'. In other words $\triangleright_{\overrightarrow{r}}$ explicates singular inductive explanation and inductive determination. In a rather obvious sense it then also, accepting (*11.52*), explicates the inductive conditional 'if an *F*-event occurred then with inductive strength *r* a *G*-event would occur'.[1]

The singular probabilistic conditional $\triangleright_{\overline{r}}$ also serves to codify what the inference scheme (*11.48*) expresses, as we interpreted it. Thus, we may, given that *a* and *b* occurred, read '*F*(*a*) $\triangleright_{\overline{r}}$ *G*(*b*)' as 'given *F*(*a*) one may inductively infer *G*(*b*)'.

In giving the above interpretations for $\triangleright_{\overline{r}}$, we naturally assumed that the inductive relation I involved in the definition (*11.44*) of ρ is explicated by means of the positive relevance condition and the Leibniz-condition, viz. (ii'). The formal properties of the thus interpreted inductive relation I and of the relation ρ of inductive explanation were discussed in Chapter 11. Thus nothing more needs to be said here concerning the formal properties of $\triangleright_{\overline{r}}$ under that interpretation.

However, let us now consider this conditional by weakening I to represent only positive relevance, i.e., I(*Tc*, *L*) if and only if p(*L*/*Tc*) > *p*(*L*). This situation must be considered when, for instance, low-probability events are explained. For in their case we may use explanations which do not give the explananda a probability greater than one half (cf. our above slot machine example). The only cases of the explanation of such low probability events that we are considering will be probabilistically *caused* events. Their explanation amounts to citing their cause, i.e. stating a suitable *causal* conditional with $\triangleright_{\overline{r}}$. Although the positive relevance relation obviously satisfies some conditions that positive relevance together with the Leibniz requirement do not satisfy, the new weakened relation of inductive explanation still does not gain any interesting new formal properties (cf. Rescher (1970), Niiniluoto (1972), and Niiniluoto and Tuomela (1973) on positive relevance). Thus the formal aspects of the new $\triangleright_{\overline{r}}$ do not need any further discussion.

On the basis of the new interpretation of $\triangleright_{\overline{r}}$, which will be adopted for the rest of this chapter, we may now study some formal aspects of probabilistic singular causation. Analogously with the weakening of the conditional $\triangleright\rightarrow$ to $\triangleright_{\overline{r}}$ we will now define a relation of singular probabilistic causation to correspond to the relation *C*(*a*, *b*) of "strict" or deterministic singular causation of Tuomela (1976a) and Chapter 9. This characterization concerns a relation '*PC*(*a*, *b*)' which reads 'singular event *a* is (or was) a direct probabilistic cause of singular event *b*'. We assume that *a* and *b* have occurred and define the truth of the statement

'$PC(a, b)$' in the material mode by:

(2) $PC(a, b)$ if and only if there are true descriptions D_1 and
                D_2 and a true causal probabilistic law $p(G/F) = r$ such
                that
         (1) D_1 truly describes a as an F and D_2 truly describes b as
             a G;
         (2) $D_1 \mathrel{\vartriangleright_{\!\!\overrightarrow{r}}} D_2$ (on the basis of the law $p(G/F) = r$);
         (3) a is causally prior to b;
         (4) a is spatiotemporally contiguous with b.

Several clarifying remarks are now in order concerning definition (2),
which gives a backing law analysis of singular probabilistic causation. It
assumes the existence of event descriptions D_1 and D_2 which describe a
and b, respectively, as an F and a G. For instance, we could have $D_1 = F(a)$
and $D_2 = G(b)$. Properties or generic events F and G are nomically
connected by means of a probabilistic law, viz. true lawlike statement,
$p(G/F) = r, 0 < r < 1$, it is assumed.

Conditions (1) and (2) concern the event descriptions D_1 and D_2, while
conditions (3) and (4) are concerned directly with the singular events a
and b. Conditions (1) and (2) will be regarded as sufficiently clear on the
basis of what has been said above. As for (3) causal priority seems best
analyzed in terms of "marking" method related to the generic events
F and G (cf. Reichenbach (1956)); also recall the asymmetry of ρ. The
temporal priority of cause is not analytically required, but in normal circumstances (3) indeed will involve this. The notion of spatiotemporal contiguity in (4) will not be analyzed here at all (see Kim (1973) and the
remarks in Tuomela (1976a)). Conditions (3) and (4) play no central role
in our discussion below, and hence a further analysis of them is not
needed here.

One objection that may immediately come to mind in looking at
our "definition" (2) is that it represents a circular analysis. Fair
enough. Even the word 'causal' occurs in the analysans, and twice, in
fact. Its occurrence in (3) is innocuous and non-essential, however.
But our requirement that $p(G/F)$ be a *causal* probabilistic law is
essential.

It must be said immediately that I do not attempt to give a reductive
analysis of causality at all. That is not possible, I claim (cf. Tuomela
(1976a)). Analysis (2) primarily connects singular causation with generic
causation. It does not really analyze generic causation.

What are causal probabilistic laws like as distinguished from non-causal probabilistic laws? The basic idea involved in a causal law such as $p(G/F) = r$ is that F produces G with probability r. Let F represent the inserting of a coin and G the ejection of a chocolate bar in our earlier slot machine example. When we repeatedly instantiate F, the products, i.e. instantiations of outputs, will exhibit a certain distribution of relative frequencies which in the limit converge to the probability values (such as $p(G/F) = r$), cf. the appendix to Chapter 11. The manipulability of G by means of F in roughly this sense serves as a test of a *causal* probabilistic connection, which marks a difference with respect to mere (= non-causal) probabilistic nomic coexistence of properties. We shall not attempt to give a more detailed analysis of causal probabilistic laws here. Our concerns in this chapter lie elsewhere.

Our definition of $PC(a, b)$ relies on explanation and information-giving (cf. our earlier discussion concerning ρ). What we are analyzing here is causality between two objective singular events a and b. However, this causality is to be analyzed or justified by reference to explanation and to the logical space of reasons. We can perhaps say that singular causation takes place "out there" in the external world, but much of what a philosophical analysis can say about such causation (save the existence of such a connection) is inevitably *epistemic* in nature. Definition (2) is not intended to be a mere meaning analysis but rather, it purports to say what the use of singular probabilistic causal claims, such as $PC(a, b)$, most centrally philosophically involve.

A common objection against a backing law analysis such as ours is that it seems to entail that no causal connections can obtain without there being linguistic descriptions such as D_1 and D_2. This objection is mistaken. We are discussing causality between structured events a and b here. In other words, we are concerned with a as an F probabilistically causing b as a G. What we are assuming, essentially, is the existence of *universals* of some sort. This granted, everybody can use his own favorite semantical apparatus. Thus our analysis could be given in terms of properties, generic events, concepts or just universals, as well.

Another common objection against a backing law analysis is the following related one. The analysis seems to presuppose the existence of laws. But surely causal processes exist without (linguistically formulated) laws, the objection goes. But here again it suffices for our purposes to speak about *nomic connections between universals*. In what sense that involves connections between properties *in rerum natura* and in what sense between concepts (in one's conceptual system), which somehow adequately represent aspects of reality, is not central here. Our analysis is compatible with different views in this respect. Given these remarks it is also easier to accept that our analysans is not only necessary but also sufficient for $PC(a, b)$. (See also the relevant remarks in Tuomela (1976a).)

We have been emphasizing the role of propensity-bases in accounting for the overt probabilistic behavior of chance set-ups. In our (2) this aspect is hidden in the property expressed by the predicate F, which may be a *complex* one. This property may involve both the relevant overt "stimulus" characteristics and the backing causally active propensity-basis (cf. Section 11.1).

One interesting aspect about our notion of probabilistic causation is that it is, in a sense, compatible with deterministic causation. That is, our '$PC(a, b)$' and '$C(a, b)$' may both be true (or assertable) at the same time. This could happen if, for instance, so-called "gambling device laws" governing the behavior of such chance set-ups as dice boxes and roulette wheels could function as backing laws for $PC(a, b)$. For if we now accept that the probabilistic processes in question are deterministic at some level in the "scientific image", then $C(a, b)$ would hold, too (with a suitable physical backing law). This situation is possible, even if we do not keep to a fixed nomic field $\mathscr{F}_\lambda$ and set λ, but let λ vary. I do not regard this "conflict" as paradoxical, however.

In Chapter 11 we noticed a connection between our requirement of maximal specificity and screening off. Suppose that $Tc = \{p(G/F) = r, F(a)\}$, $L = \{G(b)\}$ and that $\rho(L, Tc)$. Then there is no property in λ which screens$_0$ off (or screens$_1$ off) F from G, and F is in this sense inductively indispensable. Now we can immediately see this:

(3)   If $PC(a, b)$, then there is no property D in λ which screens$_0$ off F from G, given that F and G are the backing properties for $PC(a, b)$ and $\mathscr{F}_\lambda$ is a nomic field relative to which the backing holds.

Notice that there may be several ways of establishing a backing for $PC(a, b)$ with (3) holding for all of them.

We may now compare our approach to those of Reichenbach (1956), Suppes (1970), and Good (1960), which are perhaps the main accounts available today. Let us first consider Reichenbach's analysis of probabilistic causation, based on positive relevance. We reconstruct it as follows (cf. Reichenbach (1956), p. 204):

(4)   A singular event of kind F at time t' is *causally relevant* to a singular event of kind G at time t if and only if

(i) $t' < t$;

(ii) $p(G/F) > p(G)$;

(iii) there exists no set of singular events of kinds D_1, $D_2, \ldots, D_n$ which are earlier than or simultaneous with the above singular event of kind F such that (the conjunction of the members of) this set screens$_0$ off F from G.

The above definition of causal relevance does not, however, characterize only direct causal relevance but also indirect or mediated causal influence. We shall later return to indirect causation and to this notion of causal relevance. Presently we are interested in *direct* causal relevance only. Although Reichenbach did not himself define such a notion it is obvious enough how the above definition is to be complemented in order to implement this idea:

(5)   A singular event of kind F at time t' is *directly causally relevant* to a singular event of kind G at time t
if and only if

(1) this singular event of kind F is causally relevant to the singular event of kind G;

(2) there is no set of singular events of kinds C_1, $C_2, \ldots, C_m$ which occur before t but later than t' such that (the conjunction of the members of) this set screens$_0$ off F from G.

This definition thus excludes a causally relevant singular event from being a direct cause if some intermediate screening event is found, but even in such a case this singular event will, of course, continue to be causally relevant in the sense of definition (4).

Let us now think of the event kinds in definition (5) in terms of predicates as in Chapter 11. We can then see that the following result holds true, assuming $PC(a, b)$ to be backed by $p(G/F) = r$:

(6)   If $PC(a, b)$, then a (*as an F*) is directly causally relevant to b (*as a G*) in the sense of definition (5) (given that the events C_i, $i = 1, \ldots, m$ in its clause (2) belong to λ); but not conversely.

If a is causally prior to b, we understand that clause (i) of definition (4) is satisfied. Its clause (ii) incorporates the positive relevance requirement, which our explanatory relation ρ fulfils. We also easily see that our (1) can be taken to entail the truth of clause (iii) of definition (4). For, as the sets C_i, $i = 1, \ldots, m$, and D_j, $j = 1, \ldots, n$, are finite we can at least formally take λ to include them. Thus the given-clause of (6) also becomes satisfied. Accordingly, our relation $PC(a, b)$ is stronger than the above notion of direct causal relevance, which perhaps could be called "Reichenbachian". (That the converse of (6) does not hold should be obvious.)

Suppes has defined and discussed several notions of probabilistic causation. The following definition gives – in Suppes' terminology – the most sophisticated and presumably the most satisfactory of them (Suppes (1970), p. 28):

(7)   A singular event $F_{t'}$ is a *direct cause* of singular event G_t if and only if

    (i) $t' < t$;

    (ii) $p(F_{t'}) > 0$;

    (iii) $p(G_t/F_{t'}) > p(G_t)$;

    (iv) there is no t'' and partition $\pi_{t''}$ of the sample space such that for every $D_{t''}$ in $\pi_{t''}$

       (a) $t' < t'' < t$,

       (b) $p(F_{t'} \cap D_{t''}) > 0$

       (c) $p(G_t/D_{t''} \cap F_{t'}) = p(G_t/D_{t''})$.

In presenting this definition we have followed Suppes' notation rather closely. Thus e.g. $F_{t'}$ in the definiendum represents a singular event of the kind F occurring at t'. But $F_{t'}$ in the definiens represents rather a *kind* of event. It is assumed that in (7) an ordinary set-theoretic framework is employed with a fixed sample space X. The times of the occurrence of the events F, G etc. are included in the formal characterization of the probability space, and hence e.g. the expression '$p(F_{t'})$' makes formal sense.

Clause (i) of the above definition of a direct cause is identical with Reichenbach's analogous condition in (4). Clause (ii) is trivial and implicitly assumed both in our (2) and in Reichenbach's analysis. Clause (iii) states the positive relevance condition. So far Suppes' definition agrees with Reichenbach's and our own analysis. Condition (iv) formalizes a kind of "non-screening off" condition. It speaks of partitions of the sample space X relative to t''. It requires of *every* set $D_{t''}$ in *any* partition

$\pi_{t''}$ (all denied the existence) that $D_{t''}$ screens$_1$ off $F_{t'}$ from G_t. Although (iv) quantifies over all the partitions of X it is still rather weak in that it (somewhat oddly) concerns every $D_{t''}$ in any such $\pi_{t''}$.

If we find a partition $\pi_{t''}$ for which Suppes' condition (iv) becomes false, then also condition (2) in definition (5) becomes false. For any single element $D_{t''}$ in $\pi_{t''}$ can be taken as the one-member conjunction of the members of the set assumed to screen$_0$ off F from G in clause (2) of (5). We get:

(8)   If a singular event $F_{t'}$ is directly causally relevant to a singular event G_t in the sense of definition (5), then $F_{t'}$ is a direct cause of G_t in Suppes' sense (7); but not conversely.

This result holds even if clause (iv) of (7) were falsified by an infinite partition $\pi_{t''}$, for any element $D_{t''}$ again suffices for our purpose! The only if-part of (8) does not hold because in Suppes' definition (iv) concerns every $D_{t''}$ in $\pi_{t''}$ (one $D_{t''}$ suffices for falsification) and because, furthermore, Definition (5) concerns any (timewise appropriate) screening set and not only all the relevant partitions of some sample space X, and finally because of clause (1) of (5).

Combining (6) and (8) and appropriately transforming Suppes' framework into ours such that Suppes' partitions $\pi_{t''}$ of sets in X become partitions of predicate families in λ, we get:

(9)   If $PC(a, b)$, then a (as an F) is a direct cause in Suppes' sense of b (as a G); but not conversely.

Within our framework there is no dependence on a fixed sample space as in Suppes' original set-theoretical framework. (Neither do we depend on reference classes in Salmon's sense, of course.) This is an advantage. On the other hand, we must restrict ourselves to finite families λ in order to do anything "useful". A purely set-theoretic framework may, of course, employ an infinite class of sets (kinds of events), on the contrary. Still, these sets may include sets which are not recursively enumerable and which are not determined by place selection rules (as e.g. Salmon wants), and so on.

In view of (9), and keeping the differences in framework in mind, Suppes' results concerning his direct causes can be seen to characterize our probabilistic causes $PC(-, -)$ as well.[2]

Our conception of probabilistic causation has an important advantage over Reichenbach's and Suppes' notions as well as over Salmon's events with the screening off property. This advantage is that in our analysis cause-effect relationships can be tracked by means of *redescriptions* of events. To see what I mean let us consider the following example.

Suppes (1970), p. 41 discusses the following example about a golfer. The golfer makes a shot that hits a limb of a tree close to the green and is thereby deflected directly into the hole, for a beautiful birdie. Let the explanandum event A_t here be the event of the golfer's making a birdie, and let $B_{t'}$ be the event of the ball's hitting the limb earlier. We estimate that the probability of the golfer's making a birdie on this particular hole to be low. However, Suppes argues, we ordinarily estimate the conditional probability of the golfer's making a birdie on this hole to be still lower, given that the ball hits the branch. Yet, when we see the event happen, we realize that hitting the branch in exactly the way it did was essential for the ball's going into the cup, Suppes concludes.

Suppes says very little by way of solving this puzzle. We notice, first, that even if $p(A_t/C_{t''}) > p(A_t)$, where $C_{t''}$ is taken as the event of the golfer swinging his club, we seem to have $p(A_t/B_{t'} \& C_{t''}) < p(A_t/C_{t''})$ in our example. Suppes' main comment now seems to be that $C_{t''}$ may yet be regarded as a weak kind of cause, viz. a prima facie cause, of A_t in his technical sense (cf. Suppes (1970), p. 12). However, this does not take us very far at all.

We may then ask whether $C_{t''}$ is directly causally relevant to A_t in the sense of (5) or (7). Knowing an answer to this would be informative, but the example, unfortunately, does not tell us that (we may or may not have $p(A_t/B_{t'}) = p(A_t/B_{t'} \& C_{t''})$).

There are two interesting things to be said about this example, however. They are connected to how the situation is conceptualized. First, very much obviously depends on how the ball's hitting the branch is exactly described. We can imagine that $p(A_t/B_{t'} \& C_{t''})$ indeed is low if nothing special is said about the branch-hitting. But if B is taken to represent the generic event of the ball's hitting the branch with such and such an angle and speed, we might arrive at a characterization which makes $p(A_t/B_{t'} \& C_{t''})$ much higher. Thus, the conceptualization and (possibly) reconceptualization of the situation seems important, but in Suppes' account the causal claims are discussed relative to a fixed but rather unclearly given way of describing the case.

What is more, it seems that the whole causal process leading to this spectacular birdie is describable in deterministic (or "near-deterministic") terms. For we may imagine that under a suitable redescription we could find out that the process (causal chain) is backed by some nomic physical laws. The original description, in terms of probabilities, is then superseded by this new account, which explains why everything happened so smoothly in that situation. This shows the importance of allowing (in an explicit fashion) redescriptions of the singular events involved in the causal situation. This is built into our notion $PC(-, -)$, as we have seen.

Our definition (2) does not require that if a probabilistic causal conditional $F(a) \triangleright_{\vec{r}} G(b)$ holds that (anything like) the conditional $\sim F(a) \triangleright_{\vec{r}} \sim G(b)$ holds. The latter conditional would obviously represent the probabilistic analogue of necessary conditionship. This conditional is proportional to $p(\sim G / \sim F)$, and it satisfies the positive relevance requirement $p(\sim G / \sim F) > p(\sim G)$, which equals $p(G / \sim F) < p(G)$. Now, in the deterministic as well as in the probabilistic case, causal *overdetermination* may be possible (cf. Chapter 9). Thus, the fact that F is not exemplified does not preclude another cause of G from occurring. Therefore, we cannot make it part of our notion of a cause that the non-occurrence of the causal factor somehow "produces" the non-occurrence of the effect event.

But in most "ordinary" situations we might want to say that $p(G / \sim F)$ should be taken into account, when measuring the causal impact of F on G (cf. our slot machine example as an extreme case with $p(G / \sim F) \approx 0$). Good (1960) offers the following quantitative measure for the amount (Q) of causal support for G provided by F:

$$(10) \qquad Q(F; G) =_{df} \log \frac{p(\sim G / \sim F)}{p(\sim G / F)}$$

This measure is equivalent to

$$(11) \qquad Q(F; G) = \log p(G/F) - \log p(G / \sim F).$$

According to (11) $Q(F; G)$ is directly proportional to $p(G/F)$ and inversely proportional to $p(G / \sim F)$. From (11) we see that $Q(F; G)$ gains its maximum when F entails G (and when, hence, $p(G/F) = 1$). If $p(G/F) = 0$ (inductive independence between F and G) and if $p(G / \sim F) = 1$ (e.g. when $\sim F$ entails G), then $Q(F; G)$ obtains its minimum.

Good's measure is just one among many which are directly proportional to $p(G/F)$ and inversely proportional to $p(G / \sim F)$. Our measure

$$\text{syst}_2 (F, G) = \frac{1 - p(F \vee G)}{1 - p(G)}$$

is in fact one of those measures (see (*11.23*) of Chapter 11). For $\text{syst}_2 (F, G)$ gains its maximum when $p(G/F) = 1$ and its minimum when $p(G/\sim F) = 1$ (e.g. when $\sim F$ entails G). Thus, for cases where overdetermination is not to be expected, the measure $\text{syst}_2 (F, G)$ can be used to measure the amount of causal support for G provided by F. In other cases again, the measures $\text{syst}_1 (F, G)$ and $\text{syst}_3 (F, G)$ of Chapter 11 seem more appropriate. This is because they satisfy the likelihood condition according to which $\text{syst}_i (F', G) > \text{syst}_i (F, G)$ if and only if $p(G/F') > p(G/F)$.

There are several special problems related to overdetermination (e.g. conditional and linked overdetermination), but we will not in this context discuss them. Let us just point out that a backing law approach like ours (armed in addition with the notion of supersessance from Chapter 11) seems capable of handling such puzzles as well as any other, e.g. a "conditionalist", approach (cf. the discussion in Tuomela (1976a)).

Our next topic to be commented on in this section is *common causes* and *common effects* (cf. Reichenbach (1956)). Let us consider a simple example to illustrate the notion of a common cause. Suppose there is a significant probabilistic correlation between a sudden drop in the barometer (*A*) and the breaking of a storm (*B*). Thus we may assume:

$$(12) \qquad p(A \ \& \ B) > p(A) \cdot p(B).$$

In a methodological situation where (*12*) holds, we often start looking for a common cause of *A* and *B*. Such a common cause here might be a (temporally preceding) sudden drop in atmospheric pressure (*C*). What kind of probabilistic properties does a common cause have? One strong property that we may be willing to attribute to a common cause (at least in examples like the above) is the following

$$(13) \qquad p(A \ \& \ B/C) = p(A/C) \cdot p(B/C).$$

This condition says that *C* makes *A* and *B* independent or that *A* and *B* are independent relative to *C*. If we are willing to accept (*13*) then we immediately see that *C* screens$_1$ off *A* from *B* (and *B* from *A*). We get, provided that *C* does not occur later than *A*:

(*14*)   If *C* is a common cause of *A* and *B* and if $p(B/C) > p(B/A)$, then *A* is *not* causally relevant (in Reichenbach's sense (*4*)) to *B*.

In our barometer example we may plausibly assume that the antecedent conditions of (*14*) are fulfilled. Thus, a drop in barometer is shown not to cause a storm, which is what is desired. Condition (*14*) concerns generic events, but it can obviously be translated into talk about singular probabilistic causation as well. Thus, we may as well say, in view of (*6*), that if the generic events (predicates) *A*, *B*, and *C* satisfy the antecedent of (*17*) then no singular event *a* (which is an *A*) is causally relevant to any singular event *b* (which is a *B*), and hence not *PC(a, b)* under the backing predicates *A* and *B*.

We note that common causes may often be theoretical properties introduced to explain probabilistic connections between observables (such as the above *A* and *B*). The situation considered above is in fact typically one in which the need for explanatory theoretical properties is felt (cf. Tuomela (1973a), Chapters 6 and 7).

Next we consider common effects. What is meant here by a common effect is roughly the causal product of two independent causes. For instance, suppose that two persons are required to lift and carry a heavy table. Let the liftings be called *A* and *B* and the table's becoming elevated by *E*. Then *A* and *B* jointly caused *E*, and from *E* we may be able to induce that *A* and *B* took place. Thus we assume that a common *effect* satisfies this:

(*15*) $p(A \ \& \ B/E) > p(A) \cdot p(B)$.

However, in the present case we do not accept (*12*), which partially defines common causes. (Rather we have equality in (*12*).) Thus there is an *asymmetry* between probabilistic common causes and common effects, and in a desired manner.

There are lots of problems about probabilistic causation that we have left out of our consideration. Let us just point out two directions which were left unexplored here. There are fairly close connections between our approach and approaches using random variables. Some comments on them are made in Suppes (1970); recall that our $PC(-, -)$ entails Suppes' notion of a direct cause and his other weaker notions. Another direction, which we have not explicitly discussed, is the study of probabilistic causes under determinism (i.e. under the assumption that they approximate strict sufficient and/or necessary causes). Nowak (1974) has studied some of these questions.

Until now we have mostly been discussing *direct* probabilistic causation. Only Reichenbach's notion of causal relevance takes into account indirect probabilistic causation as well (see our definition (*4*)). We will need an account of indirect probabilistic causation for our discussion of complex actions in the next section. Let us therefore briefly consider probabilistic chains here.

In the case of deterministic causation we defined indirect singular causation simply in terms of a (deterministic) causal chain (definition (9.13)). One thing to be noticed about it is that it preserves determination. The first causal event thus is seen to determine the last event in the chain. Another aspect of this case is that the causation of any event in the chain is due to its direct cause only and thus in an important sense history-independent (or "path-independent"). We want to respect these two features below as far as possible.

In the probabilistic case we proceed as follows. We consider a sequence of *singular* events $e_1, \ldots, e_{n-1}, e_n, e_{n+1}, \ldots, e_r$. We want to explicate the idea that e_1 probabilistically causes e_r through the intermediate events so that e_1 probabilistically causes e_2, which again causes e_3, and so on until the end of the chain. As before we assume that singular causation involves causal nomic probabilistic relationships between some appropriate *generic* events $E_1, \ldots, E_{n-1}, E_n, E_{n+1}, \ldots, E_r$ which $e_1, \ldots, e_{n-1}, e_n, e_{n+1}, \ldots, e_r$ respectively exemplify.

We start by defining a relation of nomic betweenness for these generic events (predicates) E_i. Thus let $N(E_{n-1}, E_n, E_{n+1})$ be read 'the event E_n is nomically between E_{n-1} and E_{n+1}'. We define:

(16) $N(E_{n-1}, E_n, E_{n+1})$ if and only if
   (1) E_n supersedes E_{n-1} with respect to E_{n+1};
   (2) E_n supersedes E_{n+1} with respect to E_{n-1};
   (3) E_n screens$_1$ off E_{n-1} from E_{n+1}.

The notion of supersessance used here is that specified by the definiens of definition (11.12) (with the omission of reference class). Conditions (1) and (2) are clearly symmetrical with respect to E_{n-1} and E_{n+1}. It is also easily seen that the relation of screening$_1$ off is symmetrical with respect to its second and third arguments. Thus (3) entails that E_n also screens$_1$ off E_{n+1} from E_{n-1}. It follows that the relation $N(E_{n-1}, E_n, E_{n+1})$ is symmetrical with respect to its first and third arguments. ($N(E_{n-1}, E_n, E_{n+1})$ can be shown to be formally equivalent to Reichenbach's relation of causal betweenness (see Reichenbach (1956), p. 190).)

Furthermore, $N(E_{n-1}, E_n, E_{n+1})$ is easily seen to be unique in the sense that if $N(E_{n-1}, E_n, E_{n+1})$, then neither $N(E_{n+1}, E_{n-1}, E_n)$ nor $N(E_{n-1}, E_{n+1}, E_n)$, for otherwise the screening$_1$ off condition would be violated.

Another property of $N(E_{n-1}, E_n, E_{n+1})$ is that it is not transitive. If

$N(E_{n-1}, E_n, E_{n+1})$ and $N(E_n, E_{n+1}, E_{n+2})$, then it need not be the case that $N(E_{n-1}, E_n, E_{n+2})$. This is hardly a surprising observation.

What kind of properties of $N(E_{n-1}, E_n, E_{n+1})$ are philosophically most central? One feature is that conditions (1) and (2) tell us that probabilistic relevance diminishes with increasing distance in the "causal net" defined by the relation N. (Note: N admits of bifurcating causal trees; thus, it does not specify a linear chain.) The probabilistic relevance between two adjacent generic events E_n and E_{n+1} is never less than that between E_n and E_{n+j} $(j > 1)$. If the sequence $E_1, \ldots, E_{n-1}, E_n, E_{n+1}, \ldots, E_r$ could be taken to represent a causal sequence, the analogous remark would go for the strength of causal influence.

Another philosophically interesting feature of the relation N is that its condition (3) specifies a *Markov property* for the sequence $E_1, \ldots, E_{n-1}, E_n, E_{n+1}, \ldots, E_r$. The sequence becomes probabilistically ahistorical in the sense that the causal (or nomic) influence directed towards E_n is entirely represented in E_{n-1}. No previous events are relevant to the probabilistic determination of E_n in the presence of E_{n-1}.

The fact that $N(E_{n-1}, E_n, E_{n+1})$ is symmetrical entails that not all sequences $E_1, \ldots, E_{n-1}, E_n, E_{n+1}, \ldots, E_r$ defined by means of this relation are going to be causal sequences. Causal sequences are asymmetrical, we assume. How do we introduce asymmetry, then? Temporal asymmetry can hardly be taken as defining causal asymmetry (cf. Mackie (1974)). Causal priority thus has to be characterized otherwise, although it may turn out to covary with temporal priority. "Marking" conditions (Reichenbach), as well as various other "interventionistic" conditions, have been offered to resolve this problem about causal priority. We cannot here discuss this problem at all (but recall our comments on the asymmetry of common causes and effects).

We assume, however, that the following formal requirement can be established on the basis of any tenable solution to the problem of causal priority (for a sequence $e_1, \ldots, e_{n-1}, e_n, e_{n+1}, \ldots, e_r$ of the kind discussed above:

(*17*)   If $i > j$, then event e_i (of kind E_i) cannot causally influence event e_j (of kind E_j).

(For our present purposes we can take the subindices i and j to represent moments of time.)

One problem to be answered in discussing probabilistic causal sequences is what kind of a causal relation should obtain between two singular events e_n and e_{n+1} in the sequence. Obviously we want it to be a probabilistic relation, but perhaps *direct* probabilistic causation would be too strict in general. On the other hand, if we are given e_1 and e_r and asked to explicate how e_1 indirectly caused e_r, we presumably will always be able *in principle* to conceive of and find a sequence $e_1, \ldots, e_{n-1}, e_n, e_{n+1}, \ldots, e_r$, where direct probabilistic causation holds between the elements.

On the basis of our above discussion we now propose the following definition for the indirect probabilistic causation of e_r by e_1 (in symbols, $IPC(e_1, e_r)$), given that a sequence $e_1, \ldots, e_{n-1}, e_n, e_{n+1}, \ldots, e_r$ of singular events occurred:

(18) $IPC(e_1, e_r)$ if and only if there are generic events (represented by the predicates) $E_1, \ldots, E_{n-1}, E_n, E_{n+1}, \ldots, E_r$ such that e_n exemplifies E_n, $n = 1, \ldots, r$, and
   (1) $N(E_{n-1}, E_n, E_{n+1})$ for all n, $1 < n < r$;
   (2) e_1 (*as an* E_1) is causally relevant (in the sense of (4)) to e_r (*as an* E_r);
   (3) for all n, $1 \leq n < r$, $PC(e_n, e_{n+1})$.

In this definition condition (1) guarantees that the sequence $e_1, \ldots, e_{n-1}, e_n, e_{n+1}, \ldots, e_r$ is a potential causal sequence in the sense of being generated by the relation N of nomic betweenness. Condition (2) guarantees that the first event e_1 is causally relevant to e_r in the sense of definition (4), i.e. in Reichenbach's sense. This requirement entails that e_1 is positively relevant to e_r in the sense that $p(E_r/E_1) > p(E_r)$ and that the causal power of e_1, therefore, does not fade away in the process but does have an effect on e_r. Furthermore, condition (2) guarantees that e_1 is not merely spuriously relevant to e_r as no earlier events screen off e_1 from e_r.

Condition (3) is rather strong in that it requires each e_n ($1 \leq n < r$) to be a direct probabilistic cause (in the sense of definition (2)) of e_{n+1}. This obviously introduces the required causal asymmetry into the whole sequence. Condition (3) also guarantees that the sequence $e_1, \ldots, e_{n-1}, e_n, e_{n+1}, \ldots, e_r$ is *linear*. An event e_n cannot have two or more direct causes belonging to the sequence, nor can it be a direct cause of more than one event (viz. e_{n+1}) in the sequence.

For some purposes it is useful to make *IPC* cover chains $e_1, \ldots, e_r$ where some links are deterministic (non-probabilistic). Then we require in clause (3) that *either PC or C* holds between e_n and e_{n+1}, $1 \leq n < r$. In the next section this liberalization will in general be made. We denote the liberalized condition by (3').

Instead of condition (3) one might also consider the following two requirements. First, one might only require that each e_n be causally prior to e_{n+1}. (Perhaps a requirement of spatiotemporal contiguity could still be added.) Another possibility would be to employ either the notion of direct causal relevance of definition (5) or to use Suppes' notion of direct cause to replace $PC(-, -)$ in (3). We shall not, however, here consider the possible merits of these weakenings.

2. ACTIONS, PROPENSITIES, AND INDUCTIVE-PROBABILISTIC EXPLANATION

2.1. We have been emphasizing that human actions are causal products of their agent's activated powers in suitable circumstances. A causal theory of action was constructed earlier in this book as an elaboration of this general idea. Now we are in a position to "probabilify", so to speak, the basic concepts of this theory of action. Instead of dispositions to act, we now come to speak of propensities to act; instead of strict causation or generation of actions we use probabilistic causation and generation, and so forth.

Below, we shall not attempt to solve the metaphysical problem of whether human actions involve some inherently probabilistic elements. It will suffice for our purposes that there be stable probabilistic generalizations related to actions. It should also be emphasized that there is a need for a probabilistic framework of action concepts for the purposes of behavioral scientists. The employment of probabilistic learning theories, for instance, presuppose such a framework. Currently existing probabilistic behavior theories do not, for instance, make even the most basic conceptual distinctions related to behavior (cf. Bush and Mosteller (1956), Atkinson *et al.* (1965)). Thus, actions are not even distinguished from responses and reflexes in general, not to speak of such finer distinctions as basic versus generated actions, simple versus compound actions, and so on. Still, our general approach seems to accord well with most of what is being done in current mathematical behavior theory. It is to be hoped that our framework, to be developed in more detail below, will clarify this agreement better, although no more direct references to, and detailed comparisons with, mathematical behavior theories will be made.

We start by defining the basic probabilistic concepts for action theory;

this will be done directly on the basis of our results in Chapter 11 and in Section 1 of the present chapter.

In Chapter 9 we employed direct deterministic causation in analyzing trying-overt action causation. This notion $C(t, u_o)$ was also given a "purposive" counterpart $C^*(t, u_o)$. The probabilistic counterparts of these notions are obvious. Corresponding to $C(t, u_o)$ we obviously have $PC(t, u_o)$ as characterized by definition (2). On the basis of $PC(t, u_o)$ and definition (9.12) we define $PC^*(t, u_o)$, which reads 't is a direct purposive probabilistic cause of u_o'. (For stylistic convenience we use the semantic phrase "'$PC^*(t, u_o)$' is true" rather than the "material mode".)

(19)   '$PC^*(t, u_o)$' is true (with respect to the behavior situation
            conceptualized by $\mathcal{M}$)
            if and only if
        (a) '$PC(t, u_o)$' is true (with respect to $\mathcal{M}$); and
        (b) there is a complete conduct plan description K satisfied$_2$
            by $\mathcal{M}$ such that in K t is truly described as A's trying (by
            means of his bodily behavior) to exemplify that action
            which $u(= \langle t, u_o \rangle)$ is truly describable as tokening
            according to K, and u_o is represented in K as a maximal
            overt component-event of u; t and u_0 belong to the domain
            of $\mathcal{M}$.

Clause (b) of the definiens is identical with clause (b) of definition (9.12). Therefore the above definition does not need any further clarification. $PC^*(t, u_o)$ simply represents, so to speak, intention-relative singular probabilistic causation.

Analogously with our definition of $\triangleright\!\rightarrow$ in Chapter 9, we may now define a conditional $\triangleright_{\overrightarrow{r}}$, which is relativized to a conduct plan K. We can do it on analogy with definition (9.15) as follows:

(20)   '$D_1 \triangleright_{\overrightarrow{r}} D_2$' is true (with respect to $\mathcal{M}$)
            if and only if
            '$PC^*(t, u_o)$' is true (with respect to $\mathcal{M}$) such that D_1 truly
            describes t and D_2 truly describes u_o as required by clause (b)
            of definition (19).

Before we try to deal with complex actions, we have to define probabilistic action generation. We shall not attempt here to be more precise in characterizing the backing law (here probabilistic) than in the earlier deterministic case. Thus, we arrive by obvious analogy with definitions

(9.17) and (9.19), at the following definitions for $PR_1(e_1, e_2)(= e_1$ prob-abilistically factually generates e_2) and $PR_2(e_1, e_2)(= e_1$ probabilistically conceptually generates e_2):

(21)   '$PR_1(e_1, e_2)$' is true (in a given situation $\mathcal{M}$)
         if and only if
         (with respect to $\mathcal{M}$) there are true descriptions D_1 of e_1 and
         D_2 of e_2 and a suitable true factual probabilistic law T such
         that $\rho(D_2, D_1 \& T)$.

(22)   '$PR_2(u, D_1, D_2)$' is true (in a given situation $\mathcal{M}$)
         if and only if
         (with respect to $\mathcal{M}$) there is a suitable true probabilistic
         meaning postulate T such that $\rho(D_2, D_1 \& T)$, where D_1 and
         D_2 truly describe u.

In definition (21) we understand the inductive explanatory relation ρ in the weak sense in which I (the inducibility relation) equals positive probabilistic relevance. The probabilistic generation of a singular event e_2 by a singular event e_1 then basically amounts to subsuming e_1 and e_2 under a probabilistic law $p(E_2/E_1) = r$ such that $p(E_2) < r$.

Probabilistic conceptual generation may seem to be a peculiar notion. What could probabilistic meaning postulates be? One answer to this might be obtained from taking our ordinary concepts to be fuzzy (cf. 'rich', 'bald') and using probabilistic tools for analyzing this fuzziness. We do not, however, want to defend such an analysis here. $PR_2(u, D_1, D_2)$ does not play any important role in our treatment, and hence the problem of the meaningfulness and the exact content of $PR_2(e_1, e_2)$ can be left open.

We still have to characterize indirect probabilistic causation and generation in action contexts. As to indirect probabilistic causation, we will simply adopt our $IPC(e_1, e_r)$ of definition (18) to characterize it. In the typical case we take $e_1 = t, e_2 = b$, and $e_r = r$ in the sequence of singular events presupposed by definition (18).

Indirect purposive probabilistic causation can be defined on the basis of $IPC(e_1, e_r)$ in complete analogy with definition (9.14) as follows:

(23)   '$IPC^*(t, r)$' is true (with respect to $\mathcal{M}$)
             if and only if
         (a) '$IPC(t, r)$' is true (with respect to $\mathcal{M}$) such that t corre-
             sponds to e_1, b to e_2, and r to e_r;

(b) there is a complete conduct plan description K satisfied$_2$
by $\mathscr{M}$ such that in K t is truly described as A's trying to do
by his bodily behavior what u tokens according to K,
and u_o (composed of b and r) is truly described as a
maximal overt part of an action which is (directly or
indirectly) probabilistically and factually (and thus not
merely conceptually) generated by some of A's actions in
this behavior situation; t and u_o are assumed to belong to
the domain of $\mathscr{M}$.

Here we use the liberalized clause (3′) in defining $IPC(t, r)$, as was said
at the end of Section 1 of this chapter.

Corresponding to our IPC and IPC^*, we can define notions of indirect
probabilistic generation (IPG) and indirect purposive probabilistic
generation (IPG^*) simply by using R_1 instead of PC in their definientia.
This is in complete analogy with the deterministic case (see Chapter 9),
and therefore no further discussion is needed here.

Likewise, corresponding to the deterministic case we may technically
define probabilistic conditionals $\vartriangleright_{\vec{r}}^{\rightarrow}$, $\bigcirc_{\vec{r}}^{\rightarrow}$, $\bigcirc_{\vec{r}}^{\rightarrow}$, $|\bigcirc_{\vec{r}}^{\rightarrow}$ to represent,
respectively, indirect purposive probabilistic causation, direct probabilistic
generation (i.e. by means of R_1 or R_2), direct purposive probabilistic
generation, and indirect purposive probabilistic generation (cf. Chapter 9).
As it is obvious how to do this, we need not write out the definitions
explicitly here.

Given the above concepts of probabilistic causation and generation we
can "probabilify" some of the key concepts of our action theory. This
probabilification technically simply amounts to augmenting the earlier
deterministic concepts by the probabilistic ones. Thus, whenever we earlier
spoke about deterministic causation or generation we now speak about
causation and generation in general, irrespective of its being deterministic
or probabilistic. (Thus, we preserve the old deterministic concepts as
special cases of their generalized counterparts.)

Among the action concepts that can be fruitfully probabilified are the
concepts of *basic action type* (cf. definition (10.1)), *compound basic action
type* (definition (10.4)), *action token₁* (definition (10.5)), *compound action
type* (definition (10.8)), *complex action type* (definition (10.9)), *power*
(cf. definitions (10.3), (10.11), and (10.12)), the concept of an *intentional
action token* (definition (10.14)); and the concepts of *conduct plan*, and
probabilistic action law. We shall not here make a special investigation

of these probabilistic counterpart concepts, but only make a few comments.

Our concept of a basic action type is probabilified simply by using $\rhd_{\vec{r}}$ instead of $\rhd_{\rightarrow}$ (in its clause (1)). Similarly, a probabilistic compound basic action type differs from its deterministic counterpart only in this same respect. If we think of arm raisings or ear wigglings, we might want to require that the probability r be high (i.e. close to 1). In any case, we must now impose the Leibniz-condition on the inductive relation I indicating that $r > \frac{1}{2}$.

In the case of action tokens$_1$, we get the probabilistic counterpart by allowing probabilistic (direct and indirect) causes in clause (1) of definition (10.5). Here I do not see strong reasons for requiring r to be close to 1 nor even for its being greater than $\frac{1}{2}$.

In the case of the rest of the action concepts mentioned above, the probabilification concerns indirect generation. Technically we need no syntactic changes in our definitions of a compound action type, complex action type, or power (definition (10.11)). We simply reinterpret the notion of generation (i.e. the word 'generated' in clause (3) of both definitions (10.8) and (10.11). Thus, it now comes to mean deterministic or probabilistic generation (instead of merely deterministic generation).

What does the probabilification of compound and complex actions then involve? If an agent builds a house, travels around the world or runs 5000 meters in fourteen minutes, chance elements may be involved in the performances of these complex actions. Given suitable qualifications (cf. appendix of Chapter 11), we may perhaps apply a relative frequency interpretation here and say that as a matter of nomic law the agent has an $s\%$ chance of succeeding in bringing about the required end result of such a complex action as, say, travelling around the world in a sail boat, given that he sets himself to do it. If our agent succeeded, his action (i.e. the end result r) was only probabilistically generated (i.e. $IPG^*(t, r)$). To the extent this kind of probabilistic considerations are acceptable, we then need probabilistic action concepts as well. (I only mean to say that *some* of our (complex) doings require probabilistic action concepts for their conceptualization and for speaking about them.)

Our concept of power still needs some extra remarks as it is connected to our notion of propensity as defined in Chapter 11. That is, an agent's probabilistic power to do something should of course fall under our general concept of propensity. Let us see what the connections here exactly are.

Our definition (11.9) analyzes the propensity (of some strength, say s) of a chance set-up x_1 at t_1, to do G, given F. The analysis assumes that x_1 has a basis property B which, together with F, inductively explains x_1's G-behavior. This inductive explanation assumes that there is a probabilistic backing law of the type $p(G/F \ \& \ B) = s$ to serve as a covering law.

Let us assume that G represents a simple or complex action type. We now want to analyze an agent A's active propensity ($=$ probabilistic power) to do G, given some suitable circumstances F. Obviously the agent A here corresponds to the chance set-up x_1 above. While definition (11.9) uses substance-ontology, we here prefer to use event-ontology.

Thus, we simply start with the general idea that A has the (active) propensity (of strength s) to do G in circumstances F just in case if A tried to do G in those circumstances he would succeed with probability s. This is taken to involve (within our event-framework) that A's tryings t generate the end results r to come about in circumstances of the type F.

To be more explicit, any action token (of A) *manifesting* A's propensity to do G, given F, will be of the general form $u = \langle t, \ldots, b, \ldots, r \rangle = \langle t, \ldots, b_1, \ldots, b_m, \ldots, r_1, \ldots, r_k, r \rangle$, as discussed in Chapter 10. Here t will instantiate or "partially" instantiate a propensity basis B. (In fact, as we saw in Chapter 11, we may even think that t cum b, or $t + b$, has to satisfy a propensity-basis predicate understood in a broader sense.) What this propensity basis B is we do not necessarily have to know. However, we do know, for instance, that t is describable (at least in part) as A's trying (by his bodily behavior) to bring about whatever is required for G to be exemplified. Still, it is not necessary that B be a psychological predicate; it might be, for instance, a neurophysiological one as well.

F will, in the present context, just represent various "normal conditions" that t and b ($= b_1 + \ldots + b_m$) have to satisfy. We do not have to add anything here to what was said about such conditions in Chapters 7 and 10.

Our definition (11.9) requiring that the backing law $p(G/F \ \& \ B) = s$ inductively explain its instances must, in the present situation, be explicated in a strong sense. For we must require that $IPG^*(t, r)$ be the case. We recall from our definition (23) defining IPC^* and from our later comments related specifically to IPG^* that PG will in the general case have to obtain between single elements of the generational chain. Hence t (as a B or a part of B) will only indirectly inductively explain b cum r as a G. In other words, instances of B and F here only indirectly explain instances of G (but still so that B & F is positively relevant to G). In any case, our definition of A's probabilistic power to do G, given F, has been shown to fit our earlier more general characterization of a chance set-up's propensity to do G, given F.

Above, we silently assumed that the generation of action tokens of complex actions takes place according to the generational principle (P**) of Chapter 10 with IPG replacing IG. Thus, we have also assumed that generation may take place between composed events. It may, however, seem problematic whether composite events (subsumed under appropriate universals) can and do instantiate nomic probabilistic laws as required by our characterization of IPG (cf. my building a house). Perhaps, it would be better to think of generation between composite events as somehow summed up from the *simple event generations*, we may think. However, until definitive counterexamples of this kind against our new (P) and (P**) as well as IPG are produced, I do not see any strong reasons to think that our new probabilistic machinery will not work.

How about our characterization of intentionality? Do our present probabilistic considerations apply to this problem? They do. Let us consider an example. Suppose Smith kills Jones "in a normal way" by shooting him. Suppose for the sake of our argument that it is a nomic fact that the probability of Smith's lethally wounding Jones by shooting at him in such and such circumstances is r. Let us assume that r is non-negligible, e.g. $r > \frac{1}{2}$. Now, in a certain sense it was possible that Smith intended to kill Jones, even if r is clearly less than 1 (cf. our earlier remarks in Chapter 6). Given this, we can say that Smith in fact killed Jones (as everything went according to Smith's plan). Still the firing of the gun only probabilistically generated the death of Jones.

What does this entail concerning our earlier characterization of intentionality? We can put it as follows. Our earlier discussion implicitly involved the assumption that determinism is a true doctrine. If that were the case there would in principle be a way of analyzing Smith's killing Jones in a fully deterministic (non-probabilistic) fashion. Thus, our earlier technical criteria of intentionality in Chapter 10 could stay intact.

If, however, we are willing to accept the possibility that indeterminism is true in the realm of human affairs, we can make the following obvious changes in our earlier characterization. The basic definition of intentionality, i.e. definition (*10.14*), stays syntactically the same, and so does its main explicate (I) in Chapter 10. However, we now allow the because of-relation and the notion of generation in them to contain probabilistic generation exactly parallel with out earlier "probabilifications". Concerning the specific explicates (a), (b), (c), and (d) of (I) we now get their more general probabilified counterparts by using PC^*, IPG^*, IPC^*, and IPG^* to replace C^*, IG^*, IC^*, and IG^*, respectively. Recall that, because of the liberalized clause (3′) in our definition of IPC and IPG, the deterministic case is included as a special case here. Practically everything else we have said about intentionality remains intact.

We must again emphasize here that our characterization of intentionality concerns the *causal source* of intentionality, so to speak. Effective intendings (= tryings) thus are the causal source of intentional actions in that the *truth conditions* for action tokens being intentional involve reference to them. We recall from Chapters 3, 7 and 10 that the *semantical* or *conceptual source* of intentionality in our analysis is, in a sense, meta-linguistic (semantical) discourse. In this conceptual sense we say that an item (e.g., an action description) is intentional in case this item is the kind of thing which makes *reference* to something (or is *about* something). Given our analogy theory of thought-episodes (such as tryings), these two senses of intentionality become as intimately connected as one can possibly demand (recall our discussion in Chapter 10).

In Chapter 9 we briefly discussed action-laws, viz. potential laws specifying how agents' wants and beliefs determine their actions. It should be emphasized now that at least in actual scientific practice such laws will be probabilistic rather than strict (cf. various probabilistic learning theories). A very simple type of such a probabilistic action law would be:

$$(24) \qquad p(\text{Verbs}(A, o, z)/W(A, p_i, x) \ \& \ B(A, q_j, y) \ \& \ F(A, x, y)) = r.$$

The generalization (or possible law) (24) employs the symbolism of Chapter 7 and specifies that the agent's want-state W and belief-state B, together with some suitable additional factors F, determine his "Verbing" with a propensity probability of strength r. The propensity in question may be either "active" (when the agent intentionally acts "on" his want) or "passive" (when the agent acts nonintentionally "because of" his want).

More complex examples of potential probabilistic action laws are provided by probabilistic learning theories. They usually satisfy an independence of path property such as the Markov property discussed in the previous section. Our present conceptual and technical tools could easily be extended to handle them. We shall not here do it, however. Let us only mention, as an example, one such probabilistic law that is a theorem of both Bush's and Mosteller's linear theory and Estes' and Suppes' stimulus sampling theory. It is the so-called *probability matching law*, which is found to be true for a variety of circumstances. This law says that in a choice experiment with a certain type of probabilistic reinforcement schedule subjects will asymptotically start choosing the choice alternatives in direct proportion to their reinforcement probability. This asymptotic action law is formally a special case of the simple law (*24*), although it usually does not explicitly mention the agent's wants and beliefs at all. (I have elsewhere, in Tuomela (1968), discussed the conditions of validity of the probability matching law.)

2.2. Our probabilistic notions developed above can be built into a number of other methodologically central notions in our system, although we shall not try to do it here in detail. Let us however emphasize some such cases.

Our *conduct plans* can be "probabilified" in an obvious way by allowing our various notions of probabilistic causation and generation to be used in them. In fact, we are going to need such probabilistic conduct plans as soon as we use probabilistic concepts in speaking about actions. This is because of the symmetry between conduct plans and actions, with the notion of trying (willing) as the "reflection point" of this symmetry (cf. our discussion in Chapter 7 and below).

Notice here that as practical syllogisms, in fact, are conduct plans we also get *probabilistic practical syllogisms*. In other words, we may, for instance, have practical syllogisms in whose second premises the concept of probabilistic generation is employed. (Such practical syllogisms are what we have called liberal practical syllogisms, and they, in general, are not "logically conclusive" in the sense discussed earlier.)

To illustrate what we have said about probabilistic generation and probabilistic conduct plans, we may reconsider our example discussed in Section 1 of Chapter 7 and earlier in Chapter 2. In fact, our example brings together all of our notions of generation, illustrating them, summarily, at the same time. With some new details added to the earlier example and viewing the agent's behavior *ex post facto*, we get Figure 12.1.

Our flip switching example here has some new nodes (the singular events a_1, a_5, a_6, and b_1^*) which have to be explained. Let us assume that our agent A has an evening free with nothing particular to do. He then forms an intention to read the newspaper (a_1). Right after forming this intention he realizes that he must also find something else to do in the

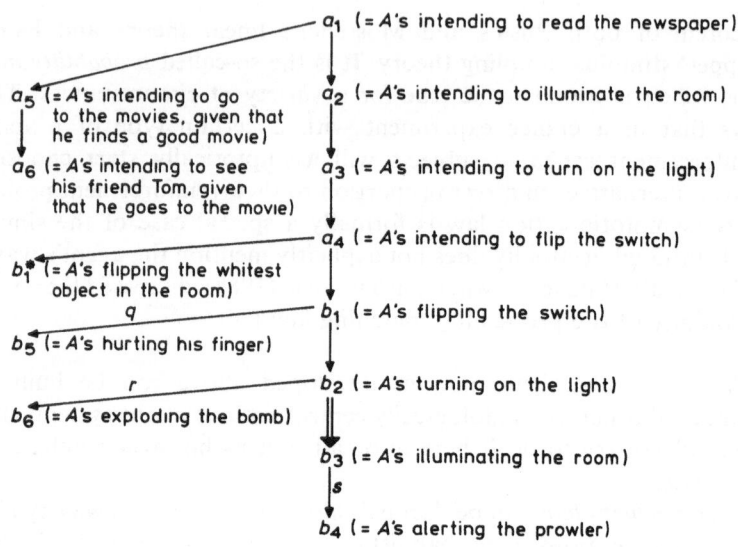

Figure 12.1

evening, and he forms the intention to go to the movies that night, provided he finds something worth seeing (a_5). This intention again leads him to form the conditional intention to go and see his old friend Tom after the movie (a_6).

In our Figure 12.1 the a_i-events, $i = 1, \ldots, 6$, are mental events or episodes. They are the singular episodes that took place before A acted and which activated A's conduct plan (cf. our description of a part of it in Chapter 7). The b_j-events, $j = 1, \ldots, 6$, are overt behavioral events which are obviously related to A's carrying out his conduct plan. At least b_1, b_2, b_3, and b_4 are assumed to satisfy$_2$ A's conduct plan. (Note the symmetry between these events and their mental counterparts). The event b_1^* is meant to be iden ical with b_1. Now we recall from Chapter 9 that if A actively and correctly believed that the switch is identical with the whitest object in the room, then also b_1^* satisfies$_2$ A's conduct plan.

Let us now discuss the generational relationships, starting from intention-generation. A's intending to read the newspaper (a_1) is assumed to lead to his intending to illuminate the room via his belief that illuminating the room is necessary for his being able to read the newspaper. Thus generation can be formulated as an ordinary kind of practical syllogism with the exception that the conclusion is an intention rather than an action or an initiation of an action (cf. Chapter 7). Similarly, we have intention transferral via the practical syllogism from a_2 to a_3 and from a_3 to a_4.

Our diagram uses single arrows for factual generation and the double arrow for conceptual generation, as before. The letters q, r, and s attached to the arrows mean generation probabilities (see Chapter 7). When no such letter is attached, deterministic generation is in question.

Earlier, we have tentatively assumed that the factual generations involved between the b_j-events are causal. We also assumed, still more tentatively, that the process leading from a_1 to a_4 is causal (cf. Chapters 6 and 8). Presently, we do not have the psychological theories and the psychological evidence needed for making a "conclusive" rational evaluation of the situation.

What can then be said about the claims that a_1 generates a_5 and that a_5 generates a_6? Concerning the first, we may consider the following principle:

(25)   If A intends that q, if p, and if A believes that p is true, then A intends that q.

This principle seems to be needed to back the generation of a_5 by a_1. Is (25) a conceptual truth? If we assume that A is a rational agent, then (25) cannot fail. But if A is, say, a moron who fails to connect the two antecedents of (25), both of which he nevertheless separately satisfies, he might not intend that q. It is thus not necessary to think that (25), without qualifications, is a conceptually true principle.

In fact, it is not at all important for our purposes whether (25) is a conceptual truth or not. If a_1 causally generated a_5 and if, furthermore, this generation was purposive (and thus description-relative) vis-à-vis (25), that suffices in this context. Meaning analysis is not very helpful here, we feel. The "interesting" transformation principles for intendings, etc. are likely to come from psychology. Thus, for instance, the "free-association-like" generation of a_6 by a_5 is something for a psychologist rather than for a philosopher to give an account of.

Let us summarize what now can be said about the generational relationships in Figure 12.1 (we call generation by means of a practical syllogism PS-generation):

(26) a_i PS-generated a_{i+1}, $i = 1, 2, 3$;

(27) a_1 conditionally generated a_5;

(28) a_5 conditionally generated a_6;

(29) $C^*(a_4, b_1)$;

(30) $C^*(a_4, b_1^*)$, provided A actively and correctly believed that the whitest object in the room was identical with the switch;

(31) $IG^*(b_1, b_2)$;

(32) $PC(b_1, b_5)$ (but not $PC^*(b_1, b_5)$);

(33) $R_2^*(b_2, b_3)$;

(34) $IPG(b_2, b_6)$ (but not $IPG^*(b_2, b_6)$);

(35) $IPG(b_3, b_4)$ (but not $IPG^*(b_3, b_4)$).

On the basis of this list, couched in our familiar symbolism, all the other generational relationships, holding between nonadjacent nodes in our Figure 12.1, can be seen.

Our example does not mention whether A realized his intentions expressed in a_5 and a_6. In any case, his plan involved a gap in that he had to find out whether there are good movies playing. When this gap is filled (due to A's feedback from his searching action), the carrying out of that part of A's plan could go on. Perhaps A did not find a good movie. This would explain the fact that the corresponding behavior-events did not occur.

We have been claiming that causal processing takes place when an agent forms and carries out his conduct plan. Some of the processes in question are mental. Thus, for instance, intention–intention generation and intention–belief generation, as we have earlier discussed them, seem to be causal mental processes. So far we have said next to nothing about how an agent's conduct plans come about. Presumably, the agent's stabler conative and cognitive structures are centrally involved here (cf. our remarks in Chapter 6).

How all this takes place is up to psychologists to clarify. Indeed, there are many psychological motivation theories concerned with these matters. Thus, Freudians may refer to intention generation due to repressed desires and intentions. Cognitive balance theorists think that agents tend to form their intentions so as to make their conative and cognitive attitudes form balanced wholes (cf. Heider (1958)). Lewinians think in terms of resultants of suitable instigating and consummative forces (cf. our remarks in Chapter 6).

We will next briefly discuss an example related to intention generation and to our discussion of probabilistic causal powers — Atkinson's account of motivation (cf. Atkinson (1964)). It may be called an expected-value type of motivation theory, although it, on the other hand, is clearly Lewinian.

Suppose that in the present situation there are two actions that an agent A can do. A can either continue writing a scientific paper (X_1) or A can go to the cafeteria and have a cup of coffee and socialize with his collegues (call this X_2). According to Atkinson's theory, A will do X_1 rather than X_2 if and only if A's tendency to do X_1 is greater than A's tendency to do X_2. Let us denote the first tendency by $T_{x_1,s}$ and the second by $T_{x_2,aff}$. Here 's' stands for success and 'aff' for affiliation. In other words, X_1 is here associated with the motive for achievement or success and X_2 with the motive for affiliation. The theory now gives us this formula to compare the tendencies:

(36) $T_{x_1,s}$ is stronger than $T_{x_2,aff}$ if and only if
$$(M_S \times E_{x_1,s} \times I_s) + T_{S,i} > (M_{AFF} \times E_{x_2,aff} \times I_{aff}) + T_{AFF,i}.$$

Here M_S represents the strength of the agent's general motive for success or achievement and M_{AFF} that of his general motive for affiliation. $E_{x_1,s}$ represents the degree of expectancy that action X_1 has for success (e.g., A's finishing his paper), and $E_{x_2,aff}$ in our example represents the degree of expectancy that A's going to the cafeteria has for satisfying his motive of affiliation. I_s and I_{aff} stand for the incentive value of success and affiliation, respectively. Finally $T_{S,i}$ and $T_{AFF,i}$ stand for the general initial tendency for success and affiliation, respectively. Here 'general' means, roughly, independent of a particular action type and situation.

Now, we proceed with the following technical reinterpretation of the situation. We take the tendencies $T_{x_1,s}$ and $T_{x_2,aff}$ to represent A's propensities to act (to do X_1 and X_2, respectively, in the given situation). Thus, formally, $T_{x_1,s}$ and $T_{x_2,aff}$ are taken to be probabilities of the sort our definition (11.9) specifies. They are conditional probabilities to do X_i, $i = 1, 2$, viz. conditional on the situation, say C, and on some suitable basis property, say B, of the agent. The expectancies $E_{x_1,s}$ and $E_{x_2,aff}$ are taken to be conditional probabilities of getting to the goals s and aff by doing X_1 and X_2 in C, respectively. M_S and I_s are taken to be utilities for A, M_S being the general utility of success and I_s the specific utility obtained from a consequence of an action, of A's finishing his paper. Analogous remarks go for M_{AFF} and I_{aff}. $T_{S,i}$ and $T_{AFF,i}$ simply stand for initial propensities to do actions leading to achievement and affiliation, respectively.

With this new interpretation (36) can be regarded as a kind of principle of "propensity kinematics". Incidentally, as Atkinson's theory has been rather well supported by empirical evidence, it is likely that it would continue to be supported under our slightly different reinterpretation.

We shall not try to draw any further philosophical conclusions from (36) and our reinterpretation of it. What is central is that we have a clearly contingent psychological principle here which is supposed to determine *intention formation*. For the strongest propensity to act in the case of intentional action just develops into the agent's intending, given our analysis in Chapter 6. (If it would not, then we would not in general speak of intentional action.)

2.3. Our final topic in this chapter is the probabilistic explanation of actions, where the explanandum, in particular, is an intentional action (as in Chapters 8 and 9). As we have already discussed action-explanations at length in this book, we can below rely on our previous discussion and be rather brief and sketchy. This is especially so since our main interest here is (practically) to "probabilify" the main points of our previous discussion. Almost everything except this probabilification will remain as in Chapters 8 and 9. Another way of putting the matter is to say that whereas action-explanations were discussed in Chapters 8 and 9 as if determinism were true, now we will relax this assumption and consider indeterminism to be a viable factual possibility.

Explanations under determinism may in principle give us deductively *conclusive* arguments, whereas explanations under indeterminism only may give us, in that sense, formally *inconclusive* arguments (cf. Chapter 11). Probabilistic action-explanations will thus be (parts of) inconclusive arguments, viz. p-arguments (cf. (11.49)). This is the formal side of the matter. As for the contents of proper probabilistic action-explanations we again claim that they, as well as their deterministic counterparts, will be IT-explanations (intentional-teleological explanations). Accordingly, our theses (T1)–(T4) of Chapter 8 will remain as they are except that probabilistic notions (such as purposive probabilistic causation and generation) will now be allowed and used in their explication in addition to deterministic ones.

We have taken action-explanations to typically be (complete or incomplete) answers to why-questions. We did, however, mention some types of explanation which are not answers to why necessary-questions (cf. Chapter 8). For instance, some explanations answer a how possible-question. Thus, explanations in terms of necessary conditions as well as certain probabilistic explanations fall into this category. Still, it seems that all nomological probabilistic-inductive explanations, as we have construed them, can also be regarded as answers to (widely understood) reason-asking why-questions (cf. Chapter 11).

We shall below concentrate on *causal* probabilistic explanations. They, in any case, may be regarded as answers to widely understood why-questions. For instance, in the example with which this chapter started we may explain the chocolate bar's coming out of the machine (a low-probability event) by saying that it was causally produced by the inserting of the coin. A more dramatic example is obtained if the output is taken to be the firing of a revolver which killed Smith. Now Jones' inserting the coin probabilistically caused Smith's death. We ask: Why did Smith get shot? The answer: He got shot because Jones inserted the coin in the machine. (Notice that we of course also have a how possible-explanation here. Smith's death was made possible by Jones inserting the coin.)

Accordingly, concentrating on explanations in terms of probabilistic causes and thus in terms of answers to why-questions, may now characterize them as inconclusive explanatory answers (cf. (*11.49*)). Recall that inconclusive explanatory answers only require making the explanandum nomically expectable to some degree rather than guaranteeing the occurrence of the explanandum-episode. In examples like the above, we do not require that the inductive relation I satisfies more than the positive relevance condition.

As we saw in Chapter 11, not all inconclusive answers (as defined by (*11.49*)) are inductive-probabilistic explanations. We may note, furthermore, that not all inductive-probabilistic explanations are causal, for we may answer why-questions in terms of non-causal laws. Consider the following example. Why is this bird white? It is a swan and all swans are very likely to be white. Here the explanatory law is not a causal law.

All IT-explanations (at least indirectly) involve purposive causation (though, perhaps, imbedded in purposive *generation*). (Cf. Chapters 8, 9, and 10.) Let us accept this claim here. Now, if indeterminism happens to be true (in some strong sense), we must accept *probabilistic* purposive causation to be our basic pattern of explanation. Thus, instead of C^*,

IC^*, and IG^* we typically have to make use (implicitly or explicitly) of PC^*, IPC^* and IPG^* in our action-explanations (cf. Chapter 9). Notice, too, that in explaining actions by citing their reasons we accordingly use IPG^* instead of IG^* (cf. Chapter 8). As an example of a probabilistic action-explanation we may recall the golfer-example from Section 1. There, the golfer's intentionally swinging his club indirectly, and perhaps even purposively, caused his making a birdie. We might add to the example that the golfer's effective *intending* to get the ball into the hole caused his swinging the club, and thus it indirectly explains the birdie.

When we use singular explanatory conditionals, we now typically employ $\triangleright_{\vec{r}}$, $\triangleright_{\vec{r}}$, and $\mid\odot_{\vec{r}}$ instead of their deterministic counterparts. Thus, we might conceivably explain A's going to Stockholm over the weekend by citing his wanting to see a friend there. A's wanting might only probabilistically entail his going to Stockholm, and technically we might employ $\triangleright_{\vec{r}}$ in our explanation. (Presumably we would now require I to satisfy the Leibniz-condition.) There is no need to go into further details as we would mainly be repeating, *mutatis mutandis*, what has been said earlier about action-explanations and probabilistic explanations.

There is one feature of probabilistic action-explanations which is worth emphasizing here, however. It is that stable wants and other stable conative and doxastic attitudes or "cognitive structures", as well as perhaps stable emotions and traits, may play the role of "common causes" in action-explanations (cf. Section 1). Wants and analogous attitudes are complex dispositions. Here we think of them in probabilistic terms and regard them as *propensities*. Any such stable propensity will become activated from time to time and generate intentions to act (cf. our earlier discussion of Atkinson's motivation theory). The intentions produce overt actions, which may appear very different from each other. However, the general want (or some other propensity) which generated the overt actions is a feature which connects these actions (by being their common cause) and thus also serves to explain them.

Let us consider a simple example to illustrate this. Suppose that an agent, who is known to often lie or "verbally hide" his real attitudes, speaks in an appraising way about orthodox communists. Let us call this verbal action X_1. Suppose, furthermore, that the agent gives his vote to the orthodox communists in an election (X_2). Now we may be able to find a common probabilistic cause for X_1 and X_2 from the fact that the agent is almost fanatically conservative (C). We may assume that such conservatives normally vote in an extreme way, either for the extreme right or extreme left. Now, it is not implausible to think that X_1 and X_2 are connected (by being actions of the same consistent agent) and explainable by C. Thus, we have (cf. Section 1):

(37) $p(X_1 \& X_2) > p(X_1) \cdot p(X_2)$

(38) $p(X_1 \& X_2/C) = p(X_1/C) \cdot p(X_2/C)$.

Formula (38) in effect says that C is a probabilistic common cause of X_1 and X_2, and it serves to explain both X_1 and X_2. C then screens$_1$ off X_1 from X_2. If, furthermore $p(X_2/C) > p(X_2/X_1)$ (our agent often lies!) then X_1 is *not* even causally relevant to X_2 (cf. Section 1).

Our hypothetical example shows that underlying causal factors may profitably be searched for, since such factors may serve to unify and connect several distinct explanations given in terms of intentions (and thus in terms of probabilistic practical syllogisms). It is partly in this sense that reference to underlying stable cognitive structures can be said to have more explanatory power than explanations merely in terms of particular intentions.

NOTES

[1]   We are now in the position to formally define a notion of probabilistic implication for generic events (and other universals). Let F and G be monadic predicates, for simplicity. We define a probabilistic nomological of strength r, say $\exists_{\overrightarrow{r}}$, as follows:

$F \exists_{\overrightarrow{r}} G$ if and only if for every instantiation i,
$F(i) \vartriangleright_{\overrightarrow{r}} G(i)$.

However, as a philosophical characterization of probabilistic lawlikeness this definition is clearly circular, for $\vartriangleright_{\overrightarrow{r}}$ presupposes that notion.

[2]   Rosen (1975) presents a slight modification of Suppes' (1970) definition of a direct cause. The essential difference concerns clause (iv) (of definition (5)). Rosen requires that for all times t and partitions $\pi_{t''}$ we must have, for all $D_{t''}$ in $\pi_{t''}$:

$p(G_t/D_{t''} \cap F_{t'}) > p(G_t/D_{t''})$.

Clearly, Rosen's notion of a direct (or genuine) cause then is included in Suppes notion above. However, the converse is not true. A direct cause in Suppes' sense need not be a genuine cause in Rosen's above sense.

Our notion $PC(-, -)$ neither entails nor is entailed by Rosen's interesting concept (not even if λ were infinite).

REFERENCES

Ackermann, R.: 1965, 'Deductive Scientific Explanation', *Philosophy of Science* **32**, 155–167.

Ackermann, R. and A. Stenner: 1966, 'A Corrected Model of Explanation', *Philosophy of Science* **33**, 168–171.

Anscombe, G.: 1957, *Intention*, Blackwell, Oxford.

Armstrong, D.: 1969, 'Dispositions Are Causes', *Analysis* **30**, 23–26.

Armstrong, D.: 1973, *Belief, Truth and Knowledge*, Cambridge University Press, Cambridge.

Atkinson, J.: 1964, *An Introduction to Motivation*, Van Nostrand, Princeton.

Atkinson, J. and D. Birch: 1970, *The Dynamics of Action*, Wiley, New York.

Atkinson, R., Bower, G. and E. Crothers: 1965, *An Introduction to Mathematical Learning Theory*, Wiley, New York.

Audi, R.: 1972, 'The Concept of Believing', *The Personalist* **53**, 43–62.

Audi, R.: 1973a, 'The Concept of Wanting', *Philosophical Studies* **24**, 1–21.

Audi, R.: 1973b, 'Intending', *The Journal of Philosophy* **70**, 387–403.

Baier, A.: 1971, 'The Search for Basic Actions', *American Philosophical Quarterly* **8**, 161–170.

Bandura, A.: 1969, *Principles of Behavior Modification*, Holt, Rinehart and Winston, New York.

Becker, W., Iwase, K., Jürgens, R. and H. Kornhuber: 1973, 'Brain Potentials Preceding Slow and Rapid Hand Movements', paper presented at the 3rd International Congress on Event Related Slow Potentials of the Brain, held in Bristol, August 13–18, 1973. Printed in McCallum, C. and J. Knott (eds.): 1976, *The Responsive Brain*, Wright, Bristol.

Berofsky, B.: 1971, *Determinism*, Princeton University Press, Princeton.

Block, N. and J. Fodor: 1972, 'What Psychological States Are Not', *Philosophical Review* **31**, 159–181.

Brand, M.: 1970, 'Causes of Actions', *The Journal of Philosophy* **67**, 932–947.

Bruner, J., Goodnow, J. and G. Austin: 1956, *A Study of Thinking*, Wiley, New York.

Bush, R. and F. Mosteller: 1956, *Stochastic Models for Learning*, Wiley, New York.

Carnap, R.: 1950, *The Logical Foundations of Probability*, University of Chicago Press, Chicago.

Chisholm, R. D.: 1966, 'Freedom and Action', in K. Lehrer (ed.), *Freedom and Determinism*, Random House, New York, pp. 11–44.

Chomsky, N.: 1965, *Aspects of the Theory of Syntax*, M.I.T. Press, Cambridge.

Chomsky, N.: 1970, 'Remarks on Nominalization', in Jacobs, R. and P. Rosenbaum (eds.), *Readings in English Transformational Grammar*, Ginn and Co., Waltham. Reprinted in Davidson, D. and G. Harman (eds.), 1972: *The Logic of Grammar*, Dickenson, Encino, pp. 262–289.

Chomsky, N.: 1971, 'Deep Structure, Surface Structure, and Semantic Interpretation', in Steinberg, D. and L. Jakobovits (eds.), *Semantics*, Cambridge University Press, Cambridge, pp. 183–216.

Churchland, P. M.: 1970, 'The Logical Character of Action-Explanations', *The Philosophical Review* **79**, 214–236.

Clark, R.: 1970, 'Concerning the Logic of Predicate Modifiers', *Noûs* **4**, 411–335.

Coffa, A.: 1974, 'Hempel's Ambiguity', *Synthese* **28**, 141–163.

Collingwood, R.: 1940, *An Essay on Metaphysics*, Clarendon Press, Oxford.

Cornman, J.: 1971, *Materialism and Sensations*, Yale University Press, New Haven.

Cornman, J.: 1972, 'Materialism and Some Myths about Some Givens', *The Monist* **56**, 215–233.

Cresswell, M.: 1974, 'Adverbs and Events', *Synthese* **28**, 455–481.

Davidson, D.: 1963, 'Actions, Reasons, and Causes', *The Journal of Philosophy* **60**, 685–700.

Davidson, D.: 1965, 'Theories of Meaning and Learnable Languages', in Bar-Hillel, 4. (ed.), *Proceedings of the 1964 International Congress for Logic, Methodology, and Philosophy of Science*, North-Holland, Amsterdam, pp. 390–396.

Davidson, D.: 1966, 'The Logical Form of Action Sentences', in Rescher, N. (ed.), *The Logic of Decision and Action*, Univ. of Pittsburgh Press, Pittsburgh, pp. 81–95.

Davidson, D.: 1967, 'Causal Relations', *The Journal of Philosophy* **64**, 691–703.

Davidson, D.: 1969, 'The Individuation of Events', in Rescher, N. *et al.* (eds.), *Essays in Honor of Carl G. Hempel*, Reidel, Dordrecht, pp. 216–234.

Davidson, D.: 1970, 'Mental Events', in Foster, L. and J. Swanson (eds.), *Experience and Theory*, Univ. of Massachusetts Press, Amherst, pp. 79–101.

Davidson, D.: 1971, 'Agency', in Binkley, *et al.* (eds.), *Agent, Action and Reason*, Toronto Univ. Press, Toronto, pp. 1–25.

Davidson, D.: 1973, 'Freedom to Act', in Honderich, T. (ed.), *Essays on Freedom of Action*, Routledge and Kegan Paul, London, pp. 139–156.

Davidson, D.: 1975, 'Semantics for Natural Languages', in Davidson, D. and G. Harman (eds.), *The Logic of Grammar*, Dickenson, Encino, pp. 18–24.

Dulany, D.: 1968, 'Awareness, Rules, and Propositional Control: A Confrontation with S-R Behavior Theory', in Dixon, T. and D. Horton (eds.), *Verbal Behavior and General Behavior Theory*, Prentice Hall, Englewood Cliffs, pp. 340–387.

Fenstad, J.: 1968, 'The Structure of Logical Probabilities', *Synthese* **18**, 1–27.

Feyerabend, P.: 1963, 'How to be a Good Empiricist', in Baumrin, B. (ed.), *Philosophy of Science, The Delaware Seminar*, Vol. 2, Wiley, New York, pp. 3–39.

Fishbein, M. and I. Ajzen: 1975, *Belief, Attitude, Intention, and Behavior*, Addison-Wesley, Reading.

Fodor, J.: 1975, *The Language of Thought*, Crowell, New York.

Fodor, J., Bever, T. and M. Garrett: 1974, *The Psychology of Language*, McGraw-Hill, New York.

Frankfurt, H.: 1971, 'Freedom of the Will and the Concept of a Person', *The Journal of Philosophy* **68**, 5–20.

Giere, R.: 1973, 'Objective Single-Case Probabilities and the Foundations of Statistics', in Suppes, P. *et al.* (eds.), *Logic, Methodology and Philosophy of Science IV*, North-Holland, Amsterdam, pp. 467–483.

Goldman, A.: 1970, *A Theory of Human Action*, Prentice-Hall, Englewood Cliffs.

Good, I.: 1960, 'A Causal Calculus I' *British Journal for the Philosophy of Science* **11**, 305–318.

Goodman, N.: 1955, *Fact, Fiction, and Forecast*, Harvard University Press, Cambridge, Mass.

Green, T.: 1885, 'Introduction to Hume's Treatise', reprinted in Lemos, R. (ed.), *Thomas Hill Green's Hume and Locke*, New York, sec. 24.

Greeno, J.: 1970, 'Evaluation of Statistical Hypotheses Using Information Transmitted' *Philosophy of Science* **37**, 279–293.

Hacking, I.: 1965, *The Logic of Statistical Inference*, Cambridge Univ. Press, Cambridge.

Harman, G.: 1972, 'Logical Form', *Foundations of Language* **9**, 38–65. Reprinted in Davidson, D. and G. Harman (eds.), *The Logic of Grammar*, Dickenson, Encino, pp. 289–307.

Harman, G.: 1973, *Thought*, Princeton University Press, Princeton.

Heider, F.: 1958, *The Psychology of Interpersonal Relations*, Wiley, New York.

Heidrich, C.: 1975, 'Should Generative Semantics Be Related to Intensional Logic?', in Keenan, E. (ed.), *Formal Semantics of Natural Language*, Cambridge University Press, Cambridge, pp. 188–204.

Hempel, C.: 1965, *Aspects of Scientific Explanation*, Free Press, New York.

Hempel, C.: 1968, 'Maximal Specificity and Lawlikeness in Probabilistic Explanation', *Philosophy of Science* **35**, 116–134.

Hintikka, J.: 1965, 'Distributive Normal Forms in First-Order Logic', in Crossley, J. and M. Dummett (eds.), *Formal Systems and Recursive Functions*, North-Holland, Amsterdam, pp. 47–90.

Hintikka, J.: 1966, 'A Two-Dimensional Continuum of Inductive Methods', in Hintikka, J. and P. Suppes (eds.), *Aspects of Inductive Logic*, North-Holland, Amsterdam, pp. 113–132.

Hintikka, J.: 1973, *Logic, Language-Games, and Information*, Clarendon Press, Oxford.

Hintikka, J.: 1974, 'Questions about Questions', in M. Munitz and P. Unger (eds.), *Semantics and Philosophy*, New York University Press, New York, pp. 103–158.

Hintikka, J.: 1976, 'A Counterexample to Tarski-Type Truth-Definitions as Applied to Natural Languages', in Kasher, A. (ed.), *Language in Focus*, Reidel, Dordrecht and Boston, pp. 107–112.

Hintikka, J. and J. Pietarinen: 1966, 'Semantic Information and Inductive Logic', in Hintikka, J. and P. Suppes (eds.), *Aspects of Inductive Logic*, North-Holland, Amsterdam, pp. 96–112.

Hintikka, J. and R. Tuomela: 1970, 'Towards a General Theory of Auxiliary Concepts and Definability in First-Order Theories', in Hintikka, J. and P. Suppes (eds.), *Information and Inference*, Reidel, Dordrecht, pp. 298–330.

Jenkins, J. and D. Palermo: 1964, 'Mediation Processes and the Acquisition of Linguistic Structure', in Bellugi, U. and R. Brown (eds.), *The Acquisition of Language, Monographs of the Society for Research on Child Development*, **29**, pp. 79–92.

Kenny, A.: 1965, *Action, Emotion and Will*, Routledge and Kegan Paul, London.

Kim, J.: 1969, 'Events and Their Descriptions: Some Considerations', in Rescher, N. et al., *Essays in Honor of Carl G. Hempel*, Reidel, Dordrecht, pp. 198–215.

Kim, J.: 1973, 'Causation, Nomic Subsumption, and the Concept of Event', *The Journal of Philosophy* **70**, 217–236.

Kim, J.: 1976, 'Intention and Practical Inference', in Manninen, J. and R. Tuomela (eds.), *Essays on Explanation and Understanding*, Synthese Library, Reidel, Dordrecht, pp. 249–269.

Kuhn, T.: 1969, *The Structure of Scientific Revolutions*, University of Chicago Press, 2nd Ed., Chicago.

Lakoff, G.: 1971, 'On Generative Semantics' in Steinberg, D. and L. Jakobovits (eds.), *Semantics*, Cambridge University Press, Cambridge, pp. 232–296.

Lakoff, G.: 1972, 'Linguistics and Natural Logic', in Davidson, D. and G. Harman (eds.), *Semantics of Natural Language*, Reidel, Dordrecht and Boston, 2nd Edition, pp. 545–665.

Lakoff, G.: 1975, 'Pragmatics in Natural Logic', in Keenan, E. (ed.), *Formal Semantics of Natural Language*, Cambridge University Press, Cambridge, pp. 253–286.

Lewis, D.: 1970, 'How to Define Theoretical Terms', *The Journal of Philosophy* **67**, 427–446.

Lewis, D.: 1972, 'Psychophysical and Theoretical Identifications', *The Australasian Journal of Philosophy* **50**, 249–258.

Lewis, D.: 1973a, *Counterfactuals*, Blackwell, Oxford.

Lewis, D.: 1973b, 'Causation', *Journal of Philosophy* **70**, 556–567.

Lewis, D., 1974, ' 'Tensions', in M. Munitz and P. Unger (eds.), *Semantics and Philosophy*, New York University Press, New York, pp. 49–61.

Lindsay, P. and D. Norman: 1972, *Human Information Processing*, Academic Press, New York and London.

Łoś, J.: 1963a, 'Remarks on Foundations of Probability', *Proceedings of the International Congress of Mathematicians 1962*, Stockholm, pp. 225–229.

Łoś, J.: 1963b, 'Semantic Representation of the Probability of Formulas in Formalized Theories'. Reprinted in Hooker, C. A. (ed.): 1975, *The Logico-Algebraic Approach to Quantum Mechanics*, Reidel, Dordrecht and Boston, pp. 205–219.

Low, M., Wada, F. and M. Fox: 1973, 'Electroencephalographic Localization of Conative Aspects of Language Production in the Human Brain', *Transactions of the American Neurological Association* **98**, 129–133.

MacCorquodale, K. and P. Meehl: 1948, 'On a Distinction between Hypothetical Constructs and Intervening Variables', *Psychological Review* **55**, 97–107.

Mackie, J.: 1965, 'Causes and Conditions', *American Philosophical Quarterly* **2**, 245–264.

Mackie, J.: 1973, *Truth, Probability, and Paradox*, Clarendon Press, Oxford.

Mackie, J.: 1974, *The Cement of the Universe*, Clarendon Press, Oxford.

Malcolm, N.: 1971, 'The Myth of Cognitive Processes and Structures', in Mischel, T. (ed.), *Cognitive Development and Epistemology*, Academic Press, New York, pp. 385–392.

Maltzman, I.: 1968, 'Theoretical Conceptions of Semantic Conditioning and Generalization', in Dixon, T. and D. Horton (eds.), *Verbal Behavior and General Behavior Theory*, Prentice-Hall, Englewood Cliffs, pp. 291–339.

Manna, Z.: 1974, *Mathematical Theory of Computation*, McGraw-Hill, New York.

Manninen, J. and R. Tuomela: 1976, *Essays on Explanation and Understanding*, Reidel, Dordrecht and Boston.

Marras, A.: 1973a, 'Sellars on Thought and Language', *Noûs* **7**, 152–163.

Marras, A.: 1973b, 'On Sellars' Linguistic Theory of Conceptual Activity', *Canadian Journal of Philosophy* **2**, 471–483.

Marras, A.: 1973c, 'Reply to Sellars', *Canadian Journal of Philosophy* **2**, 495–501.

Martin, R.: 1972, 'Singular Causal Explanations', *Theory and Decision* **2**, 221–237.

Martin, R.: 1976, 'The Problem of the 'Tie' in von Wright's Schema of Practical Inference: A Wittgensteinian Solution', in Hintikka, J. (ed.), *Essays on Wittgenstein*, *Acta Philosophica Fennica*, **28**, pp. 326–363.

McAdam, D. and Whitaker, H.: 1971, 'Language Production: Electroencephalographic Localization in the Human Brain', *Science* **172**, 499–502.

McCann, H.: 1972, 'Is Raising One's Arm a Basic Action', *Journal of Philosophy* **69**, 235–249.

McCann, H.: 1975, 'Trying, Paralysis, and Volition', *The Review of Metaphysics* **28**, 423–442.

McCawley, J.: 1972, 'A Program for Logic', in Davidson, D. and G. Harman (eds.), *Semantics of Natural Language*, 2nd Edition, Reidel, Dordrecht, pp. 498–544.

McGuigan, F. and R. Schoonover (eds.): 1973, *The Psychophysiology of Thinking*, Academic Press, New York.

Melden, A.: 1961, *Free Action*, Routledge and Kegan Paul, London.

Miller, G., Galanter, E. and K. Pribram: 1960, *Plans and the Structure of Behavior*, Holt, Rinehart and Winston, New York.

Morton, A.: 1975, 'Because He Thought He Had Insulted Him', *Journal of Philosophy* **72**, 5–15.

414   REFERENCES

Neisser, U.: 1967, *Cognitive Psychology*, Appleton Century Crofts, New York.
Newell, A. and H. Simon: 1972, *Human Problem Solving*, Prentice-Hall, Englewood Cliffs.
Niiniluoto, I.: 1972, 'Inductive Systematization: Definition and a Critical Survey', *Synthese* **25**, 25–81.
Niiniluoto, I.: 1976, 'Inductive Explanation, Propensity, and Action', in Manninen, J. and R. Tuomela (eds.), *Essays on Explanation and Understanding*, Reidel, Dordrecht, pp. 335–368.
Niiniluoto, I. and R. Tuomela: 1973, *Theoretical Concepts and Hypothetico-Inductive Inference*, Synthese Library, Reidel, Dordrecht and Boston.
Nordenfelt, L.: 1974, *Explanation of Human Actions*, Philosophical Studies published by the Philosophical Society and the Department of Philosophy, University of Uppsala, No. 20.
Nowak, S.: 1974, 'Conditional Causal Relations and Their Approximations in the Social Sciences', in Suppes, P. *et al.* (eds.), *Logic, Methodology and Philosophy of Science IV*, North-Holland, Amsterdam, pp. 765–787.
Omer, I.: 1970, 'On the D-N Model of Scientific Explanation', *Philosophy of Science* **37**, 417–433.
Osgood, C.: 1953, *Method and Theory in Experimental Psychology*, Oxford University Press, New York.
Osgood, C.: 1968, 'Toward a Wedding of Insufficiencies', in Dixon, R. and D. Horton (eds.), *Verbal Behavior and General Behavior Theory*, Prentice-Hall, Englewood Cliffs, pp. 495–519.
Osgood, C.: 1970, 'Is Neo-behaviorism up a Blind Alley?', preprint.
Osgood, C.: 1971, 'Where Do Sentences Come From?', in Steinberg, D. and L. Jakobovits (eds.), *Semantics*, Cambridge University Press, Cambridge, pp. 497–529.
Osgood, C. and M. Richards: 1973, 'From Yang and Yin to *And* or *But*', *Language* **49**, 380–412.
O'Shaughnessy, B.: 1973, 'Trying (as the Mental "Pineal Gland")', *The Journal of Philosophy* **70**, 365–386.
Pap, A.: 1958, 'Disposition Concepts and Extensional Logic', in Feigl, H., Scriven, M., and G. Maxwell (eds.), *Minnesota Studies in the Philosophy of Science II*, University of Minnesota Press, Minneapolis, pp. 196–224.
Parsons, T.: 1970, 'Some Problems Concerning the Logic of Grammatical Modifiers', *Synthese* **21**, 320–334.
Pribram, K.: 1969, 'The Neurophysiology of Remembering', *Scientific American*, January Issue, 73–86.
Przełecki, M.: 1969, *The Logic of Empirical Theories*, Routledge and Kegan Paul, London.
Przełecki, M. and R. Wójcicki: 1969, 'The Problem of Analyticity', *Synthese* **19**, 374–399.
Putnam, H.: 1960, 'Minds and Machines', in Hook, S. (ed.), *Dimensions of Mind*, New York Univ. Press, New York, pp. 138–164.
Pörn, I.: 1974, 'Some Basic Concepts of Action', in Stenlund, S. (ed.), *Logical Theory and Semantic Analysis*, Reidel, Dordrecht and Boston, pp. 93–101.
Reichenbach, H.: 1956, *The Direction of Time*, University of California Press, Berkeley.
Rescher, N.: 1970, *Scientific Explanation*, Free Press, New York.
Rorty, R.: 1970, 'Incorrigibility as the Mark of the Mental', *The Journal of Philosophy* **67**, 399–424.
Rorty, R.: 1974, 'More on Incorrigibility', *Canadian Journal of Philosophy* **4**, 195–197.
Rorty, R.: 1976, *Philosophy and the Mirror of Nature*, manuscript.
Rosen, D.: 1975, 'An Argument for the Logical Notion of a Memory Trace', *Philosophy of Science* **42**, 1–10.

Rosenberg, A.: 1974, 'On Kim's Account of Events and Event-Identity', *The Journal of Philosophy* **71**, 327–336.
Rosenberg, J.: 1974, *Linguistic Representation*, Reidel, Dordrecht and Boston.
Rosenthal, D.: 1976, 'Mentality and Neutrality', *The Journal of Philosophy* **73**, 386–415.
Rosenthal, D. and W. Sellars: 1972, 'The Rosenthal–Sellars Correspondence on Intentionality', in Marras, A. (ed.), *Intentionality, Mind, and Language*, University of Illinois Press, Urbana, pp. 461–503.
Rozeboom, W.: 1963, 'The Factual Content of Theoretical Concepts', in Feigl, H. and G. Maxwell (eds.), *Minnesota Studies in the Philosophy of Science III*, Univ. of Minnesota Press, Minneapolis, pp. 273–357.
Rozeboom, W.: 1973, 'Dispositions Revisited', *The Philosophy of Science* **40**, 59–74.
Ryle, G.: 1949, *The Concept of Mind*, Hutchinson, London.
Salmon, W.: 1970, 'Statistical Explanation', in Colodny, R. (ed.), *Nature and Function of Scientific Theories*, University of Pittsburgh Press, Pittsburgh, pp. 173–231. Reprinted in Salmon (1971), to which the page references are made.
Salmon, W.: 1971, *Statistical Explanation and Statistical Relevance*, University of Pittsburgh Press, Pittsburgh.
Scott, D.: 1967, 'Some Definitional Suggestions for Automata Theory', *Journal of Computer and System Sciences* **1**, 187–212.
Scott, D. and P. Krauss: 1966, 'Assigning Probabilities to Logical Formulas', in Hintikka, J. and P. Suppes (eds.), *Aspects of Inductive Logic*, North-Holland, Amsterdam, pp. 219–264.
Scriven, M.: 1965, 'An Essential Unpredictability in Human Behavior', in Wolman, B. (ed.), *Scientific Psychology, Principles and Approaches*, Basic Books, New York, pp. 411–425.
Scriven, M.: 1972, 'The Logic of Cause', *Theory and Decision* **2**, 49–66.
Sellars, W.: 1956, 'Empiricism and the Philosophy of Mind', in Feigl, H. and M. Scriven (eds.), *Minnesota Studies in the Philosophy of Science I*, Univ. of Minnesota Press, Minneapolis, pp. 253–329.
Sellars, W.: 1963, *Science, Perception and Reality*, Routledge and Kegan Paul, London.
Sellars, W.: 1965, 'The Identity Approach to the Mind-Body Problem'. Reprinted in Sellars, W.: (1967), *Philosophical Perspectives*, Charles C. Thomas, Springfield, pp. 370–388.
Sellars, W.: 1966, 'Fatalism and Determinism', in Lehrer, K. (ed.), *Freedom and Determinism*, Random House, New York, pp. 141–202.
Sellars, W.: 1967, *Philosophical Perspectives*, Charles C. Thomas, Springfield.
Sellars, W.: 1968, *Science and Metaphysics*, Routledge and Kegan Paul, London.
Sellars, W.: 1969a, 'Metaphysics and the Concept of Person', in Lambert, K. (ed.), *The Logical Way of Doing Things*, Yale University Press, New Haven, pp. 219–252.
Sellars, W.: 1969b, 'Language as Thought and as Communication', *Philosophy and Phenomenological Research* **29**, 506–527.
Sellars, W.: 1969c, 'Are There Non-Deductive Logics?', in Rescher, N. *et al.* (eds.), *Essays in Honor of Carl G. Hempel*, Reidel, Dordrecht, pp. 83–103.
Sellars, W.: 1973a, 'Actions and Events', *Noûs* **7**, 179–202.
Sellars, W.: 1973b, 'Reply to Marras', *Canadian Journal of Philosophy* **2**, 485–493.
Sellars, W.: 1974, *Essays in Philosophy and Its History*, Reidel, Dordrecht and Boston.
Sellars, W.: 1975, 'Reply to Alan Donagan', *Philosophical Studies* **27**, 149–184.
Sherwood, M.: 1969, *The Logic of Explanation in Psychoanalysis*, Academic Press, New York.
Shoemaker, S.: 1975, 'Functionalism and Qualia', *Philosophical Studies* **27**, 291–315.

Stegmüller, W.: 1969, *Wissenschaftliche Erklärung und Begründung, Probleme und Resultate der Wissenschaftstheorie und Analytischen Philosophie*, Band I, Springer, Berlin, Heidelberg, and New York.

Stegmüller, W.: 1973, *Personelle und Statistische Wahrscheinlichkeit, Probleme und Resultate der Wissenschaftstheorie und Analytischen Philosophie*, Band IV, Springer, Berlin, Heidelberg, and New York.

Stenius, E.: 1967, 'Mood and Language-Game', *Synthese* 17, 254–274.

Stoutland, F.: 1976a, 'The Causal Theory of Action', in Manninen, J. and R. Tuomela (eds.), *Essays on Explanation and Understanding*, Synthese Library, Reidel, Dordrecht and Boston, pp. 271–304.

Stoutland, F.: 1976b, 'The Causation of Behavior', in Hintikka, J. (ed.), *Essays on Wittgenstein, Acta Philosophica Fennica*, 28, pp. 286–325.

Suppes, P.: 1966, 'Probabilistic Inference and the Concept of Total Evidence', in Hintikka, J. and P. Suppes (eds.), *Aspects of Inductive Logic*, North-Holland, Amsterdam, pp. 49–65.

Suppes, P.: 1970, *A Probabilistic Theory of Causality, Acta Philosophica Fennica*, North-Holland, Amsterdam.

Taylor, C.: 1964, *The Explanation of Behaviour*, Routledge and Kegan Paul, London.

Taylor, C.: 1970, 'Explaining Action', *Inquiry* 13, 54–89.

Tuomela, R.: 1968, *The Application Process of a Theory; with Special Reference to some Behavioral Theories, Annales Academiae Scientiarum Fennicae*, ser. B, tom, 154; 3, Helsinki.

Tuomela, R.: 1972, 'Deductive Explanation of Scientific Laws', *Journal of Philosophical Logic* 1, 369–392.

Tuomela, R.: 1973a, *Theoretical Concepts*, Library of Exact Philosophy, vol. 10, Springer, Vienna and New York.

Tuomela, R.: 1973b, 'Theoretical Concepts in Neobehavioristic Theories', in Bunge, M. (ed.), *The Methodological Unity of Science*, Reidel, Dordrecht and Boston pp. 123–152.

Tuomela, R.: 1973c, 'A Psycholinguistic Paradox and Its Solution', *Ajatus* 35, 124–139.

Tuomela, R.: 1974a, *Human Action and Its Explanation*, Reports from the Institute of Philosophy, University of Helsinki, No. 2.

Tuomela, R.: 1974b, 'Causality, Ontology, and Subsumptive Explanation', forthcoming in the proceedings of the Conference for Formal Methods in the Methodology of Empirical Sciences, held in Warsaw, Poland in 1974.

Tuomela, R.: 1975, 'Purposive Causation of Action', forthcoming in Cohen, R. and M. Wartofsky (eds.), *Boston Studies in the Philosophy of Science*, vol. 31, Reidel, Dordrecht and Boston.

Tuomela, R.: 1976a, 'Causes and Deductive Explanation', in A. C. Michalos and R. S. Cohen (eds.), *PSA 1974*, Boston Studies in the Philosophy of Science, vol. 32, Reidel, Dordrecht and Boston, pp. 325–360.

Tuomela, R.: 1976b, 'Morgan on Deductive Explanation: A Rejoinder', *Journal of Philosophical Logic* 5, 527–543.

Tuomela, R.: 1976c, 'Explanation and Understanding of Human Behavior', in Manninen, J. and R. Tuomela (eds.), *Essays on Explanation and Understanding*, Reidel, Dordrecht and Boston, pp. 183–205.

Wallace, J.: 1971, 'Some Logical Roles of Adverbs', *The Journal of Philosophy* 68, 690–714.

von Wright, G. H.: 1968, *An Essay in Deontic Logic and the General Theory of Action, Acta Philosophica Fennica*, North-Holland, Amsterdam.

von Wright, G. H.: 1971, *Explanation and Understanding*, Cornell University Press, Ithaca.

von Wright, G. H.: 1972, 'On So-Called Practical Inference', *Acta Sociologica* **15**, 39–53.
von Wright, G. H.: 1974, *Causality and Determinism*, Columbia University Press, New York and London.
von Wright, G. H.: 1976a, 'Replies', in Manninen, J. and R. Tuomela (eds.), *Essays on Explanation and Understanding*, Reidel, Dordrecht and Boston, pp. 371–413.
von Wright, G. H.: 1976b, 'Determinism and the Study of Man' in Manninen, J. and R. Tuomela (eds.), *Essays on Explanation and Understanding*, Reidel, Dordrecht and Boston, pp. 415–435.

INDEX OF NAMES

INDEX OF SUBJECTS

SYNTHESE LIBRARY

Monographs on Epistemology, Logic, Methodology,
Philosophy of Science, Sociology of Science and of Knowledge, and on the
Mathematical Methods of Social and Behavioral Sciences

Managing Editor:
JAAKKO HINTIKKA (Academy of Finland and Stanford University)

Editors:

ROBERT S. COHEN (Boston University)
DONALD DAVIDSON (University of Chicago)
GABRIËL NUCHELMANS (University of Leyden)
WESLEY C. SALMON (University of Arizona)

1. J. M. Bocheński, *A Precis of Mathematical Logic.* 1959, X + 100 pp.
2. P. L. Guiraud, *Problèmes et méthodes de la statistique linguistique.* 1960, VI + 146 pp.
3. Hans Freudenthal (ed.), *The Concept and the Role of the Model in Mathematics and Natural and Social Sciences, Proceedings of a Colloquium held at Utrecht, The Netherlands, January 1960.* 1961, VI + 194 pp.
4. Evert W. Beth, *Formal Methods. An Introduction to Symbolic Logic and the Study of Effective Operations in Arithmetic and Logic.* 1962, XIV + 170 pp.
5. B. H. Kazemier and D. Vuysje (eds.), *Logic and Language. Studies Dedicated to Professor Rudolf Carnap on the Occasion of His Seventieth Birthday.* 1962, VI + 256 pp.
6. Marx W. Wartofsky (ed.), *Proceedings of the Boston Colloquium for the Philosophy of Science, 1961-1962,* Boston Studies in the Philosophy of Science (ed. by Robert S. Cohen and Marx W. Wartofsky), Volume I. 1973, VIII + 212 pp.
7. A. A. Zinov'ev, *Philosophical Problems of Many-Valued Logic.* 1963, XIV + 155 pp.
8. Georges Gurvitch, *The Spectrum of Social Time.* 1964, XXVI + 152 pp.
9. Paul Lorenzen, *Formal Logic.* 1965, VIII + 123 pp.
10. Robert S. Cohen and Marx W. Wartofsky (eds.), *In Honor of Philipp Frank,* Boston Studies in the Philosophy of Science (ed. by Robert S. Cohen and Marx W. Wartofsky), Volume II. 1965, XXXIV + 475 pp.
11. Evert W. Beth, *Mathematical Thought. An Introduction to the Philosophy of Mathematics.* 1965, XII + 208 pp.
12. Evert W. Beth and Jean Piaget, *Mathematical Epistemology and Psychology.* 1966, XII + 326 pp.
13. Guido Küng, *Ontology and the Logistic Analysis of Language. An Enquiry into the Contemporary Views on Universals.* 1967, XI + 210 pp.
14. Robert S. Cohen and Marx W. Wartofsky (eds.), *Proceedings of the Boston Colloquium for the Philosophy of Science 1964-1966, in Memory of Norwood Russell Hanson,* Boston Studies in the Philosophy of Science (ed. by Robert S. Cohen and Marx W. Wartofsky), Volume III. 1967, XLIX + 489 pp.

15. C. D. Broad, *Induction, Probability, and Causation. Selected Papers.* 1968, XI + 296 pp.
16. Günther Patzig, *Aristotle's Theory of the Syllogism. A Logical-Philosophical Study of Book A of the Prior Analytics.* 1968, XVII + 215 pp.
17. Nicholas Rescher, *Topics in Philosophical Logic.* 1968, XIV + 347 pp.
18. Robert S. Cohen and Marx W. Wartofsky (eds.), *Proceedings of the Boston Colloquium for the Philosophy of Science 1966-1968,* Boston Studies in the Philosophy of Science (ed. by Robert S. Cohen and Marx W. Wartofsky), Volume IV. 1969, VIII + 537 pp.
19. Robert S. Cohen and Marx W. Wartofsky (eds.), *Proceedings of the Boston Colloquium for the Philosophy of Science 1966-1968,* Boston Studies in the Philosophy of Science (ed. by Robert S. Cohen and Marx W. Wartofsky), Volume V. 1969, VIII + 482 pp.
20. J.W. Davis, D. J. Hockney, and W. K. Wilson (eds.), *Philosophical Logic.* 1969, VIII + 277 pp.
21. D. Davidson and J. Hintikka (eds.), *Words and Objections: Essays on the Work of W. V. Quine.* 1969, VIII + 366 pp.
22. Patrick Suppes, *Studies in the Methodology and Foundations of Science. Selected Papers from 1911 to 1969.* 1969, XII + 473 pp.
23. Jaakko Hintikka, *Models for Modalities. Selected Essays.* 1969, IX + 220 pp.
24. Nicholas Rescher *et al.* (eds.), *Essays in Honor of Carl G. Hempel. A Tribute on the Occasion of His Sixty-Fifth Birthday.* 1969, VII + 272 pp.
25. P. V. Tavanec (ed.), *Problems of the Logic of Scientific Knowledge.* 1969, XII + 429 pp.
26. Marshall Swain (ed.), *Induction, Acceptance, and Rational Belief.* 1970, VII + 232 pp.
27. Robert S. Cohen and Raymond J. Seeger (eds.), *Ernst Mach: Physicist and Philosopher,* Boston Studies in the Philosophy of Science (ed. by Robert S. Cohen and Marx W. Wartofsky), Volume VI. 1970, VIII + 295 pp.
28. Jaakko Hintikka and Patrick Suppes, *Information and Inference.* 1970, X + 336 pp.
29. Karel Lambert, *Philosophical Problems in Logic. Some Recent Developments.* 1970, VII + 176 pp.
30. Rolf A. Eberle, *Nominalistic Systems.* 1970, IX + 217 pp.
31. Paul Weingartner and Gerhard Zecha (eds.), *Induction, Physics, and Ethics: Proceedings and Discussions of the 1968 Salzburg Colloquium in the Philosophy of Science.* 1970, X + 382 pp.
32. Evert W. Beth, *Aspects of Modern Logic.* 1970, XI + 176 pp.
33. Risto Hilpinen (ed.), *Deontic Logic: Introductory and Systematic Readings.* 1971, VII + 182 pp.
34. Jean-Louis Krivine, *Introduction to Axiomatic Set Theory.* 1971, VII + 98 pp.
35. Joseph D. Sneed, *The Logical Structure of Mathematical Physics.* 1971, XV + 311 pp.
36. Carl R. Kordig, *The Justification of Scientific Change.* 1971, XIV + 119 pp.
37. Milič Čapek, *Bergson and Modern Physics,* Boston Studies in the Philosophy of Science (ed. by Robert S. Cohen and Marx W. Wartofsky), Volume VII. 1971, XV + 414 pp.

38. Norwood Russell Hanson, *What I Do Not Believe, and Other Essays* (ed. by Stephen Toulmin and Harry Woolf). 1971, XII + 390 pp.

39. Roger C. Buck and Robert S. Cohen (eds.), *PSA 1970. In Memory of Rudolf Carnap*, Boston Studies in the Philosophy of Science (ed. by Robert S. Cohen and Marx W. Wartofsky), Volume VIII. 1971, LXVI + 615 pp. Also available as paperback.

40. Donald Davidson and Gilbert Harman (eds.), *Semantics of Natural Language*. 1972, X + 769 pp. Also available as paperback.

41. Yehoshua Bar-Hillel (ed.), *Pragmatics of Natural Languages*. 1971, VII + 231 pp.

42. Sören Stenlund, *Combinators, λ-Terms and Proof Theory*. 1972, 184 pp.

43. Martin Strauss, *Modern Physics and Its Philosophy. Selected Papers in the Logic, History, and Philosophy of Science*. 1972, X + 297 pp.

44. Mario Bunge, *Method, Model and Matter*. 1973, VII + 196 pp.

45. Mario Bunge, *Philosophy of Physics*. 1973, IX + 248 pp.

46. A. A. Zinov'ev, *Foundations of the Logical Theory of Scientific Knowledge (Complex Logic)*, Boston Studies in the Philosophy of Science (ed. by Robert S. Cohen and Marx W. Wartofsky), Volume IX. Revised and enlarged English edition with an appendix, by G. A. Smirnov, E. A. Sidorenka, A. M. Fedina, and L. A. Bobrova. 1973, XXII + 301 pp. Also available as paperback.

47. Ladislav Tondl, *Scientific Procedures*, Boston Studies in the Philosophy of Science (ed. by Robert S. Cohen and Marx W. Wartofsky), Volume X. 1973, XII + 268 pp. Also available as paperback.

48. Norwood Russell Hanson, *Constellations and Conjectures* (ed. by Willard C. Humphreys, Jr.). 1973, X + 282 pp.

49. K. J. J. Hintikka, J. M. E. Moravcsik, and P. Suppes (eds.), *Approaches to Natural Language. Proceedings of the 1970 Stanford Workshop on Grammar and Semantics*. 1973, VIII + 526 pp. Also available as paperback.

50. Mario Bunge (ed.), *Exact Philosophy — Problems, Tools, and Goals*. 1973, X + 214 pp.

51. Radu J. Bogdan and Ilkka Niiniluoto (eds.), *Logic, Language, and Probability. A Selection of Papers Contributed to Sections IV, VI, and XI of the Fourth International Congress for Logic, Methodology, and Philosophy of Science, Bucharest, September 1971*. 1973, X + 323 pp.

52. Glenn Pearce and Patrick Maynard (eds.), *Conceptual Chance*. 1973, XII + 282 pp.

53. Ilkka Niiniluoto and Raimo Tuomela, *Theoretical Concepts and Hypothetico-Inductive Inference*. 1973, VII + 264 pp.

54. Roland Fraïssé, *Course of Mathematical Logic — Volume 1: Relation and Logical Formula*. 1973, XVI + 186 pp. Also available as paperback.

55. Adolf Grünbaum, *Philosophical Problems of Space and Time*. Second, enlarged edition, Boston Studies in the Philosophy of Science (ed. by Robert S. Cohen and Marx W. Wartofsky), Volume XII. 1973, XXIII + 884 pp. Also available as paperback.

56. Patrick Suppes (ed.), *Space, Time, and Geometry*. 1973, XI + 424 pp.

57. Hans Kelsen, *Essays in Legal and Moral Philosophy*, selected and introduced by Ota Weinberger. 1973, XXVIII + 300 pp.

58. R. J. Seeger and Robert S. Cohen (eds.), *Philosophical Foundations of Science. Proceedings of an AAAS Program, 1969*, Boston Studies in the Philosophy of

Science (ed. by Robert S. Cohen and Marx W. Wartofsky), Volume XI. 1974, X + 545 pp. Also available as paperback.

59. Robert S. Cohen and Marx W. Wartofsky (eds.), *Logical and Epistemological Studies in Contemporary Physics*, Boston Studies in the Philosophy of Science (ed. by Robert S. Cohen and Marx W. Wartofsky), Volume XIII. 1973, VIII + 462 pp. Also available as paperback.

60. Robert S. Cohen and Marx W. Wartofsky (eds.), *Methodological and Historical Essays in the Natural and Social Sciences. Proceedings of the Boston Colloquium for the Philosophy of Science, 1969-1972,* Boston Studies in the Philosophy of Science (ed. by Robert S. Cohen and Marx W. Wartofsky), Volume XIV. 1974, VIII + 405 pp. Also available as paperback.

61. Robert S. Cohen, J. J. Stachel and Marx W. Wartofsky (eds.), *For Dirk Struik. Scientific, Historical and Political Essays in Honor of Dirk J. Struik*, Boston Studies in the Philosophy of Science (ed. by Robert S. Cohen and Marx W. Wartofsky), Volume XV. 1974, XXVII + 652 pp. Also available as paperback.

62. Kazimierz Ajdukiewicz, *Pragmatic Logic*, transl. from the Polish by Olgierd Wojtasiewicz. 1974, XV + 460 pp.

63. Sören Stenlund (ed.), *Logical Theory and Semantic Analysis. Essays Dedicated to Stig Kanger on His Fiftieth Birthday.* 1974, V + 217 pp.

64. Kenneth F. Schaffner and Robert S. Cohen (eds.), *Proceedings of the 1972 Biennial Meeting, Philosophy of Science Association*, Boston Studies in the Philosophy of Science (ed. by Robert S. Cohen and Marx W. Wartofsky), Volume XX. 1974, IX + 444 pp. Also available as paperback.

65. Henry E. Kyburg, Jr., *The Logical Foundations of Statistical Inference.* 1974, IX + 421 pp.

66. Marjorie Grene, *The Understanding of Nature: Essays in the Philosophy of Biology*, Boston Studies in the Philosophy of Science (ed. by Robert S. Cohen and Marx W. Wartofsky), Volume XXIII. 1974, XII + 360 pp. Also available as paperback.

67. Jan M. Broekman, *Structuralism: Moscow, Prague, Paris.* 1974, IX + 117 pp.

68. Norman Geschwind, *Selected Papers on Language and the Brain*, Boston Studies in the Philosophy of Science (ed. by Robert S. Cohen and Marx W. Wartofsky), Volume XVI. 1974, XII + 549 pp. Also available as paperback.

69. Roland Fraïssé, *Course of Mathematical Logic* – Volume 2: *Model Theory.* 1974, XIX + 192 pp.

70. Andrzej Grzegorczyk, *An Outline of Mathematical Logic. Fundamental Results and Notions Explained with All Details.* 1974, X + 596 pp.

71. Franz von Kutschera, *Philosophy of Language.* 1975, VII + 305 pp.

72. Juha Manninen and Raimo Tuomela (eds.), *Essays on Explanation and Understanding. Studies in the Foundations of Humanities and Social Sciences.* 1976, VII + 440 pp.

73. Jaakko Hintikka (ed.), *Rudolf Carnap, Logical Empiricist. Materials and Perspectives.* 1975, LXVIII + 400 pp.

74. Milič Čapek (ed.), *The Concepts of Space and Time. Their Structure and Their Development*, Boston Studies in the Philosophy of Science (ed. by Robert S. Cohen and Marx W. Wartofsky), Volume XXII. 1976, LVI + 570 pp. Also available as paperback.

75. Jaakko Hintikka and Unto Remes, *The Method of Analysis. Its Geometrical Origin and Its General Significance*, Boston Studies in the Philosophy of Science (ed. by Robert S. Cohen and Marx W. Wartofsky), Volume XXV. 1974, XVIII + 144 pp. Also available as paperback.

76. John Emery Murdoch and Edith Dudley Sylla, *The Cultural Context of Medieval Learning. Proceedings of the First International Colloquium on Philosophy, Science, and Theology in the Middle Ages – September 1973*, Boston Studies in the Philosophy of Science (ed. by Robert S. Cohen and Marx W. Wartofsky), Volume XXVI. 1975, X + 566 pp. Also available as paperback.

77. Stefan Amsterdamski, *Between Experience and Metaphysics. Philosophical Problems of the Evolution of Science*, Boston Studies in the Philosophy of Science (ed. by Robert S. Cohen and Marx W. Wartofsky), Volume XXXV. 1975, XVIII + 193 pp. Also available as paperback.

78. Patrick Suppes (ed.), *Logic and Probability in Quantum Mechanics*. 1976, XV + 541 pp.

79. H. von Helmholtz, *Epistemological Writings*. (A New Selection Based upon the 1921 Volume edited by Paul Hertz and Moritz Schlick, Newly Translated and Edited by R. S. Cohen and Y. Elkana), Boston Studies in the Philosophy of Science, Volume XXXVII. 1977 (forthcoming).

80. Joseph Agassi, *Science in Flux*, Boston Studies in the Philosophy of Science (ed. by Robert S. Cohen and Marx W. Wartofsky), Volume XXVIII. 1975, XXVI + 553 pp. Also available as paperback.

81. Sandra G. Harding (ed.), *Can Theories Be Refuted? Essays on the Duhem-Quine Thesis*. 1976, XXI + 318 pp. Also available as paperback.

82. Stefan Nowak, *Methodology of Sociological Research: General Problems*. 1977, XVIII + 504 pp. (forthcoming).

83. Jean Piaget, Jean-Blaise Grize, Alina Szeminska, and Vinh Bang, *Epistemology and Psychology of Functions*. 1977 (forthcoming).

84. Marjorie Grene and Everett Mendelsohn (eds.), *Topics in the Philosophy of Biology*, Boston Studies in the Philosophy of Science (ed. by Robert S. Cohen and Marx W. Wartofsky), Volume XXVII. 1976, XIII + 454 pp. Also available as paperback.

85. E. Fischbein, *The Intuitive Sources of Probabilistic Thinking in Children*. 1975, XIII + 204 pp.

86. Ernest W. Adams, *The Logic of Conditionals. An Application of Probability to Deductive Logic*. 1975, XIII + 156 pp.

87. Marian Przełęcki and Ryszard Wójcicki (eds.), *Twenty-Five Years of Logical Methodology in Poland*. 1977, VIII + 803 pp. (forthcoming).

88. J. Topolski, *The Methodology of History*. 1976, X + 673 pp.

89. A. Kasher (ed.), *Language in Focus: Foundations, Methods and Systems. Essays Dedicated to Yehoshua Bar-Hillel*, Boston Studies in the Philosophy of Science (ed. by Robert S. Cohen and Marx W. Wartofsky), Volume XLIII. 1976, XXVIII + 679 pp. Also available as paperback.

90. Jaakko Hintikka, *The Intentions of Intentionality and Other New Models for Modalities*. 1975, XVIII + 262 pp. Also available as paperback.

91. Wolfgang Stegmüller, *Collected Papers on Epistemology, Philosophy of Science and History of Philosophy*, 2 Volumes, 1977 (forthcoming).

92. Dov M. Gabbay, *Investigations in Modal and Tense Logics with Applications to Problems in Philosophy and Linguistics.* 1976, XI + 306 pp.
93. Radu J. Bogdan, *Local Induction.* 1976, XIV + 340 pp.
94. Stefan Nowak, *Understanding and Prediction: Essays in the Methodology of Social and Behavioral Theories.* 1976, XIX + 482 pp.
95. Peter Mittelstaedt, *Philosophical Problems of Modern Physics,* Boston Studies in the Philosophy of Science (ed. by Robert S. Cohen and Marx W. Wartofsky), Volume XVIII. 1976, X + 211 pp. Also available as paperback.
96. Gerald Holton and William Blanpied (eds.), *Science and Its Public: The Changing Relationship,* Boston Studies in the Philosophy of Science (ed. by Robert S. Cohen and Marx W. Wartofsky), Volume XXXIII. 1976, XXV + 289 pp. Also available as paperback.
97. Myles Brand and Douglas Walton (eds.), *Action Theory. Proceedings of the Winnipeg Conference on Human Action, Held at Winnipeg, Manitoba, Canada, 9-11 May 1975.* 1976, VI + 345 pp.
98. Risto Hilpinen, *Knowledge and Rational Belief.* 1978 (forthcoming).
99. R. S. Cohen, P. K. Feyerabend, and M. W. Wartofsky (eds.), *Essays in Memory of Imre Lakatos,* Boston Studies in the Philosophy of Science (ed. by Robert S. Cohen and Marx W. Wartofsky), Volume XXXIX. 1976, XI + 762 pp. Also available as paperback.
100. R. S. Cohen and J. Stachel (eds.), *Leon Rosenfeld, Selected Papers.* Boston Studies in the Philosophy of Science (ed. by Robert S. Cohen and Marx W. Wartofsky), Volume XXI. 1977 (forthcoming).
101. R. S. Cohen, C. A. Hooker, A. C. Michalos, and J. W. van Evra (eds.), *PSA 1974: Proceedings of the 1974 Biennial Meeting of the Philosophy of Science Association,* Boston Studies in the Philosophy of Science (ed. by Robert S. Cohen and Marx W. Wartofsky), Volume XXXII. 1976, XIII + 734 pp. Also available as paperback.
102. Yehuda Fried and Joseph Agassi, *Paranoia: A Study in Diagnosis,* Boston Studies in the Philosophy of Science (ed. by Robert S. Cohen and Marx W. Wartofsky), Volume L. 1976, XV + 212 pp. Also available as paperback.
103. Marian Przełęcki, Klemens Szaniawski, and Ryszard Wójcicki (eds.), *Formal Methods in the Methodology of Empirical Sciences.* 1976, 455 pp.
104. John M. Vickers, *Belief and Probability.* 1976, VIII + 202 pp.
105. Kurt H. Wolff, *Surrender and Catch: Experience and Inquiry Today,* Boston Studies in the Philosophy of Science (ed. by Robert S. Cohen and Marx W. Wartofsky), Volume LI. 1976, XII + 410 pp. Also available as paperback.
106. Karel Kosík, *Dialectics of the Concrete,* Boston Studies in the Philosophy of Science (ed. by Robert S. Cohen and Marx W. Wartofsky), Volume LII. 1976, VIII + 158 pp. Also available as paperback.
107. Nelson Goodman, *The Structure of Appearance,* Boston Studies in the Philosophy of Science (ed. by Robert S. Cohen and Marx W. Wartofsky), Volume LIII. 1977 (forthcoming).
108. Jerzy Giedymin (ed.), *Kazimierz Ajdukiewicz: Scientific World-Perspective and Other Essays, 1931–1963.* 1977 (forthcoming).
109. Robert L. Causey, *Unity of Science.* 1977, VIII+185 pp.
110. Richard Grandy, *Advanced Logic for Applications.* 1977 (forthcoming).

111. Robert P. McArthur, *Tense Logic*. 1976, VII + 84 pp.
112. Lars Lindahl, *Position and Change: A Study in Law and Logic*. 1977, IX + 299 pp.
113. Raimo Tuomela, *Dispositions*. 1977 (forthcoming).
114. Herbert A. Simon, *Models of Discovery and Other Topics in the Methods of Science*, Boston Studies in the Philosophy of Science (ed. by Robert S. Cohen and Marx W. Wartofsky), Volume LIV. 1977 (forthcoming).
115. Roger D. Rosenkrantz, *Inference, Method and Decision*. 1977 (forthcoming).
116. Raimo Tuomela, *Human Action and Its Explanation. A Study on the Philosophical Foundations of Psychology*. 1977 (forthcoming).
117. Morris Lazerowitz, *The Language of Philosophy*, Boston Studies in the Philosophy of Science (ed. by Robert S. Cohen and Marx W. Wartofsky), Volume LV. 1977 (forthcoming).
118. Tran Duc Thao, *Origins of Language and Consciousness*, Boston Studies in the Philosophy of Science (ed. by Robert S. Cohen and Marx. W. Wartofsky), Volume LVI. 1977 (forthcoming).
119. Jerzy Pelc, *Polish Semiotic Studies, 1894–1969*. 1977 (forthcoming).
120. Ingmar Pörn, *Action Theory and Social Science. Some Formal Models*. 1977 (forthcoming).
121. Joseph Margolis, *Persons and Minds*, Boston Studies in the Philosophy of Science (ed. by Robert S. Cohen and Marx W. Wartofsky), Volume LVII. 1977 (forthcoming).

SYNTHESE HISTORICAL LIBRARY

Texts and Studies
in the History of Logic and Philosophy

Editors:

N. KRETZMANN (Cornell University)
G. NUCHELMANS (University of Leyden)
L. M. DE RIJK (University of Leyden)

1. M. T. Beonio-Brocchieri Fumagalli, *The Logic of Abelard.* Translated from the Italian. 1969, IX + 101 pp.
2. Gottfried Wilhelm Leibniz, *Philosophical Papers and Letters.* A selection translated and edited, with an introduction, by Leroy E. Loemker. 1969, XII + 736 pp.
3. Ernst Mally, *Logische Schriften,* ed. by Karl Wolf and Paul Weingartner. 1971, X + 340 pp.
4. Lewis White Beck (ed.), *Proceedings of the Third International Kant Congress.* 1972, XI + 718 pp.
5. Bernard Bolzano, *Theory of Science,* ed. by Jan Berg. 1973, XV + 398 pp.
6. J. M. E. Moravcsik (ed.), *Patterns in Plato's Thought. Papers Arising Out of the 1971 West Coast Greek Philosophy Conference.* 1973, VIII + 212 pp.
7. Nabil Shehaby, *The Propositional Logic of Avicenna: A Translation from al-Shifā: al-Qiyās,* with Introduction, Commentary and Glossary. 1973, XIII + 296 pp.
8. Desmond Paul Henry, *Commentary on De Grammatico: The Historical-Logical Dimensions of a Dialogue of St. Anselm's.* 1974, IX + 345 pp.
9. John Corcoran, *Ancient Logic and Its Modern Interpretations.* 1974, X + 208 pp.
10. E. M. Barth, *The Logic of the Articles in Traditional Philosophy.* 1974, XXVII + 533 pp.
11. Jaakko Hintikka, *Knowledge and the Known. Historical Perspectives in Epistemology.* 1974, XII + 243 pp.
12. E. J. Ashworth, *Language and Logic in the Post-Medieval Period.* 1974, XIII + 304 pp.
13. Aristotle, *The Nicomachean Ethics.* Translated with Commentaries and Glossary by Hypocrates G. Apostle. 1975, XXI + 372 pp.
14. R. M. Dancy, *Sense and Contradiction: A Study in Aristotle.* 1975, XII + 184 pp.
15. Wilbur Richard Knorr, *The Evolution of the Euclidean Elements. A Study of the Theory of Incommensurable Magnitudes and Its Significance for Early Greek Geometry.* 1975, IX + 374 pp.
16. Augustine, *De Dialectica.* Translated with Introduction and Notes by B. Darrell Jackson. 1975, XI + 151 pp.